Powell's Chemical Principles Applied to

OP / 9.98   PC

Science & Natural History   118572

CHEMICAL PRINCIPLES APPLIED TO SPACECRAFT OPERATIONS

by
Ross E. Dueber
and
Darren S. McKnight

with Foreword by
Raymond O. Rantanen

AN ORBIT SERIES BOOK

KRIEGER PUBLISHING COMPANY
MALABAR, FLORIDA
1993

Original Edition 1993

Printed and Published by
KRIEGER PUBLISHING COMPANY
KRIEGER DRIVE
Malabar, Florida 32950

Copyright © 1993 by
KRIEGER PUBLISHING COMPANY

All rights reserved. No part of this book may be reproduced in any form or by any means, electronic or mechanical, including information storage and retrieval systems without permission in writing from the publisher.
No liability is assumed with respect to the use of the information contained herein.

Printed in the United States of America

Library of Congress Cataloging-in-Publication Data

Dueber, Ross E., 1960–
    Chemical principled applied to spacecraft operations / Ross E. Dueber and Darren S. McKnight.
       p. cm. — (Orbit, a foundation series)
    Includes bibliographical references and index.
    ISBN 0-89464-036-4 (acid-free paper)
    1. Space vehicles—Design and construction. 2. Space vehicles—Materials. 3. Space environment. 4. Astrodynamics. 5. Chemistry. I. McKnight, Darren S. II. Title.
TL875.D84 1991
629.47—dc20
                                                                        90-28269
                                                                             CIP

10 9 8 7 6 5 4 3 2

Series editor
Edwin F. Strother, Ph.D.

Acknowledgments

The authors wish to acknowledge those organizations and individuals who contributed to the successful completion of this book. Thanks to the Departments of Chemistry and Physics at the United States Air Force Academy and to Kaman Sciences Corporation for their support. At the Academy special thanks to: Dr. Ron Furstenau and Dr. Stephen Novicki for laying the groundwork for Chemistry 325 – Space Chemistry, DFSIV for photography, and Lolita Young and Ike Sleighter for typing the manuscript. And thanks to Dr. Raymond Rantanen for providing the Foreword and many useful comments. We would also like to thank our wives, Sandee and Alison, for their patience and love during the compilation of this book.

Contents

Foreword ix
Preface xi
1. **Review of General Chemistry** 1
2. **Review of Astrodynamics** 23
3. **Near Earth Space Environment** 37
4. **Planetary Environments** 55
5. **Spacecraft Materials** 73
6. **Spacecraft Power** 103
7. **Propulsion** 135
8. **Thermal Control and Protection Systems** 155
9. **Manned Space Flight** 177
   Appendixes
      A: Effects of Atomic Oxygen on Selected Materials 195
      B: Effects of Rocket Exhaust Contaminants on Selected Materials 198
   Index 199

Foreword

All the systems on a spacecraft contribute to an induced environment around the spacecraft and themselves may be susceptible to degradation from this environment. The interaction of this induced environment with the ambient environment is a complex process which alters the chemistry involved when compared to the induced and ambient environment considered separately. **Observations on Shuttle flights, the returned Long Duration Exposure Facility (LDEF), and satellites indicate we do not completely understand this complex process.** Ground based simulation of the interaction environment is difficult at best and to date had not been achieved.

The interaction volume around a spacecraft is asymmetric for several reasons. An asymmetry exists at altitudes between 200 and 1,000 km in the density variation of the neutral gases because the spacecraft has speeds near 8 km/sec and the average thermal velocity of the ambient gases, in the same reference frame, is only about 1 km/sec. The net result is that the ambient gases strike the side of the vehicle facing along the velocity vector. These are then reemitted from the surface where they collide with incoming undisturbed ambient gases and with previously reemitted gases that have been scattered back toward the surfaces. The density of gases near the ram facing surfaces is almost two orders of magnitude greater than the undisturbed ambient. Spacecraft induced sources or atomic oxygen erosion products add to this pressure. It is this energetic gas cloud that produces the surface and far field glow around low Earth orbiting systems that raises the background brightness levels.

Because about one in 10^5 or 10^4 of the neutral gases is ionized by solar flux, the concentration of ions is asymmetric and quite different from the ambient. The Earth's geomagnetic field picks up some of these ions and transfers them to the wake region behind the spacecraft. During night portions of the orbit the vehicle's distribution of ions relative to the daylight portion is significantly different.

These and other asymmetrical effects (such as surface charging) result in a systems impact on the different components of the spacecraft. Understanding this is paramount to performing vulnerability and survivability assessments. Manned systems further complicate the issue because of the additional sources of effluents that accompany these systems.

The phenomena presented in this text are required inputs to models that predict the external environment around a space system. Theses interaction models must be capable of producing predictions of gases along experiment lines-of-sight; the absorption or emissions by these gases; deposition on critical surfaces (thermal control and solar arrays) and optical surfaces (mirrors, lenses, baffles); and the reflectance, and surface conductivity changes.

The chemistry of the phenomena presented in this text is required to begin to unravel the complex gas phase and surface chemistry synergism between the source and the ambient environment. The authors are to be congratulated for compiling this much needed information in a single volume.

RAYMOND O. RANTANEN

Raymond O. Rantanen has a Ph.D. in surface physics and is a principal scientist for ROR Enterprises in Atholl, Idaho. He has studied and modeled the interaction of spacecraft with the orbital ambient environment. His work on Skylab, the Space Shuttle, and LDEF has spanned three decades.

Preface

The space environment poses challenges to scientists and engineers unlike any encountered here on Earth. Human conquest of space has been possible only through the overcoming of many diverse and unique technological hurdles. Take, for example, the Space Shuttle's ceramic tiles which replaced the earlier ablation shield as a means of thermal protection. Were it not for them, the idea of a reusable space transportation system would be only a dream.

Space exploration requires the blending of many disciplines in order to understand the effects of the space environment and successfully operate spacecraft within it. This book focuses upon the chemistry of spacecraft operations and a satellite's interaction with the environment. Many books and articles are available to aerospace students and personnel on the subjects of space environment and spacecraft design. Often, however, these sources are widely scattered and not in a readily usable form. The compilation of this information into a unified approach provides a unique insight into the complex world of satellite design and operations. In *Chemical Principles Applied to Spacecraft Operations*, the authors have brought these sources together under one cover.

The book begins with reviews of general chemistry and astrodynamics in Chapters 1 and 2. These chapters are by no means an exhaustive review of the subjects, but rather serve as the framework for material discussed in later chapters. The chemistry review outlines all basic concepts necessary to assimilate the more advanced environmental concepts covered in later chapters. For the scientist experienced with chemistry, this chapter provides the nomenclature for the remainder of the text and a convenient source of basic reference data. Yet, for the scientist not as familiar with chemistry, this review is an invaluable introduction to concepts and techniques applied in later chapters. The review of astrodynamics is necessary to bridge a gap often present in space-related texts. The chapter describes how a satellite orbits a central body and is influenced by its atmosphere. Depending on its orbital parameters, a satellite may wander through a wide range of altitudes and latitudes which in turn will determine the types of chemical principles that dominate operations. A thorough, accurate representation of the impact of chemical reactions on satellite operations requires knowledge of the spacecraft's position and velocity over time. The inclusion of these chapters makes this a unique text which can stand on its own. The interdisciplinary aspects of space exploration will require that this balanced approach be taken by more texts in the future.

Chapters 3 and 4 discuss the chemical makeup of our solar system. While the reviews of chemistry and astrodynamics provide the basic tools for the text, Chapters 3 and 4 describe the environments in which spacecraft must operate. They present the boundary conditions within which the chemical principles described in the remaining chapters are applied. Particular emphasis is placed upon the near Earth environment due to the focus of our space program on Earth orbiting systems. Yet, recent interplanetary probes and a call for a manned mission to Mars have begun to broaden the frontiers on which this text will be tested.

Chapter 5 discusses the chemical interaction between spacecraft materials and the space environment. The development of data in this section is driven by the building of the international Space Station Freedom. The processes of corrosion, atomic oxygen attack, outgassing, condensation, and radiation degradation on materials used in space are developed. A section on LDEF provides the most current research into spaceborne materials analysis. In Chapter 6, chemical principles of spacecraft power systems (e.g., solar cells, batteries, fuel cells) are covered. This review of present and proposed power systems highlights the challenge to the aerospace designer to develop efficient, cost-effective systems to fulfill future space program needs. Chapter 7 deals with the chemistry of rockets used for propulsion and attitude control. New technologies are being developed which are necessary for such "leading edge of technology" programs as the National Aero-Space Plane (NASP) and the Strategic Defense Initiative (SDI). Chapter 8 examines thermal control system operation and materials. Chemical reactions on the surface of the spacecraft may adversely affect the inner workings of the satellite due to inadequate temperature control. This chapter, for the first time, integrates nomenclature for thermal control *and* protection, including atmospheric reentry. This new approach is necessitated by the advent of reusable spacecraft such as the operational Space

Shuttle and the proposed National Aero-Space Plane. And finally Chapter 9 considers the important chemistry related to humans in space which has received renewed interest lately. The safe transport of people through the harsh space environment has always been a demanding task and will be made more challenging as the duration of manned missions increases.

Chemical Principles Applied to Spacecraft Operations began as a compilation of references and notes used in teaching Chemistry 325, Space Chemistry, at the United States Air Force Academy. The course is designed for those students with an interest in space and satellite operations. It fills a void left by traditional space environment and spacecraft design courses which only superficially cover chemistry. In light of the extended periods of time future spacecraft and space stations must operate, aerospace personnel cannot be ignorant of chemistry and expect their designs to be successful. *Chemical Principles Applied to Spacecraft Operations* is of tremendous use to the beginner because it provides all the basic material in one book. It also is of use to the expert looking to gain additional information on a particular subject since there is an unprecedented coverage of concepts and a complete set of references is included at the end of each chapter. Discussion questions and workout problems are at the end of each chapter to help organize key aspects of each chapter. Those readers possessing a good understanding of chemistry and astrodynamics can skip the review chapters, Chapters 1 and 2, and move right into the heart of the book. Few individuals are well-versed in both fields, which was one of the driving factors behind the creation of this book. Yet, it will become increasingly important for space chemist to be comfortable with astrodynamics while aerospace engineers will have to understand chemical principles affecting spacecraft operations. Whether used as a course text or aerospace design aid reference, *Chemical Principles Applied to Spacecraft Operations* provides the reader with necessary and vital information pertaining to chemistry's role in spacecraft design and operations.

The experience of former Soviet Union (FSU) space stations to the application and importance of space chemistry is an enlightening preface to the examination of this topic. Nicholas L. Johnson, an advisory scientist for Kaman Sciences Corporation in Colorado Springs, Colorado, has written extensively on the FSU space program. His presentation, "Contamination Experience and Countermeasures on Soviet Space Stations," provides an overview of concerns that span a wide variety of space chemistry concepts, while also providing us with valuable lessons learned for use on Space Station Freedom [1].

To date FSU has amassed more than 18 man-years of space station experience. Figure I.1 shows the development of the stable of space stations used over the last 20 years. Figure I.2 is the present workhorse of the FSU manned space program, Mir.

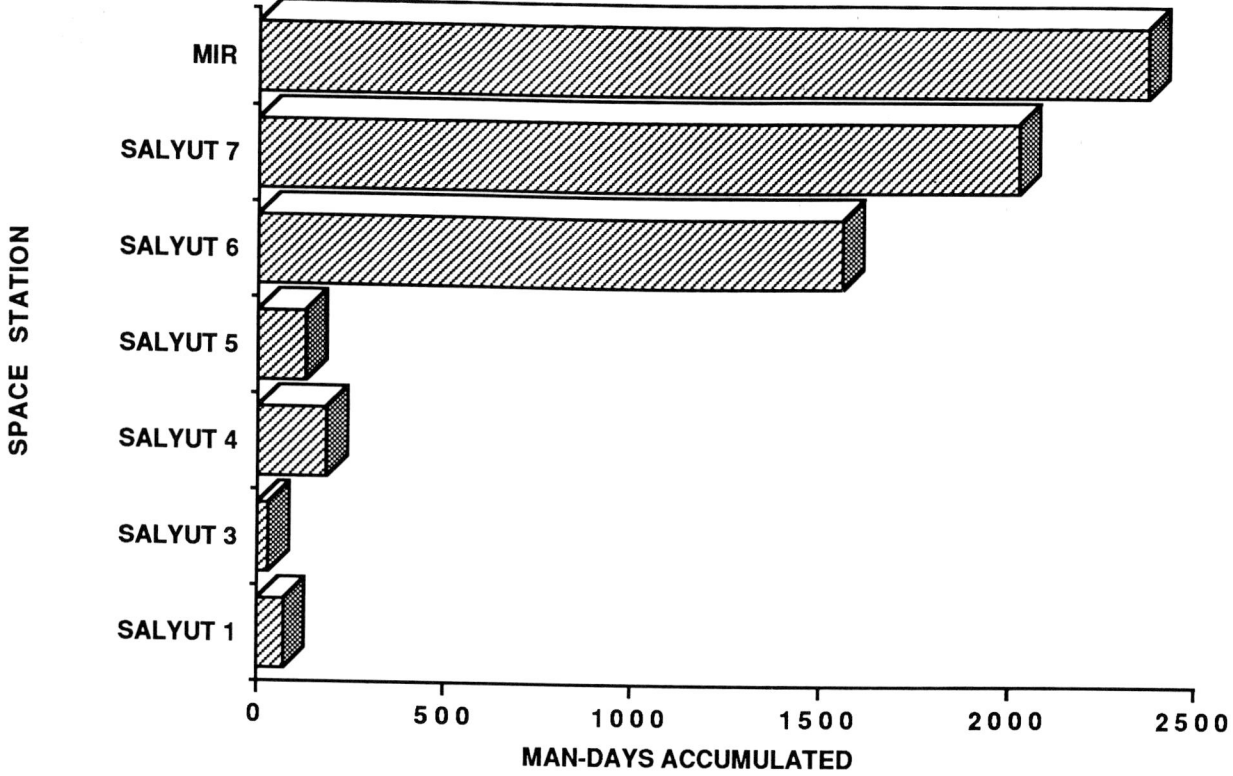

Figure I.1 A comparison of man-days accumulated on various former Soviet Union space stations.

Preface   xiii

Figure I.2 Former Soviet Union space station Mir. *Compliments of Teledyne Brown Engineering.*

Johnson describes some potential causes of exterior contamination. These include the following:

1. *Engine plumes.* Chemicals from engine exhaust produced during rendezvous, docking, orbital adjustment, and attitude control may react with the surfaces of solar panels, windows, and sensors. This contamination may be debilitating if not counteracted by changes to design and operations. These may include altering solar panel orientations, use of spring-assisted undocking, shielding of windows and sensors during rocket firings, and placing engines in remote locations.

2. *Propellant leakages and purges.* Simply the presence of propellant, without a rocket firing, may produce a potential for contamination via leakage and spillage. The introduction of raw fuel may disrupt spacecraft operations and make it hazardous for manned operations.

3. *Micrometeoroids and artificial space debris.* The particulate hazard while in orbit poses a direct and an indirect threat to a spacecraft. The direct impact by micrometeoroids or space debris may crater windows, perforate thermal insulation, splinter solar cells, and generally cause surface erosion. These events will indirectly allow some chemical reactions to have a greater effect, thus shielding or replacement of components of the space station is necessary.

4. *Waste removal and ejection.* The removal and ejection of waste from the space station must be done carefully. Storage of waste within the living quarters may cause problems for its inhabitants unless it is chemically treated. Ejectors of waste may produce contamination of the surface of the space station or actually present a collision hazard at a later time.

5. *Radiation.* Radiation may cause exposed areas to change physical properties leading to increased absorptivity, loss of transparency for windows, fatigue cracks, and erosion of metallic surfaces. Coatings resilient to this phenomenon may be breached by impacts from meteoroids and space debris rendering the coating ineffective.

6. *Extravehicular activity (EVA).* EVAs may produce hazards such as inadvertent physical contact with sensitive surfaces, release of cabin atmosphere constituents, and release of maneuvering unit effluents, all of which may contribute to adverse chemical reactions on the space station's surface.

7. *Near station environment.* Near station concerns include the generation of an altered environment near the station which remains with it for the duration of its mission. Substances that stay near may include space station outgassing, waste ejections, engine exhaust, and EVA contaminants.

The Soviets have now recognized the need to control adverse chemical effects in the vicinity of their space stations and they are actively evaluating a variety of countermeasures to mitigate their impact on operations. Spacecraft designers in the United States can learn a considerable amount from the Soviet space station experience.

Thus far, only the negative aspects of space chemistry have been highlighted, yet chemistry plays a role in the propulsion, power, and material selection for all spacecraft. Improvements in these areas will allow us to expand our horizons in aerospace activities into the 21st century. Space missions planned for the 1990s highlight this progress and accentuate the need for understanding the chemical principles which affect spacecraft operations.

In 1957, two space launches took place, but within 7 years the annual international launch rate had grown to over 110. For the last 25 years an average of over 115 space launches have taken place each year for a total of over 3,300 excursions into the harsh environment of space. This robust history of space exploration and exploitation has been made possible by advances in materials, propulsion, power systems, and thermal control. All the planets of our solar system (except Pluto) and seven comets have been visited while our Moon has been the subject of over 80 scientific missions.

The advancement of chemical sciences will propel the space age into the 21st century. Interplanetary missions will continue with encounters with

- Saturn: Cassini (ESA)
- Jupiter: Galileo
- Mars: Mars Observer (U.S.) and Mars '94 (U.S.S.R.)
- Venus: Galileo and Magellan
- Comet Kopff: CRAF
- The Sun: Ulysses

These programs will shed light on the chemistry of the solar system which can help to explain processes that we must deal with on and around Earth. Due to the large distances these spacecraft must travel, their designs will tax propulsion, power, material, and thermal control technologies.

The interaction of the solar system environment with fragile Earth is another major thrust for the decade of the 90s. Missions are being designed to monitor and quantify global environmental changes. Several of these have already been deployed. Specific areas being studied are

- Oceanography: Topex/Poseidon and NASA Scatterometer (NSCAT)
- Earth's magnetic field interactions: Global Geospace Science (GGS), Collaborative Solar Terrestrial Research Satellite (COSTR), and Combined Release and Radiation Effects Satellite (CRRES)
- Geodesy: Lageos II, Geostar

Other satellites such as the Hubble Space Telescope, Gamma Ray Observatory (GRO), Astro-02/BBXRT-02, Extreme Ultraviolet Explorer (EUVE), and the Advanced X-Ray Astrophysics Facility (AXAF) will look outward to detect energy headed toward Earth in a variety of wavelengths.

Future space missions will include a variety or orbiting laboratories built to study the Earth environment. Expansion of the Soviet Mir station and construction of the international Space Station Freedom will provide general venues for space research. A variety of specialized systems will also be deployed to monitor the environment (Earth Observation System (EOS) and Atmospheric Laboratory for Applications and Science (ATLAS)). The Upper Atmosphere Research Satellite (UARS) was deployed by the Space Shuttle in 1991 to start the collection of data on Earth's atmosphere in earnest. These satellite systems will not only provide data to better quantify the space environment and its effect on spacecraft but will also provide unique challenges for the designer. The complex sensing devices necessary for these missions must be maintained within stringent operational limits of temperature, temperature gradients, etc.

Other deployable laboratories will examine the effects of microgravity on humans, vegetation and manufacturing processes (Spacelab, Spacelab Life Sciences (SLS), and Eureca). Results from these programs will assist in the future construction and manning of space systems.

The advent of the reusable Space Shuttle ushered in a new era of spacecraft. The use of new materials was key to this innovation and will also be an integral aspect of the development of a new breed of spaceplane. A revolution of technology will be required in a variety of areas (e.g., materials, power, and propulsion) for the proposed horizontal takeoff and landing spaceplanes to become reality. This text lays the foundation for the research which will prove fundamental to most of the critical areas addressed.

The space projects of the 1990s will help mankind reach out to the cosmos with orbiting telescopes and interplanetary probes. At the same time, a continued manned presence on Space Station Freedom will be developed. The Soviet manned programs will continue over the next decade. We will look inward from these outposts to examine global environmental concerns in tandem with satellites such as UARS and EOS. The chemical principles used by all of these space systems will be examined in this text.

References:

1. Johnson, N. L., "Contamination Experience and Countermeasures on Soviet Space Stations," Presented at AIAA Spacecraft/Space Station Contamination Seminar, 10 July 1988.
2. "Space in the 1990s," *Spaceflight*, Vol. 32, January 1990.

Chapter 1
Review of General Chemistry

An understanding of the chemical environment of space and its effect upon spacecraft design and operation cannot be achieved without first knowing the fundamentals of chemistry. The absence of such a foundation will only lead to frustration and misunderstanding by the engineer or scientist whose task is to design and manufacture systems capable of functioning in this final frontier. With this idea in mind, a review of the chemical principles applicable to space operations is appropriate. Many of the topics discussed are covered in any freshman level college chemistry text, but not to the same extent. For additional information, consult the reference at the end of the chapter.

Atomic Structure

As we know it today, the universe consists of approximately 109 natural and man-made elements, each possessing its own unique physical and chemical properties. The differences and similarities between these elements can be explained by examining their atomic structure. Atoms are the smallest, single particles possessing all the characteristics of an element. They combine in numerous ways to form the millions of compounds which make up our universe.

A simple model used to depict the atom, as shown in Figure 1.1, consists of a nucleus of protons and neutrons surrounded by concentric spheres or shells containing electrons. By comparison, the mass of a proton or neutron is three orders of magnitude greater than that of an electron. The nucleus possesses a positive electrostatic charge due to its protons that attracts the negatively charged electrons. The electrical force between two charges is described by Coulomb's law as follows:

$$F = kQ_1Q_2/r^2 \qquad (1.1)$$

where

F = force in newtons
k = Coulomb's constant
$= \dfrac{1}{4\pi\epsilon_0} = 8.9875 \times 10^9 \dfrac{N \cdot m^2}{C^2}$
Q_1, Q_2 = charge in coulombs
r = distance between charges in meters.

In S. I. units charge is measured in coulombs (C) with the proton and electron possessing $+1.602 \times 10^{-19}$ C and -1.602×10^{-19} C of charge, respectively. In a neutral atom, the number of protons equals the number of electrons.

Like the planets around the Sun, electrons move about the nucleus within mathematically defined regions. Atomic orbitals have varying shapes represented by complex mathematical functions which describe the probability of the electron's location. The energies possessed by the electrons at specific locations are dependent upon the numerous electrostatic interactions between themselves and with the nucleus. Heisenberg's Uncertainty Principle stipulates the position and momentum of an electron cannot be precisely known simultaneously with any degree of certainty. However, atomic orbitals describe the areas of highest probability for locating an electron, which is the most precisely an electron's position can be determined. Atomic orbitals can hold a maximum of two electrons with the spins of the electrons being in opposite directions. Lowest energy orbitals (those closest to the nucleus) are filled with electrons first. Due to electrostatic repulsion, electrons fill unoccupied orbitals of the same energy before pairing with another electron of opposite spin.

Periodic Table

Each of the 109 elements possesses a distinct atomic structure which gives it its unique character. Discovery of these elements was accomplished over the course of history. Metals such as copper, silver, and gold trace their discovery back hundreds of years while many of the actinide series elements are the products of more recent endeavors.

Arrangement of the elements based upon similar physical and/or chemical properties is a fairly recent accomplishment for chemists. The greatest recognition for development of our current periodic table is given to the Russian chemist Demitri Mendeleev. Mendeleev published his periodic table of the known elements in 1869. He organized the elements based upon their chemical properties and left gaps in the table for undiscovered elements. Grouping of the elements in columns or "groups" and rows or "periods" enabled

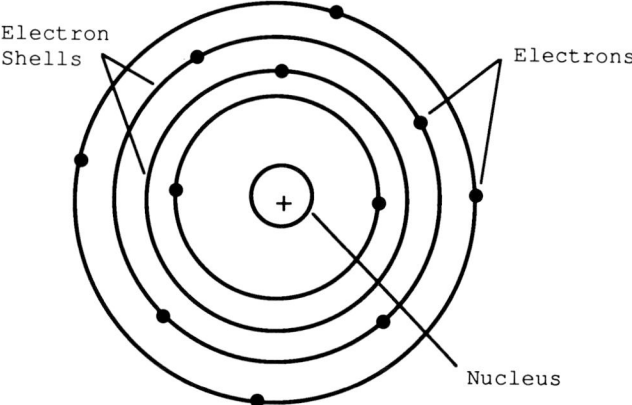

Figure 1.1 Shell model of atom showing the nucleus surrounded by concentric shells containing electrons.

Mendeleev and other chemists to recognize similarities and trends never before suspected. Arrangement of today's periodic table, Figure 1.2, is in ascending order of an element's atomic number or the number of protons in its nucleus.

Since all naturally occurring elements have zero charge, the number of electrons must equal the atomic number (number of protons). An atom's approximate total mass is the sum of the masses of its protons, neutrons, and electrons. Due to the relatively small mass of the electrons in comparison to the nucleus, the difference between atomic mass and atomic number is generally considered to be the number of neutrons in the nucleus. By convention, one atomic mass unit is equal to 1/12 the mass of ^{12}C. For example, the element aluminum (Al) has an atomic number of 13 and an average atomic mass of 26.98 amu. Aluminum atoms, therefore, contain an average of 13 protons, 13 electrons, and 14 neutrons. This relationship can be expressed as

Atomic Mass $\cong$ Atomic Number + Number of Neutrons. (1.2)

It is imperative to mention at this point that the atomic mass shown in the periodic table is the weighted average of all of the isotopes of a given element. Isotopes of an element have the same atomic number, but differ by the number of neutrons. The various atomic masses for a particular element are each multiplied by their relative abundance in nature with the sum yielding an average atomic mass; that is,

$$\text{Average Atomic Mass} = \sum_{i=1}^{N} P_i M_i \quad (1.3)$$

where

P_i = relative abundance (note: $\sum P_i = 1.00$)
M_i = atomic mass
N = number of isotopes.

The usefulness of the periodic table comes not only from its ordering of elements by ascending atomic number, but also by its elemental groupings. Several divisions exist which are useful in classifying elemental properties. The bold, stairstep line on the table's right side separates elements to the left of the line possessing metallic properties from those to the right with nonmetallic properties. Those elements directly adjacent to the line are termed metalloids or semimetals because they exhibit the properties of both classes. Semiconductors like silicon and germanium are included in this category.

Further division is evident from the horizontal periods and vertical groups. Specific names have been attached to several of the groups due to the similar nature of their elements:

Group IA—Alkali Metals
Group IIA—Alkaline Earth Metals
Group VIA—Chalcogenides
Group VIIA—Halogens
Group VIIIA—Noble Gases

The Mole

Due to the atom's extremely small size, it cannot be used as a unit of measure in any real situation. Instead a large group of them form the basic unit called the mole. A mole of atoms or molecules is the number of particles contained in the gram atomic mass or gram molecular mass of the substance. For instance, 12.011 grams of carbon contain 1 mole of atoms or 6.023×10^{23} atoms (Avogadro's number). This relationship can be represented by the following unit factors:

$$\frac{12.011 \text{ grams of carbon}}{1 \text{ mole of carbon}} = \frac{12.011 \text{ grams of carbon}}{6.023 \times 10^{23} \text{ carbon atoms}} \quad (1.4)$$

Conversions between mass and moles are accomplished using unit factors. For example, 30.0 grams of carbon is 2.50 moles since

$$30.0 \text{ grams of carbon} \times \frac{1 \text{ mole of carbon}}{12.011 \text{ grams of carbon}} = 2.50 \text{ moles} \quad (1.5)$$

and

$$3.00 \text{ moles of aluminium} \times \frac{26.98 \text{ grams of aluminium}}{1 \text{ mole of aluminium}} = 80.9 \text{ grams} \, . \quad (1.6)$$

Atomic Behavior

Atoms in their unexcited or ground state have their electrons distributed throughout the shells in an arrangement of lowest energy for the atom. Elements in their natural states possess ground state electron configurations. Excitation of

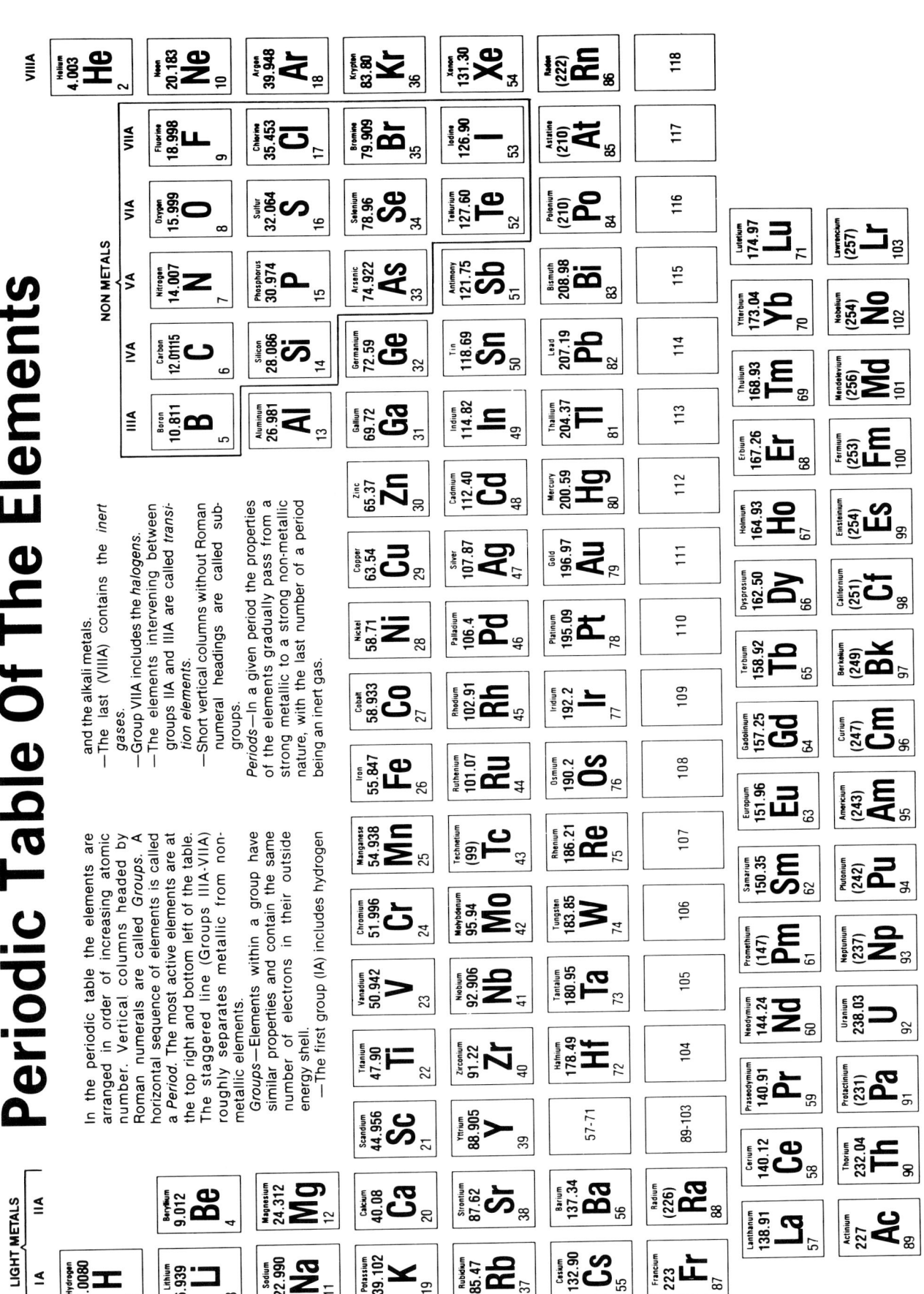

Figure 1.2 Periodic table of the elements [1].

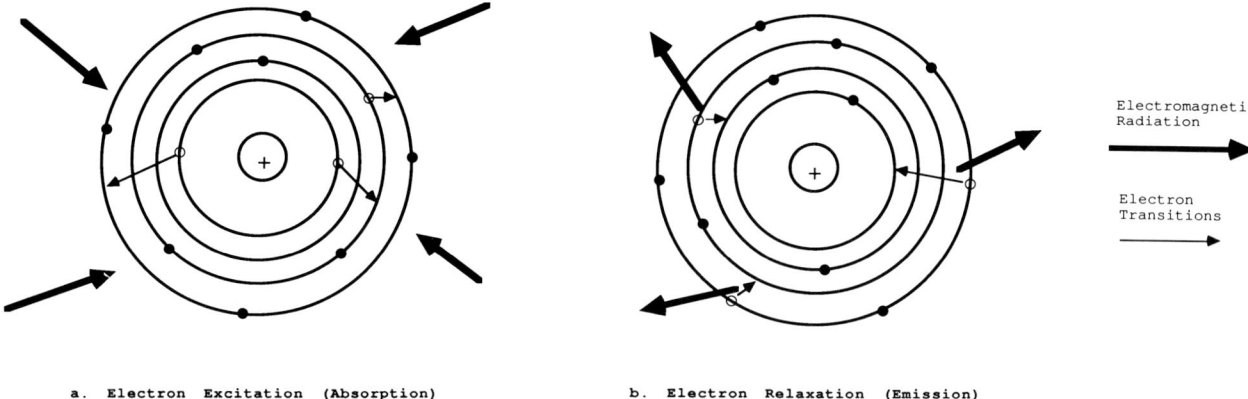

Figure 1.3 Electron transitions to higher energy levels through the absorption of electromagnetic radiation. This is followed by electron relaxation to lower energy levels through the emission of electromagnetic radiation.

atoms by exposure to heat or light, Figure 1.3a, causes the electrons to transition from their ground state shell to higher energy shells. If the energy of excitation is great enough, electrons are completely removed and the atom is said to be ionized. Outer shell electrons (those most distant from the nucleus) will be removed first because they are at the highest energy level. Short of actual removal, the electrons will lose energy through emission, Figure 1.3b, and relax back down to lower energy shells.

These electron transitions, which occur through the absorption and emission of energy, give distinct spectral patterns. The fundamental equation relating the amount of energy absorbed or emitted in the form of photons and the frequency of light observed in a spectrum is

$$E = h\nu = hc/\lambda \quad \text{(Planck's equation)} \quad (1.7)$$

where

E = transition energy
h = Planck's constant
c = speed of light
ν = frequency
λ = wavelength

Energy separation between the shells varies not only within an atom, but also between atoms of different elements. Electron transitions produce energy at wavelengths characteristic of a particular element or compound. As a result, the emission and absorption spectrum are unique for each element and can be used to identify an unknown compound. Spectroscopy is the technique used to probe the universe in search of its elemental composition. Spectral studies of the Sun, for example, showed it to consist mainly of H and He. Various regions of the electromagnetic spectrum are utilized due to the great number of possible transitions.

Chemical Bonding

Chemical bonding is the attractive force holding atoms together in a compound. It involves the transfer or sharing of one or more electrons located in the outer or valence shells of atoms. The three major classes of chemical bonding are ionic, covalent, and metallic. Ionic and covalent bonding are discussed in this section, while metallic bonding will be covered in a later section.

Ionic Bonding

Ionic bonds are created by the transfer of one or more electrons between two atoms. The electrostatically charged ions attract one another to form the bond. In the formation of sodium chloride, Na transfers its lone valence shell electron to Cl to form ionically bonded NaCl. This reaction can be illustrated as follows by the use of Lewis dot structures to show the valence or outer shell electrons involved in the transfer.

$$\text{Na} \cdot + \cdot \ddot{\underset{..}{\text{Cl}}} : \rightarrow [\text{Na}]^+ [:\ddot{\underset{..}{\text{Cl}}}:]^- \quad (1.8)$$

Electron transfer and electron sharing in a bond are a function of an atom's ability to attract electrons to itself. This property is called electronegativity and the relative values are shown in Figure 1.4. Elements with the greatest electronegativity are in the upper right corner of the periodic table while those with the smallest values are in the lower left corner. Fluorine has the highest electronegativity and will attract electrons to itself when bonded with any other atom. Attainment of a full complement of outer shell electrons, as in the Group VIII Noble Gases, is the most stable configuration for an atom. Therefore, elements from Groups IA, IIA, and IIIA readily donate electrons to form ionic bonds with the electrons accepting elements of Groups VA,

Review of General Chemistry

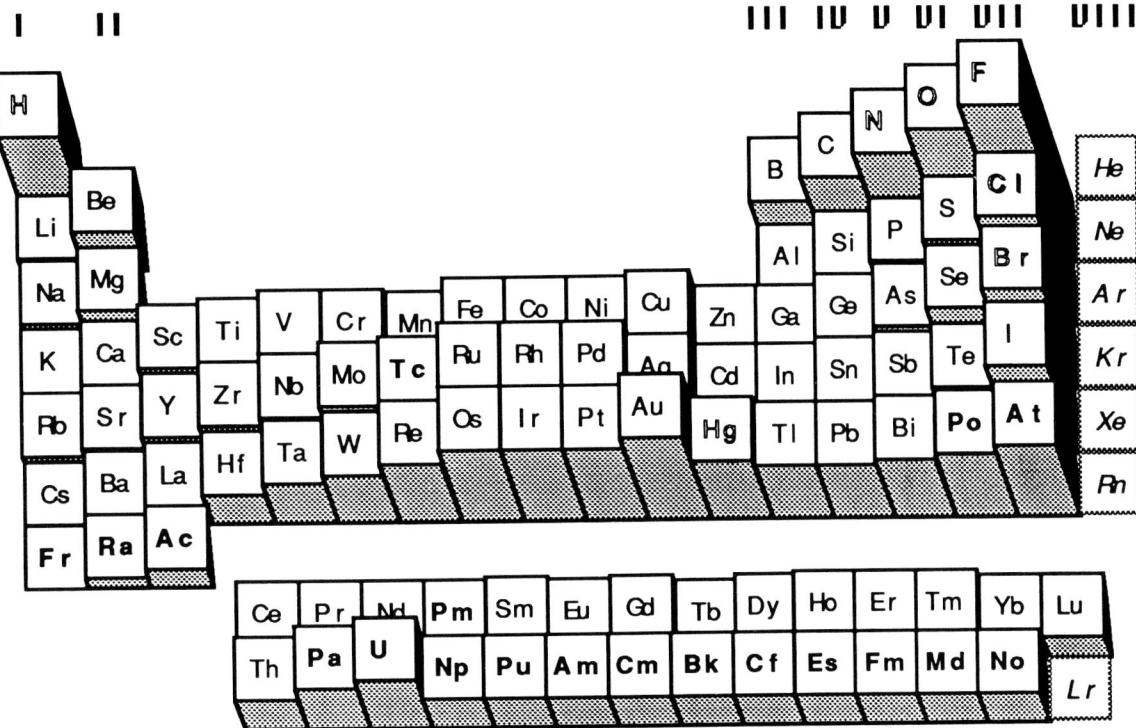

Figure 1.4 Relative electronegativities of the elements within the periodic table. The height of an element correlates to its relative electronegativity.

VIA, and VIIA. Examples of ionic compounds are given in Table 1.1.

Individual ionic molecules, however, do not exist in ionic compounds due to the isotropic nature of the electrostatic forces. For example, a single NaCl molecule could not be removed and isolated from the remaining ions. Instead, the ions are arranged in a three dimensional lattice with each ion attracting and repelling its surrounding neighbors. Figure 1.5 shows the lattice structure of NaCl.

The strong electrostatic forces between the ions are a function of each ion's size and charge. According to Coulomb's law, the force increases with decreasing ionic size and increasing ionic charge. These strong forces give ionic compounds high melting points, as is seen in Table 1.1. The presence of ions in the molten and aqueous phases also makes them excellent electrical conductors.

Table 1.1 Melting Points of Selected Ionic Compounds

Compound	Melting Point (°C)
LiBr	547
$MgCl_2$	708
CaO	2,580
Al_2O_3	2,045

Covalent Bonding

Incomplete electron transfer between two nonmetallic elements is the basis for covalent bonding. A covalent bond is the sharing of one or more pairs of electrons between two atoms. Neither atom is able to completely remove electrons from its partner so they must share in order to achieve a filled outer shell. Molecular hydrogen, oxygen, and nitrogen are three examples of compounds containing covalent bonds which can be represented as

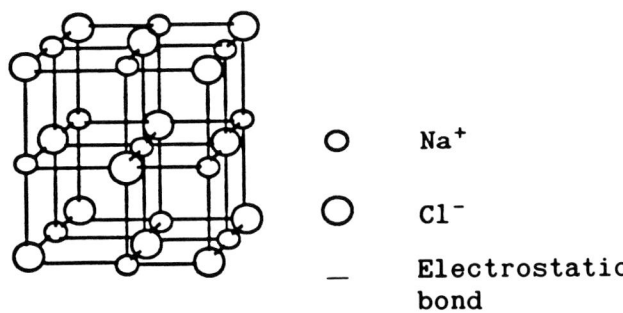

Figure 1.5 Sodium chloride lattice structure showing the three dimensional arrangement of sodium and chloride ions.

Again, Lewis dot structures are used to depict the shared and unshared outer shell electrons. Hydrogen possesses a single covalent bond from its sharing of one electron pair while oxygen and nitrogen have double covalent and triple covalent bonds from the sharing of two and three electron pairs, respectively.

Homonuclear molecules, such as those above, have equal sharing of electrons between the two atoms because there is no difference in electronegativity. These molecules are termed nonpolar because there is no separation of charge. Separation of charge does occur with heteronuclear molecules where bonding electrons are unevenly shared. An atom having a greater attraction or electronegativity for the bonding electrons will distort the bond, causing the creation of a dipole within the molecule. Figure 1.6 shows the differences in electron density in nonpolar and polar bonds. The homonuclear H_2 molecule has a symmetric electron density about the two hydrogen nuclei. On the other hand, F, with a higher electronegativity than H, pulls electron density towards itself. This distortion creates a partial positive charge ($\delta+$) on the H and a partial negative charge ($\delta-$) on the F. This is a polar, covalent bond and its importance will be discussed in a later section.

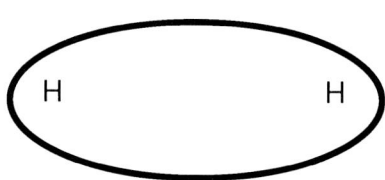

Nonpolar

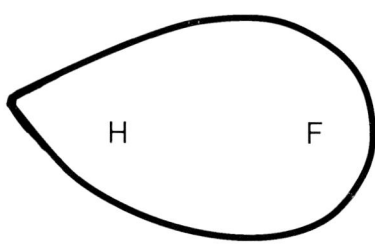

Polar

Figure 1.6 Illustrating the symmetric and asymmetric electron density distributions in the nonpolar hydrogen and polar hydrogen fluoride molecules.

Phases of Matter

To this point the focus has been upon the individual atoms and how they bond to form molecules. This, however, is only part of the picture. Now, consideration is given as to how the molecules interact with one another to form identifiable compounds that can be observed. A distinction is therefore made between intramolecular forces, which were just discussed, and intermolecular forces, which help determine the phase of matter a compound will assume under a given set of conditions.

Gases

Observations made by noted scientists like Torricelli, Boyle, Charles, and Dalton showed that gases

1. are compressible,
2. are able to exert pressure,
3. expand without limits,
4. diffuse throughout one another, and
5. can be described by their temperature, pressure, volume, and number of moles.

Several fundamental laws governing the behavior of gases under various conditions are given as follows:

Boyle's law: $PV = $ constant, $\quad P_1V_1 = P_2V_2 = $ constant
$(n, T = $ constant$)$ \hfill (1.9)

Charles's law: $V/T = $ constant, $\quad \dfrac{V_1}{T_1} = \dfrac{V_2}{T_2} = $ constant

$(n, P = $ constant$)$ \hfill (1.10)

Ideal gas law: $PV = nRT$ \hfill (1.11)

Dalton's law: $P_{TOT} = P_{gas\ A} + P_{gas\ B} + P_{gas\ C} + \ldots$
\hfill (1.12)

where

$P = $ pressure
$V = $ volume
$n = $ number of moles
$R = $ universal gas constant (8.314 J/mole · K)
$T = $ temperature (K)

Boyle first showed the inverse mathematical relationship between pressure and volume. The volume of a given mass of gas at a particular temperature is inversely proportional to the applied pressure. Picture the reaction of an inflated balloon inside an isothermal bell jar as the air is evacuated. With a smaller force exerted on the external surface of the balloon, the gas molecules inside are able to expand, thereby increasing the volume. By returning the bell jar to atmospheric pressure the balloon resumes its original size.

Variation of volume as a function of absolute temperature at constant pressure is explained by Charles's law. The directly proportional relationship between V and T is practically applied in the hot air balloon. As the gases inside the

Review of General Chemistry

balloon are heated, their kinetic energy increases, causing more frequent, forceful collisions of the air molecules with the balloon's interior surface. The result is the expansion of the gas against the constant pressure of the atmosphere.

Both Boyle's and Charles's laws make two key assumptions about the behavior of gas molecules. The first is the total volume occupied by the individual gas particles is essentially zero when compared to the volume of space the diffuse gas itself occupies. The second assumption requires no force of interaction between the gas molecules. Due to their continual motion and wide separation, gas molecules are unable to exert electrostatic forces upon each other, and they therefore exhibit simple, random motion. Under standard conditions both these assumptions are very good. The volume occupied by a gas particle is negligible in comparison to the total volume filled by the gas. Since their volume is so small, they are separated by great distances and don't feel each other's forces of attraction and repulsion. Therefore, gases behave in an ideal fashion and their behavior can be described by the ideal gas law. Algebraic manipulation of the ideal gas law is used to solve for any of the unknown parameters under a given set of conditions. Deviations from ideal behavior do occur at low temperatures or high pressures. Under these conditions the original assumptions made for this model are no longer valid and require modifications be made to the ideal gas equation. Numerous nonideal gas equations do exist, but they are not covered in this text.

Mixtures of gases behave no differently than their pure counterparts because of ideality. The total number of moles in a mixture is the sum of the moles of the individual gases.

$$n_{TOT} = \sum n_i \ . \quad (1.13)$$

Substitution of this into the ideal gas law yields

$$P_{TOT} = \sum n_i \ \{RT/V_{TOT}\} \quad (1.14)$$

or by simplifying the right-hand side of the equation

$$P_{TOT} = P_1 + P_2 + P_3 + \ldots = \sum P_i \quad (1.15)$$
$$(V, T = \text{constant}) \ .$$

This form of the equation is known as Dalton's law of partial pressures, which equates total pressure of a mixture to the sum of the partial pressures of the individual gases. Dalton's law will become quite useful when life support systems for manned spacecraft are discussed.

Intermolecular Forces

Describing the behavior of liquids and solids is much more difficult in comparison to gases due to the presence of strong intermolecular forces in these condensed phases. As a result liquids and solids can be looked at only qualitatively under different conditions. But first, the role of intermolecular forces in holding the individual molecules and atoms together must be explained.

It is important to understand the distinction between intra- and inter-molecular forces. Intramolecular forces are those like covalent bonding which occur between atoms within a molecule, while intermolecular forces are those between individual atoms and molecules in a substance. A comparison of relative strengths shows intramolecular forces to be one to two orders of magnitude greater than intermolecular.

Some confusion does exist concerning which interactions constitute intermolecular forces. Many chemistry text authors consider the ion-ion interactions of ionic bonding an intermolecular force. However, in this text intermolecular forces are limited to the two types of interactions happening among covalently bonded molecules.

Polar molecules possess partial positive and negative electrostatic charges that cause intermolecular repulsion and attraction. Figure 1.7 represents the individual polar molecules as ellipses. Lines show the various attractions and repulsions between the molecules.

As mentioned previously, these forces are weak compared to ionic and covalent bonding. However, one special case of permanent dipole-dipole interaction, hydrogen bonding, results in relatively strong interactions between the molecules. Hydrogen bonding is possible in compounds containing hydrogen atoms bonded to fluorine, oxygen, or nitrogen. The high electronegativity of these three elements creates a strong dipole within the molecule. The high partial negative charge on fluorine, oxygen, and nitrogen causes the formation of strong bonds with positively charged hydrogen atoms on adjacent molecules (Figure 1.8). Compounds containing hydrogen bonds exhibit higher values for such physical properties as melting and boiling points when compared to simple dipole-dipole interacting molecules.

The second type of intermolecular force is also present in molecules in the condensed phases, but is much weaker than dipole-dipole. Called London or dispersion forces, they arise from temporary dipoles created by instantaneous distortions of electron symmetry around the molecule by adjacent molecules. London forces are the only intermolecular force between nonpolar molecules such as CO_2, H_2, O_2, N_2, and CCl_4. On the average, with time, these molecules are nonpolar and do not have a separation of charge. At an instant in time, temporary dipoles do exist which result in attraction and repulsion among the molecules. These forces operate over a very short range, but are especially important when condensation takes place.

Properties of Liquids

Viscosity is a liquid's resistance to flow. Molecular size, shape, and interactions, as well as temperature and pressure, determine a liquid's flow characteristics.

Surface tension is a measure of the inward forces which must be overcome before a liquid can expand. Water drop-

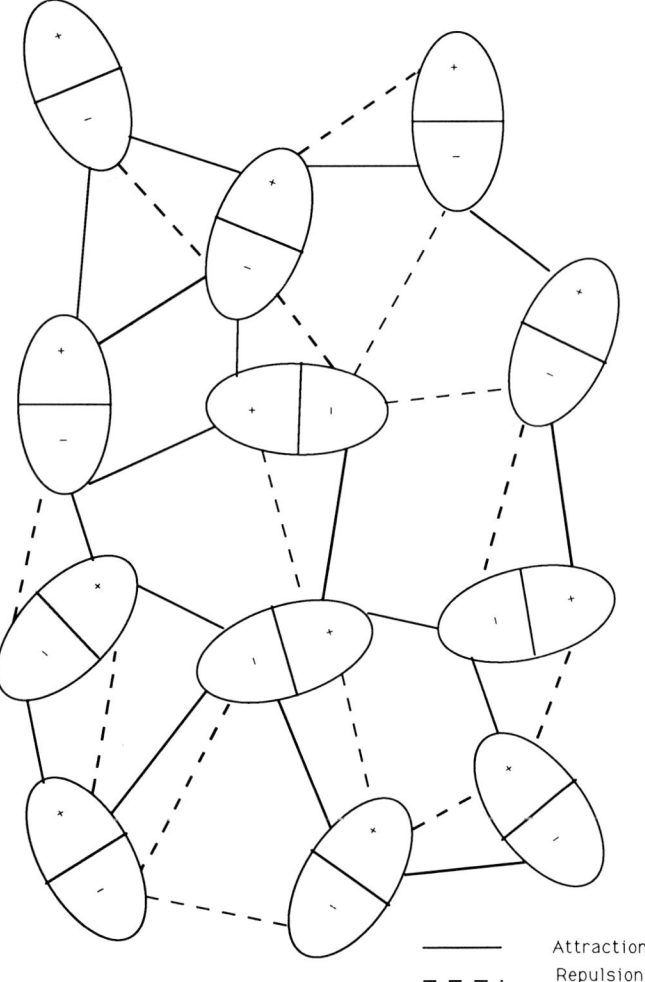

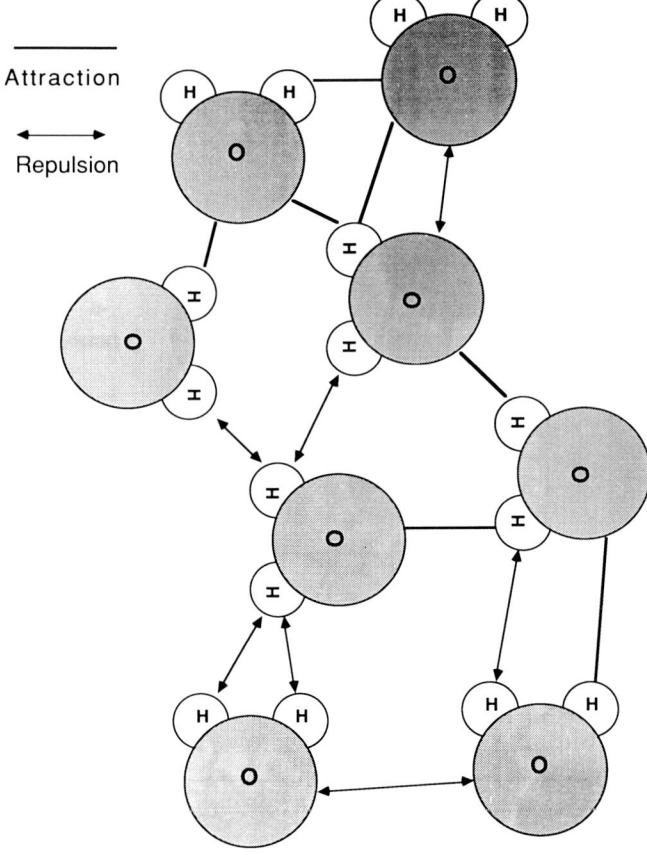

Figure 1.8 A special type of dipole-dipole interaction, hydrogen bonding, occurring between water molecules.

Figure 1.7 Attractive and repulsive forces occurring between polar molecules.

lets take their spherical shape because of attractive forces between water molecules at the surface and interior of the drop. The process of molecules leaving the surface of a liquid and entering the gas phase is commonly known as evaporation. A small portion of the liquid molecules have enough energy to escape, with that portion increasing with temperature. If the container is enclosed, gas molecules also return to the liquid phase (condensation), and thus establish a dynamic equilibrium between the two phases. The partial pressure of the gas above its own liquid at equilibrium is its vapor pressure. Liquids with small intermolecular forces are very volatile and have high vapor pressures (e.g., gasoline). These same liquids also have low boiling points. Boiling differs from evaporation due to the formation of the vapor below the liquid surface and not just at the surface. Vapor pressure equals the surrounding applied pressure at the boiling point. Vapor pressure varies inversely with applied pressure, so reducing or raising applied pressure lowers or raises the boiling point. A substance's critical temperature is that maximum temperature, regardless of pressure, above which the liquid and gas cannot exist as separate phases.

The amount of heat required to raise the temperature of 1 gram mass of liquid by 1 degree Celsius is the specific heat (J/g°C). Upon reaching the boiling point, the additional heat that must be supplied for a phase change is the molar heat of vaporization (ΔH_{vap}). The importance of specific heat and ΔH_{vap} is apparent when selecting fluids for coolants. A liquid's capacity to remove heat from a system is a function of these two properties.

Solids

Solids are the most dense and have the greatest amount of particle interaction of the three phases of matter. Two general classes of solids are amorphous and crystalline. Amorphous solids have no long-range ordered structure, but rather a random arrangement of overlapping and interlocking molecules. Differences in strength of the intermolecular forces due to the irregular molecular arrangement cause these solids to soften and melt over a wide temperature range and not at a single melting point. The additional heat that must be supplied at the melting point is expressed by the

Review of General Chemistry

Table 1.2 General Characteristics of the Four Classes of Solids Based on Their Type of Bonding [1]

	Class of Solids			
	MOLECULAR	COVALENT	IONIC	METALLIC
Basic particle	Molecules (or atoms)	Atoms	Anions, cations	Metal ions in an electron cloud
Strongest interparticle forces	London, dipole-dipole, and/or hydrogen bonds	Covalent bonds	Electrostatic	Metallic bonds (attraction between cations and electrons)
Physical properties	Soft; poor heat and electrical conductors; low melting points (-272 to $400°C$)	Very hard; poor heat and electrical conductors; high melting points ($1,200$ to $4,000°C$)	Hard; brittle; poor heat and electrical conductors; high melting points (600 to $3,000°C$)	Soft to very hard; good heat and electrical conductors; wide range of melting points (-39 to $3,400°C$)
Examples	P_4, S_8, I_2	C (diamond), SiO_2 (quartz)	NaCl, $CaBr_2$, K_2SO_4, (typical salts)	Li, K, Ca, Cu, Cr, Ni (metals)

molar heat of fusion (ΔH_{fus}). Polymers and glasses are two examples of this class of compounds.

In contrast to amorphous solids, crystalline solids possess long-range ordering in their molecular structures resulting in sharp melting points. Based upon the types of particles and their interaction, four types of crystalline solids exist. Table 1.2 outlines the various characteristics and properties of these solids.

Bonding in metals is different from ionic and covalent bonding in that the electrical attraction is between positively charged metal ions and a delocalized cloud of electrons. Overlap of atomic orbitals of adjacent atoms in the crystal creates continuous "bands" through which electrons, under an applied potential, can move among the atoms in the crystal. Movement is dependent upon vacancies in the bands formed by orbitals having less than their maximum number of electrons. The band highest in energy which is completely full is called the valence band, while the unfilled conduction band is where electron movement occurs. Metals conduct electricity because they possess partially filled conduction bands or their valence and conduction bands overlap, allowing easy electron transfer from one to the other. Insulators have a wide energy or band gap between their two bands, preventing electron flow. The semiconductors like Si, Ge, and As have small band gaps that electrons can "jump across" upon the addition of heat. Figure 1.9 shows the relative band gaps of these classes of solids.

Other metallic properties such as thermal conductivity, luster, and workability can also be explained using band theory. Addition of heat excites electrons to higher energy levels within the conduction band. These electrons return to their original state as heat is dissipated. A metal's luster is caused by the same electrons absorbing light and again becoming excited. However, the radiated energy this time takes the form of photons and not heat. Metal workability, such as malleability and ductility, comes from the ability of the metal ions to distort within the electron cloud. As bonds are broken, new ones are formed and the metal's integrity is maintained.

Phase Diagrams

Changes from one phase to another are a function of a system's temperature and pressure. Phase diagrams, like the one shown for water in Figure 1.10, show the phase or phases present as temperature and pressure are varied. Points on a solid line, like B, represent an equilibrium of the solid and gas phases at $P = 2.1$ torr and $T = -10°C$. Simultaneous presence of all three phases is at the triple point (A) where all phase boundary lines coincide. Water's critical point (C) is the highest temperature a liquid and gas can remain as separate phases. At temperatures above point C the liquid and gas phases coexist as a single phase regardless of the applied pressure. The significance of critical temperatures and pressures will be discussed in a later section on spacecraft fuel storage.

Chemical Reactions

To this point, elements and compounds have been addressed without regard to their chemical reactivity. However, chemical reactions are the basis for understanding many of the processes taking place in a spacecraft. Take, for example, the orbital maneuvering engines aboard the Space Shuttle, which are fueled by a mixture of monomethyl hydrazine and nitrogen tetroxide. It is the spontaneous ignitability of these two chemicals upon mixing that produces the force necessary to change the Space Shuttle's orbit.

Understanding what occurs in a chemical reaction is not always an easy process. In general, a reaction takes place when one or more reactants undergo a chemical change to

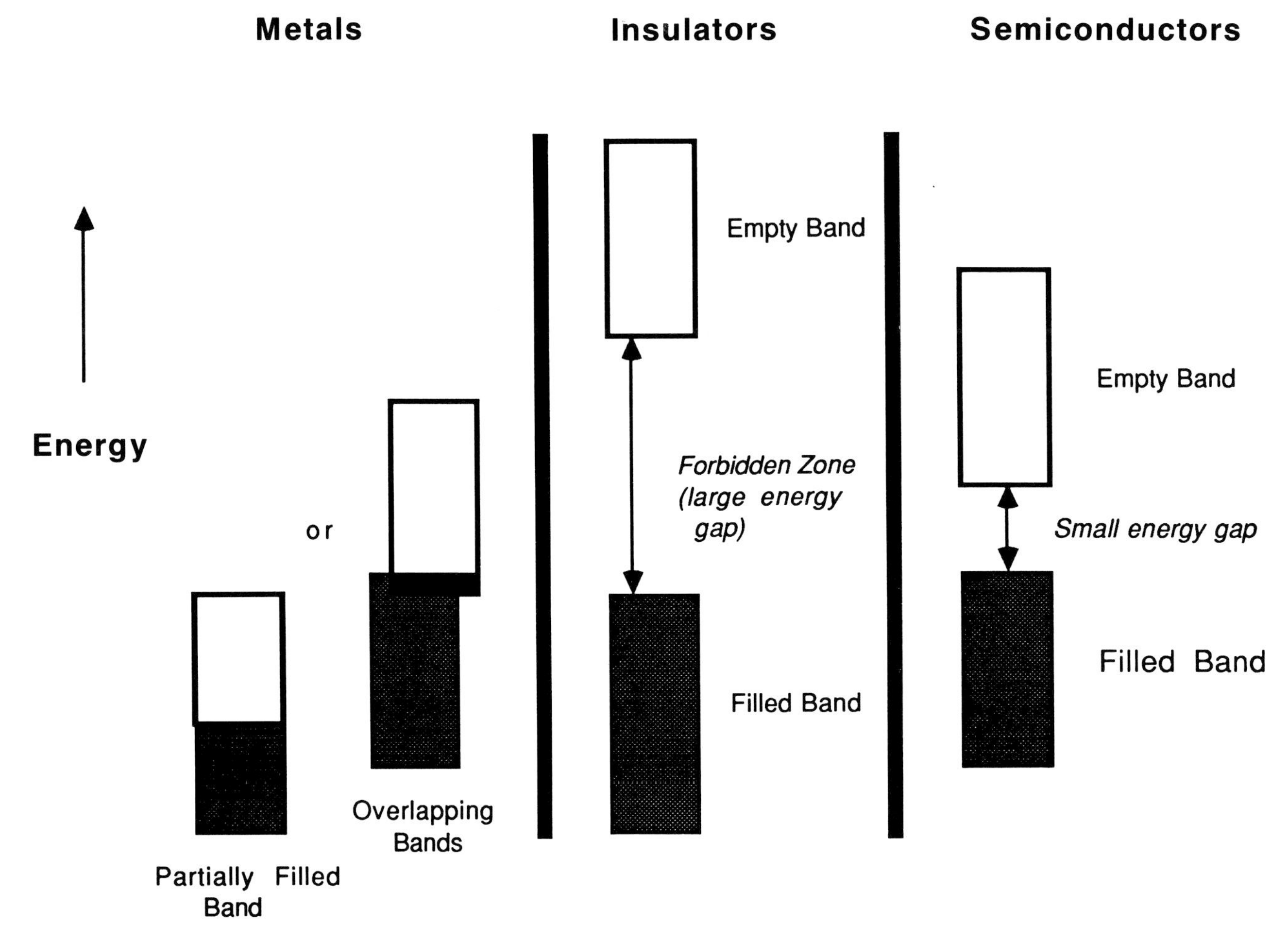

Figure 1.9 Relative band gap energies in metals, insulators, and semiconductors. In each case an unshaded area represents a conduction band [1].

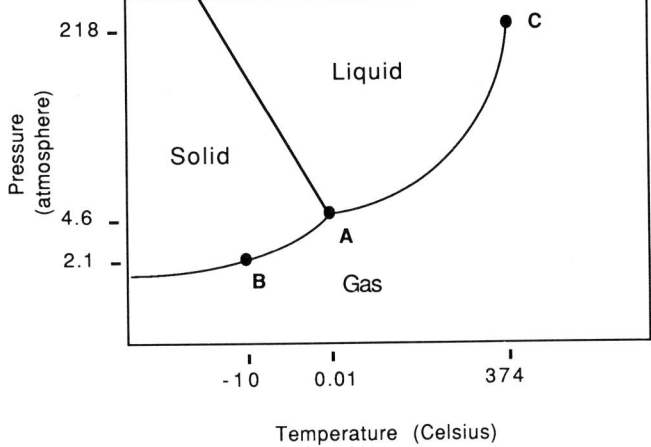

Figure 1.10 Phase diagram of water. Solid lines show the boundaries between the various phases.

form one or more products. In the mixing of hydrazine and nitrogen tetroxide the following reaction occurs:

$$5N_2H_4 + 2N_2O_4 \rightarrow 8H_2O + 2H_2 + 7N_2 . \quad (1.16)$$

The chemical reaction is apparent in this case as two highly toxic substances are converted into three relatively benign ones with the release of a large amount of energy. Equation 1.16 represents an ideal reaction in that all reactants are changed to products. Unfortunately, most chemical reactions do not proceed in such a clean fashion, but rather exhibit nonideal behavior. Depending upon the reaction conditions, other products are also produced in equation 1.16, including NO, NO_2, NH_3, and hydrazine nitrate. Such deviations from the ideal are what make predicting the outcomes of chemical reactions so difficult for chemists.

Stoichiometry

Stoichiometry deals with the relative ratios of the amounts of reactants and products in a chemical reaction. The combustion of H_2 and O_2 within the Space Shuttle's main engines proceeds as

$$2H_2(g) + O_2(g) \rightarrow 2H_2O(g) \quad (1.17)$$

in which two hydrogen molecules react with one oxygen molecule to form two water molecules. Since all reactions occur on a scale too large to use molecules as units, the stoichiometric coefficients also represent the number of moles or liters.

Ideally the fuel and oxidizer lines should supply 2 moles of H_2 and 1 mole of O_2. Any other ratio results in incomplete usage of either H_2 or O_2. Notice that both sides of the equation are balanced with respect to H and O. Multiplying coefficients and subscripts together shows four H atoms and two O atoms on each side of the equation. Chemical equations should always be checked to ensure they abide by the law of conservation of mass.

Oxidation-Reduction Reactions

Chemical reactions can be put into classes based upon the type of chemical change the reactants undergo. Four common reactions are

Combustion: $\quad (1.18)$
$\quad CH_4 + 2O_2 \rightarrow CO_2 + 2H_2O$
Decomposition: $\quad (1.19)$
$\quad 2HgO \rightarrow 2Hg + O_2$
Formation: $\quad (1.20)$
$\quad 4Al + 3O_2 \rightarrow 2Al_2O_3$
Precipitation: $\quad (1.21)$
$\quad Cd^{2+}(aq) + 2OH^-(aq) \rightarrow Cd(OH)_2(s)$

In the first three reactions, the reactants undergo chemical changes to form products, while in the precipitation reaction, solvated ions combine to form an insoluble solid. These first three reactions belong to a larger class known as oxidation-reduction reactions. Oxidation, in a specific sense, means the addition of oxygen to an element or compound. Generally speaking, however, oxidation is the apparent loss of electrons by an atom involved in a chemical reaction. The companion process by which an atom apparently gains electrons is termed reduction.

The basis for electron exchange in oxidation-reduction reactions (known as redox for short) is the previously discussed concept of electronegativity. The nonmetallic elements of the periodic table, and especially those in the upper right corner, have a large affinity for electrons. A contrasting behavior is exhibited by the metals which have a small affinity for electrons. When elements from these two groups are reacted to form a new compound, the end result is a transfer of electrons from the metal to the nonmetal. The status of an atom's electron count within a compound is denoted by an oxidation number. Oxidation numbers are assigned according to the number of electrons lost or gained by an atom with zero being the oxidation number of an uncombined element. Positive values show electron loss while negative values indicate electron gain. Table 1.3 lists the most common elemental oxidation states.

In order to better understand the concepts of oxidation and reduction, several examples are presented. The first is the reaction involving aluminum and oxygen to form alumina; namely,

$$4Al^0 + 3O_2^0 \rightarrow 2Al_2^{+3}O_3^{-2}. \quad (1.22)$$

Aluminum is oxidized as it donates its electrons to the oxygen. The oxidation of aluminium through the transfer of three electrons is shown by the change in its oxidation state from 0 to $+3$. The accompanying reduction of oxygen is apparent in its oxidation state going from 0 to -2 through the gaining of two electrons. A second example, of redox occurring is in a nickel-cadmium spacecraft battery where cadmium oxidizes $(0 \rightarrow +2)$ and nickel reduces $(+3 \rightarrow +2)$ in the following reaction:

$$Cd^0 + 2Ni^{+3}OOH + 2H_2O \quad (1.23)$$
$$\rightarrow Cd^{+2}(OH)_2 + 2Ni^{+2}(OH)_2.$$

A third example is the hydrogen-oxygen combustion reaction where hydrogen is the reducing agent $(0 \rightarrow +1)$ and oxygen is the oxidizing agent $(0 \rightarrow -2)$

$$2H_2^0 + O_2^0 \rightarrow 2H_2^{+1}O^{-2}. \quad (1.24)$$

Note that in each of these reactions, conservation of charge is observed. In other words, the amount of charge on the left side of the equation must equal the sum total on the right. This is shown in the following for each of the three above reactions respectively:

$$4(0) + (0) = 2(+3) + 3(-2) \quad (1.25)$$
$$0 + 2(+3) = 1(+2) + 2(+2) \quad (1.26)$$
$$2(0) + 0 = 4(+1) + 2(-2). \quad (1.27)$$

Equilibrium

Up to this point, chemical reactions have been treated as proceeding in only one direction, from reactants to products. However, all reactions do, in varying degrees, proceed simultaneously in the reverse direction, from products to reactants. Those reactions with a significant contribution in the reverse direction are called reversible.

$$aA + bB \rightleftarrows cC + dD \quad (1.28)$$

Chemical equilibrium is a dynamic process which takes time to establish. It is achieved when the rate of the forward reaction is the same as that in the reverse direction with the result being no net change in the concentrations of reactants and products. Take, for example, nitrogen tetroxide, which can decompose to form nitrogen dioxide.

$$N_2O_4(g) \rightleftarrows 2NO_2(g). \quad (1.29)$$

Initially, the rate of the reaction in the forward direction is very fast as the N_2O_4 decomposes to form two molecules

Table 1.3 Oxidation States of the Elements [1]

Group	IA	IIA	IIIB	IVB	VB	VIB	VIIB	VIII			IB	IIB	IIIA	IVA	VA	VIA	VIIA	0
	1 H **+1** +1																1 H **+1** −1	2 He
	3 Li **+1**	4 Be **+2**											5 B **+3**	6 C **+4** +2 −4	7 N **+5** +4 +3 +2 +1 −3	8 O −1 **−2**	9 F **−1**	10 Ne
	11 Na **+1**	12 Mg **+2**											13 Al **+3**	14 Si **+4** −4	15 P **+5** +3 −3	16 S **+6** +4 +2 −2	17 Cl **+7** +5 +3 +1 −1	18 Ar
	19 K **+1**	20 Ca **+2**	21 Sc **+3**	22 Ti **+4** +3 +2	23 V **+5** +4 +3 +2	24 Cr **+6** +3 +2	25 Mn **+7** +6 +4 +3 +2	26 Fe **+3** +2	27 Co **+3** +2	28 Ni **+2**	29 Cu **+2** +1	30 Zn **+2**	31 Ga **+3**	32 Ge **+4** −4	33 As **+5** +3 −3	34 Se **+6** +4 −2	35 Br **+7** +5 +3 +1 −1	36 Kr +4 +2
	37 Rb **+1**	38 Sr **+2**	39 Y **+3**	40 Zr **+4**	41 Nb **+5** +4	42 Mo **+6** +4 +3	43 Tc **+7** +6 +4	44 Ru **+8** +6 +4 +3	45 Rh **+4** +3 +2	46 Pd **+4** +2	47 Ag **+1**	48 Cd **+2**	49 In **+3**	50 Sn **+4** +2	51 Sb **+5** +3 −3	52 Te **+6** +4 −2	53 I **+7** +5 +3 +1 −1	54 Xe **+6** +4 +2
	55 Cs **+1**	56 Ba **+2**	57 La **+3** / 58 Ce — 71 Lu **+3**	72 Hf **+4**	73 Ta **+5**	74 W **+6** +4	75 Re **+7** +6 +4	76 Os **+8** +4	77 Ir **+4** +3	78 Pt **+4** +2	79 Au **+3** +1	80 Hg **+2** +1	81 Tl **+3** +1	82 Pb **+4** +2	83 Bi **+5** +3	84 Po **+2**	85 At −1	86 Rn

Note: The most common oxidation states are in bold typeface.

of NO_2. However, as the amount of N_2O_4 decreases, the forward reaction slows while the reverse reaction's rate increases. The point at which both rates are equal is defined as equilibrium and this dynamic interconversion will continue indefinitely as long as reaction conditions remain unchanged.

The questions now arise as to the extent a reaction proceeds in a particular direction and what is the impact upon equilibrium as reaction conditions change. In regard to the first question, the extent to which a reaction proceeds is expressed by the ratio of product amounts to reactant amounts. Using equation 1.28, the appropriate expression is

$$Q = \frac{[C]^c[D]^d}{[A]^a[B]^b} \quad (1.30)$$

where

$$Q = \text{concentration coefficient}$$
$$[X] = \text{concentration (moles/liter)}$$
$$a, b, c, d = \text{stoichiometric coefficients}$$

Q is an indicator of the relative amounts of reactant to product and it can be used to predict the direction in which a reaction will proceed in order to reach equilibrium. A special form of equation 1.30 is used when equilibrium is achieved,

$$K = \frac{[C]^c_{eq}[D]^d_{eq}}{[A]^a_{eq}[B]^b_{eq}} \quad (1.31)$$

where

$$K = \text{equilibrium constant}$$
$$[X]_{eq} = \text{concentration at equilibrium}$$

The K value for a reaction determines what the final product to reactant ratio will be under a given set of conditions. Large K values mean the reaction proceeds far to the right before equilibrium is established. Small values for K, on the other hand, mean the reverse reaction is predominant as equilibrium lies far to the left. By comparing Q to K an assessment of reaction direction can be made in the following manner:

$Q < K$ reaction proceeds to the right
$Q > K$ reaction proceeds to the left
$Q = K$ equilibrium

An example involving the reversible reaction shown in equation 1.29 will better clarify the principle of equilibrium. At 22°C the equilibrium constant for this reaction is given as

$$K = \frac{[NO_2]^2}{[N_2O_4]} = 4.66 \times 10^{-3} \quad (\text{at } 22°C). \quad (1.32)$$

Now suppose that at a given time within a 2.00 liter vessel there are 2.50 moles of N_2O_4 and 1.00 mole of NO_2. The value for Q is then

$$Q = \frac{[0.50]^2}{1.25} = 0.20$$

which is greater than K. Therefore, the reaction will proceed to the left until the value for Q equals K and the concentrations of products and reactants remain constant.

Equilibrium is, however, affected by changes to reaction conditions, such as reactant-product concentrations, pressure, and temperature. A reaction's response to these changes is explained by Le Chatlier's principle: the response by a system at equilibrium to a stress is to shift in the direction which relieves the stress, thereby reestablishing equilibrium at a different point.

Suppose the N_2O_4 system has achieved equilibrium in the vessel and now to it is added an additional 0.50 mole of NO_2. According to Le Chatlier's principle, what will happen? First, the system experiences an increase in pressure from the additional gas molecules which acts as an applied stress to the right side of the system.

$$N_2O_4 \rightleftarrows 2NO_2 + \text{stress of additional } NO_2$$

Relieving the stress requires a reduction in the number of gas molecules within the vessel so the reaction shifts to the opposite side, which in this case is to the left.

Temperature also has an important impact upon equilibrium because it affects the value of K itself. But before this effect can be understood, the role thermodynamics plays in chemical reactions must be explained.

Thermochemistry

Thermochemistry involves the study of energy changes occurring during a chemical or physical reaction. Examples of where thermochemistry is important to spacecraft operations include propellant combustion, battery cycling, fuel cell operation, material oxidation, and material phase change. What the study of thermochemistry ultimately helps to predict is whether or not a certain reaction will occur under a given set of conditions and how much energy will be absorbed or released.

Some questions a spacecraft scientist might ask are:

- Does aluminum oxidize in air at 20°C?
- Do hydrazine and nitrogen tetroxide ignite when mixed? If so, how much energy is released?
- What is an acceptable voltage for a spacecraft battery?
- How much heat does ammonia absorb per gram?

All of these questions can be answered by thermochemical analysis which involves measuring and understanding energy changes occurring during a process. The overall energy change that determines reaction spontaneity is called Gibb's free energy (ΔG). But ΔG itself depends upon two additional thermochemical parameters, enthalpy (ΔH) and entropy (ΔS). Knowledge of these two parameters, along with temperature, enables the scientist to predict the spontaneity of and available work from a given physical or chemical process.

Enthalpy

Enthalpy can be defined as the heat flow accompanying a physical or chemical reaction occurring at constant pressure. Other, more specific names for enthalpy include heat of reaction, heat of formation, heat of fusion, and heat of combustion. In each of these cases, enthalpy is the heat released or absorbed during the particular process. For example, ice must absorb a certain amount of energy, 6.02 kJ/mole, when it melts to become liquid water. The ice represents the system or that which is of interest in studying its heat flow. All else not within the system is deemed to be part of the surroundings. Together, the system and surroundings comprise the universe in which energy must always be conserved.

Enthalpy cannot, however, be measured absolutely, but rather can only be quantified when a change takes place. Thus, ice and liquid water have no absolute enthalpy values. Instead, all that can be said is that 1 mole of liquid water contains 6.02 kJ more heat than does one mole of ice.

$$H_2O(s) + 6.02 \text{ kJ} \rightarrow H_2O(\ell)$$

By definition, this reaction is endothermic because heat is absorbed by the system. On the other hand, the reverse reaction is exothermic because heat is released by the system.

$$H_2O(\ell) \rightarrow H_2O(s) + 6.02 \text{ kJ}$$

The convention is to assign a positive enthalpy value to endothermic reactions and a negative enthalpy value to exothermic reactions. Figure 1.11 shows the reactant-product heat content relationships for exothermic and endothermic reactions.

The general formula used to calculate ΔH for any reaction is:

$$\Delta H_{rxn} = \sum n\Delta H_f^{prod} - \sum n\Delta H_f^{react} \quad (1.33)$$

where

ΔH_{rxn} = enthalpy of reaction
ΔH_f^{prod} = enthalpy of formation for products
ΔH_f^{react} = enthalpy of formation for reactants
n = number of moles

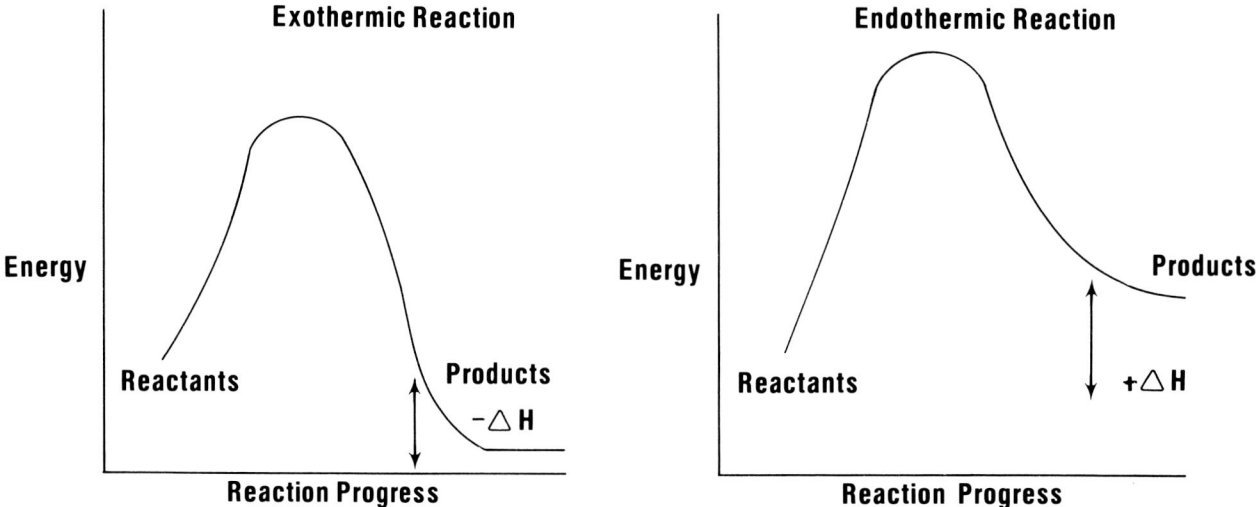

Figure 1.11 Energy diagrams for exothermic and endothermic reactions.

Enthalpy of formation is the amount of heat absorbed or released when a compound is formed from its basic elements. A table of ΔH_f values is contained in Table 1.4 which can be used to calculate the enthalpy of reaction for equation 1.16:

$$\Delta H^o_{rxn} = (8)(-241.8) + (2)(0) + (7)(0) \\ - (5)(50.6) - (2)(9.2) \quad (1.34) \\ = -2206 \text{ kJ}$$

Note that elements in their normal states are assigned ΔH_f^o values of zero. The state an element or compound is in is very important in calculating ΔH values. So too are the conditions of temperature, pressure, and concentration because they dictate the enthalpy values for both reactants and products. The "o" in thermochemistry represents the standard conditions of 25°C temperature and one atmosphere pressure. Deviations from these conditions require adjustment of standard enthalpy values.

In order to measure enthalpy values, precise measurements must be made in the laboratory under very controlled conditions. Although a great deal of empirical data is available to scientists, many reactions of interest have not been performed. Nevertheless, there are ways to determine the necessary data indirectly or at least estimate it. The first approach utilizes a fundamental principle of thermochemistry known as Hess's law. What Hess's law allows one to do is combine several chemical reactions together in order to obtain a single overall reaction. Say, a scientist needs an enthalpy value for the oxidation of Mg:

$$Mg(s) + \frac{1}{2} O_2(g) \rightarrow MgO(s) \quad (1.35)$$

Assume that this value is unknown, but enthalpy values for the following reactions are available:

$$Mg(s) + 2H^+(aq) \rightarrow Mg^{2+}(aq) + H_2(g) \quad (1.36)$$
$$\Delta H^o = 467 \text{ kJ}$$
$$MgO(s) + 2H^+(aq) \rightarrow Mg^{2+}(aq) + H_2O \quad (1.37)$$
$$\Delta H^o = 150 \text{ kJ}$$
$$H_2 + \frac{1}{2} O_2 \rightarrow H_2O \quad (1.38)$$
$$\Delta H^o = -286 \text{ kJ}$$

By reversing equation 1.37, thereby requiring a change in the sign of its ΔH^o value, and then adding the three reactions together, the scientist obtains the desired value, namely

$$Mg(s) + \frac{1}{2} O_2 \rightarrow MgO(s) \quad \Delta H^o = -603 \text{ kJ}$$

Hess's law is a powerful tool for obtaining enthalpy values of different reactions. It makes use of the fact that enthalpy is a state function. In a state function, the value of the change is independent of the reaction pathway. In the previous example, ΔH_f^o of MgO ends up being the same value regardless of going through the direct path (equation 1.35) or the indirect path (equations 1.36–1.38).

The second approach to calculating troublesome enthalpy values is based upon the bond energies of the reactants and products. Bond breaking requires that energy be input while bond formation results in the release of energy. Since enthalpy change is a measure of the relative stabilities of the reactants and products, the difference between the sums of the individual bond energies (BE) of the reactants and products should be a good estimate of ΔH^o.

$$\Delta H_{rxn} = \sum nBE_{reactants} - \sum nBE_{products} \quad (1.39)$$

The ΔH for equation 1.16 is estimated by writing the reaction using structural formulas to depict individual bonds.

Table 1.4 Thermodynamic Values for Selected Compounds

Species	$\Delta H^\circ_{f298.15}$ (kJ/mole)	$S^\circ_{298.15}$ (J/mole·K)	$\Delta G^\circ_{f298.15}$ (kJ/mole)
Al(s)	0	28.3	0
Al_2O_3(s)	−1676	50.92	−1582
Be(s)	0	9.54	0
Br_2(l)	0	152.23	0
HBr(g)	−36.4	198.59	−53.43
Ca(s)	0	41.6	0
CaH_2(s)	−189	42	−150
CaO(s)	−635.5	40	−604.2
C(s, graphite)	0	5.740	0
C(s, diamond)	1.89	2.38	2.900
C(g)	716.7	158.0	671.3
CH_4(g)	−74.81	186.2	−50.75
C_2H_2(g)	226.7	200.8	209.2
C_2H_4(g)	52.26	209.5	68.12
C_2H_6(g)	−84.86	229.5	−32.9
C_3H_8(g)	−103.8	269.9	−23.49
C_6H_6(l)	49.03	172.8	124.5
C_8H_{18}(l)	−268.8	—	—
C_2H_5OH(l)	−277.7	161	−174.9
C_2H_5OH(g)	−235.1	282.6	−168.6
CO(g)	−110.5	197.6	−137.2
CO_2(g)	−393.5	213.6	−394.4
Cl_2(g)	0	223.0	0
Cr(s)	0	23.8	0
Cu(s)	0	33.15	0
CuO(s)	−157	42.63	−130
F_2(g)	0	202.7	0
HF(g)	−271	173.7	−273
H_2(g)	0	130.6	0
H_2O(l)	−285.8	69.91	−237.2
H_2O(g)	−241.8	188.7	−228.6
H_2O_2(l)	−187.8	109.6	−120.4
I_2(s)	0	116.1	0
Fe(s)	0	27.3	0
FeO(s)	−272	—	—
Fe_2O_3(s, hematite)	−824.2	87.40	−742.2
Fe_3O_4(s, magnetite)	−1118	146	−1015
Pb(s)	0	64.81	0
Li(s)	0	28.0	0
LiOH(s)	−487.23	50	443.9
Mg(s)	0	32.5	0
MgO(s)	−601.8	27	−569.6
$Mg(OH)_2$(s)	−924.7	63.14	—
Ni(s)	0	30.1	0
NiO(s)	−244	38.6	−216
N_2(g)	0	191.5	0
NH_3(g)	−46.11	192.3	−16.5
N_2H_4(l)	50.63	121.2	149.2
NO_2(g)	33.2	240.0	51.30
N_2O_4(g)	9.16	304.2	97.82
HNO_3(l)	−174.1	155.6	−80.79
O_2(g)	0	205.0	0
O_3(g)	143	238.8	163
K(s)	0	63.6	0
KOH(s)	−424.7	78.91	−378.9
KOH(aq)	−481.2	92.0	−439.6
Si(s)	0	18.8	0
SiO(s)	−910.9	41.84	−856.7
Ag(s)	0	42.55	0
Na(s)	0	51.0	0
S(s, rhombic)	0	31.8	0
SO_2(g)	−296.8	248.1	−300.2
SO_3(g)	−395.6	256.6	371.1
Sn(s, white)	0	51.55	0
Sn(s, grey)	−2.09	44.1	0.13
W(s)	0	32.6	0
ZnO(s)	−348.3	43.64	−318.3

$$5 \begin{bmatrix} H & & H \\ & \diagdown \diagup & \\ & N\!-\!N & \\ & \diagup \diagdown & \\ H & & H \end{bmatrix} + 2 \begin{bmatrix} O & & O \\ & \diagdown\!\!\diagdown \diagup\!\!\diagup & \\ & N\!-\!N & \\ & \diagup\!\!\diagup \diagdown\!\!\diagdown & \\ O & & O \end{bmatrix} \longrightarrow$$

$$8 \begin{bmatrix} & O & \\ & \diagup \diagdown & \\ H & & H \end{bmatrix} + 2\,H\!-\!H + 7\,N\!\equiv\!N$$

Inserting bond energy values from Table 1.5 into equation 1.39 yields the following:

$$\Delta H = [7BE_{N-N} + 20BE_{N-H} + 4BE_{N-O} + 4BE_{N=O}]$$
$$\quad - [16BE_{O-H} + 2BE_{H-H} + 7BE_{N\equiv N}]$$
$$\Delta H = -2674 \text{ kJ} \quad (1.40)$$

One word of caution in using bond energies is that the values reported in Table 1.5 are an average taken from measurements of a number of different compounds. A specific bond energy is dependent upon other bonds in the molecule, so bond energy estimates of ΔH can be in error by significant amounts in some cases.

Entropy

Enthalpy, however, is not the only thermodynamic property determining reaction spontaneity. The change in order or entropy (ΔS) of the system is the other key factor. Entropy change can be exemplified by bowling pins before and after they are struck by the ball. Before the impact the pins are arranged upright in a specific pattern with each pin occupying a designated position. After the impact, the arrangement is highly disordered with pins in various orientations randomly scattered throughout the lane and pit. The net result is an increase in randomness or entropy. The second law of thermodynamics says that spontaneous changes result in greater disorder of the system and its surroundings. In other words, spontaneous reactions are favored by increased entropy in going from reactants to products. Take for example, the phase changes occurring in going from an ice cube to liquid water to steam.

$$H_2O(s) \rightarrow H_2O(l) \rightarrow H_2O(g) \quad (1.41)$$

In the progression, water goes from a highly structured crystal to a flowing liquid to an unrestrained gas. Randomness increases yielding a positive entropy value for the process. Using entropy values obtained from thermochemical tables, the general equation for calculating ΔS is

$$\Delta S_{rxn} = \sum nS_{prod} - \sum nS_{react}. \quad (1.42)$$

In the phase change $H_2O(l) \rightarrow H_2O(g)$ at 298°C the entropy change is

$$\Delta S^o = 188.7 \text{ J/mole·K} - 69.91 \text{ J/mole·K}$$
$$= 118.8 \text{ J/mole·K} \quad (1.43)$$

where ΔS^o is expressed in J/mole·K. Again, the superscript

Table 1.5 Average Bond Energies Between Selected Elements (kJ/mole)

Br—Br	193	C—H	414	Cl—N	201	H—S	339
Br—C	276	C—I	218	Cl—O	205	I—I	151
Br—Cl	218	C—N	293	Cl—S	255	I—O	201
Br—F	255	C=N	615	F—F	153	N—N	159
Br—H	368	C≡N	890	F—H	565	N=N	418
Br—I	180	C—O	351	F—I	277	N≡N	941
Br—N	243	C=O	715	F—N	272	N—O	222
Br—O	201	C≡O	1075	F—O	184	N=O	607
Br—S	213	C—S	295	F—S	285	O—O	138
C—C	347	C=S	477	H—H	436	O=O	498
C=C	612	Cl—Cl	243	H—I	297	O—S	347
C≡C	820	Cl—F	255	H—N	389	O=S	498
C—Cl	331	Cl—H	431	H—O	464	S—S	226
C—F	485	Cl—I	209				

"o" denotes standard temperature and pressure conditions of 1 atm and 25°C.

Unlike enthalpy, which measures only change and has no absolute value, absolute entropy can be calculated using statistical thermodynamics. The zero point on the entropy scale for a substance is given by the third law of thermodynamics: at absolute zero (-273°C) the entropy of a perfectly ordered crystalline solid is zero by definition.

Particles at this temperature have their lowest energy and vibrate at a minimum about fixed positions. Increasing temperature provides heat that causes the particles to vibrate and move, thus increasing entropy.

Gibbs's Free Energy

The combination of enthalpy and entropy into a mathematical equation capable of predicting reaction spontaneity was the work of J. Willard Gibbs. The Gibbs-Helmholtz equation introduced a new thermodynamic function, the Gibbs's free energy (ΔG), to express the maximum useful energy, in the form of work, yielded by a chemical reaction or any process at constant pressure and temperature. Gibbs's free energy, expressed as

$$\Delta G = \Delta H - T\Delta S \quad (T, P = constant) \quad (1.44)$$

is a balance between the net flow of heat in a reaction and the corresponding change in its disorder.

Like enthalpy, ΔG is a measure of change during a process and has no absolute value. Spontaneous processes cause a decrease in ΔG while nonspontaneous processes have positive ΔG values. Depending upon the values for ΔH and ΔS, temperature can play an important role in determining reaction spontaneity. Negative values for ΔG with different combinations of ΔH and ΔS and different temperatures are shown in Table 1.6. Obviously the best possible combination for reaction spontaneity is when the reaction is exothermic and entropy increases. Using the hydrogen/oxygen reaction,

Table 1.6 Gibbs's Free Energy Dependency upon Enthalpy, Entropy, and Temperature

$\Delta G < 0$ when $\Delta H < 0$ and $\Delta S > 0$
$\Delta G < 0$ when $\Delta H < 0$, $\Delta S < 0$, and T is low
$\Delta G < 0$ when $\Delta H > 0$, $\Delta S > 0$, and T is high

$$H_2(g) + \frac{1}{2}O_2(g) \rightarrow H_2O(g), \quad (1.45)$$

the thermodynamic values at standard temperature and pressure are

$\Delta H^o = -241.8$ kJ/mole
$\Delta S^o = -0.0444$ J/mole $\cdot$ K
$\Delta G^o = -241.8$ kJ/mole $- (298$ K$)(-0.0444$ kJ/mole$)$
$= -228.6$ kJ/mole,

Therefore, 228.6 kJ of energy are released during the formation of one mole of water from the reaction of one mole of hydrogen with one-half mole of oxygen.

Under nonstandard conditions, the expression for Gibbs's free energy change requires the following adjustment to equate ΔG and ΔG^o:

$$\Delta G = \Delta G^o + RT \ln Q. \quad (1.46)$$

The second term on the right-hand side of the equation accounts for nonstandard temperature, pressure, or concentration. At equilibrium, ΔG becomes zero because the forward and reverse reactions are occurring at the same rate. Therefore, Q can be replaced by K and equation 1.46 becomes

$$\Delta G^o = -RT \ln K$$

or

$$K = e^{-\Delta G^o/RT}. \quad (1.47)$$

Review of General Chemistry

Notice the consistency between expected values for K and ΔG^o in a spontaneous reaction. A large, negative ΔG^o yields a large, positive K value as the reaction proceeds spontaneously from left to right.

Chemical Kinetics

Thermodynamics predicts reaction spontaneity, but says nothing about how fast a reaction will proceed. In order to determine a reaction's rate, chemical kinetics must be studied.

Chemical reactions occur through the energetic collisions of reactant particles to form products. These collisions must have proper orientation and sufficient energy to break apart the reactants' chemical bonds. As is expected, a great number of molecular collisions do not possess these two characteristics and therefore produce no product. Hence, even though the production of water is spontaneous at room temperature, the reaction rate is very slow due to the low number of effective collisions between H_2 and O_2.

The need for effective collisions can be shown on an energy diagram like the one in Figure 1.12. In order to form products AB and B, reactants A and B_2 must first overcome the energy barrier represented by the activation energy (E_a). Given that the molecules are properly oriented, they must collide with sufficient energy to overcome E_a and proceed down the energy curve to the products. At the pinnacle of the curve, the reactants exist for a short time in what is known as a transition state. The dashed lines of the transition state represent the simultaneous formation and breakage of chemical bonds as reactants transition to products.

The energy available to overcome E_a is a function of the reaction temperature. Figure 1.13 shows the molecular kinetic energy distributions at temperature T_1 and T_2. At T_1, a small fraction of the molecules possess the minimum amount of energy to react. By increasing the temperature to $T_2 > T_1$ the distribution shifts to the right and now more molecules have a kinetic energy greater than E_a. This creates

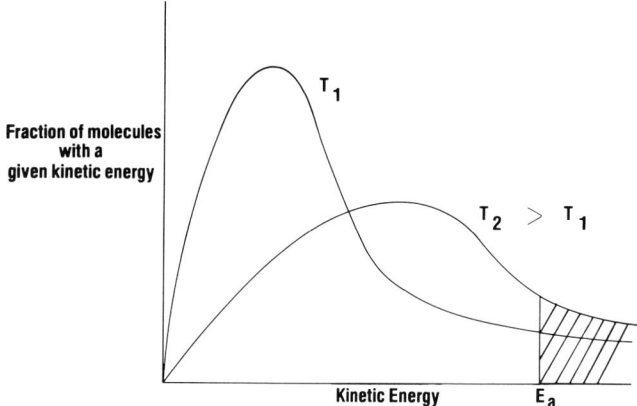

Figure 1.13 The fraction of molecules with kinetic energy greater than E_a increases as temperature increases from T_1 to T_2.

more effective collisions and leads to an increase in the reaction rate. In the case of the H_2/O_2 reaction, increasing the reaction temperature will increase the rate until rapid combustion occurs.

Reaction rates are also increased through the use of catalysts. Catalysts change the route of a chemical reaction by providing an alternate pathway with a lower E_a (Figure 1.14). True catalysts are not consumed and are chemically unchanged after completion of the reaction. Catalysts in the same phase as the reactants are homogeneous, while those in a different phase are heterogeneous.

A catalyst's effectiveness is dependent upon the absence of any inhibitor able to impede the reaction process. Platinum is a common catalyst used in automobiles to convert CO to CO_2. Adsorption of CO and O_2 on platinum's surface is crucial for the reaction to occur. Should lead be present in the exhaust it will bind to the platinum and prevent CO and O_2 from adsorbing and reacting. Exclusion of such poisons is also extremely important in the functioning of H_2/O_2 fuel cells, which will be discussed later.

Reaction Rate and Rate Law

The speed at which a chemical reaction proceeds is expressed by the change in a chemical constituent concentration with time, $d[x]/dt$. The constituent can be either a reactant or a product. In the combustion of hydrogen and oxygen, the reaction rate can be expressed as either of the following:

$$2H_2(g) + O_2(g) \rightarrow 2H_2O(g)$$

or (1.48)

$$-\frac{d}{dt}[H_2] = (-2)\frac{d}{dt}[O_2] = \frac{d}{dt}[H_2O] .$$

Negative signs denote the reactants are being consumed and therefore their concentrations are decreasing. The coeffi-

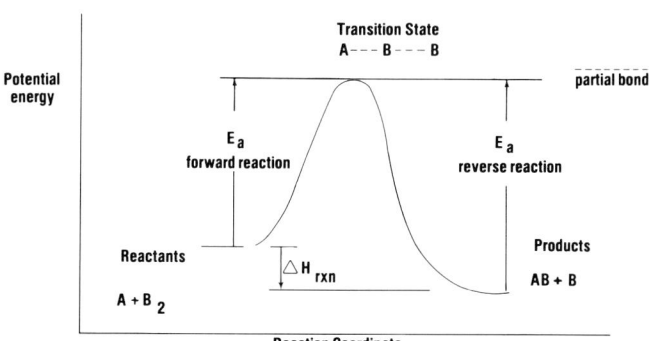

Figure 1.12 Exothermic reaction in which the activation energy E_a must be furnished to the reactants A and B_2 in order to form products AB and B.

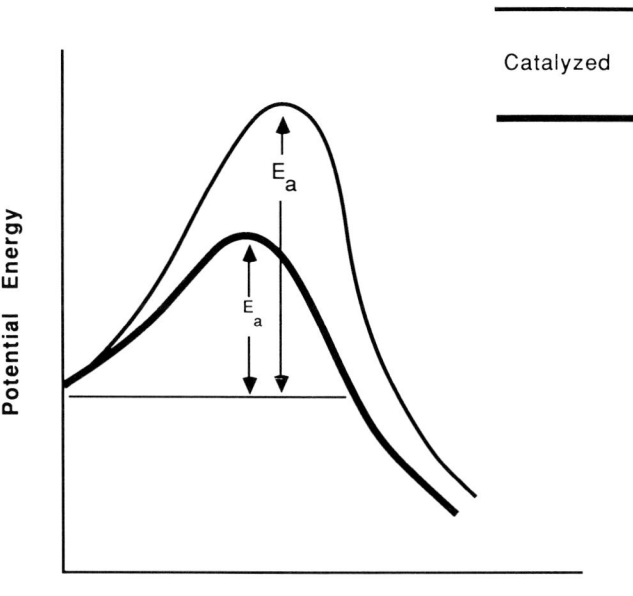

Figure 1.14 Energy diagrams of a general chemical reaction with and without the use of a catalyst.

cient of -2 for O_2 is from the balanced equation, showing that O_2 is consumed one-half as rapidly as hydrogen.

Reactant concentrations are the key determinant of reaction rate since reactant molecules must collide in order for the reaction to occur. Therefore, the reaction rate for water production can be expressed mathematically as a function of reactant concentrations.

$$\text{rate of reaction} = k[H_2]^x[O_2]^y \quad (1.49)$$

where

$$k = \text{rate constant}$$
$$x, y = \text{reactant orders}$$

Reactant order shows the degree of dependence a reaction has upon a particular reactant. A coefficient of one, for example, means the reaction is first order with respect to the reactant. An overall reaction order is obtained by summing the individual reactant orders $(x + y)$.

If the reaction takes place in a single step, then the values of the reactant orders correspond to the stoichiometric coefficients of the balanced equation $(x = 2, y = 1)$. However, many reactions must go through multiple steps from start to finish. In these reactions, the order must be determined experimentally and is dependent upon the reaction's mechanism or total steps. Rates differ for each step of the mechanism with one step generally much slower than the others. The slowest step, which is called the rate determining step, dictates the rate of the overall reaction. For example, in the decomposition of ozone (O_3) by nitrogen oxide one possible mechanism is

$$\begin{aligned}
NO + O_3 &\rightarrow NO_3 + O \quad \text{(slow)} \\
NO_3 + O &\rightarrow NO_2 + O_2 \quad \text{(fast)} \\
\hline
O_3 + NO &\rightarrow NO_2 + O_2 \quad \text{(overall)}
\end{aligned} \quad (1.50)$$

and the rate expression is

$$\text{rate} = k[NO][O_3] \, . \quad (1.51)$$

Note the concentrations and reactant orders correspond to those of the balanced rate determining step.

Polymers

Polymers are giant molecules (macromolecules) built up by the repetition of small, simple chemical units called monomers. They are a main constituent of our food (starch, protein, cheeses), our clothes (cotton, polyester, nylon), our homes (wood, cellulose, latex paint), and our bodies (polynucleic acids, proteins).

Polymers display a wide range of properties based upon the nature of the monomer, the length of the polymer molecules, and the degree of interaction between the polymer molecules. Synthetic polymers are currently classified into three general categories: (1) fibers, (2) elastomers, and (3) plastics. Fibers are flexible straight chain structures that can be woven into yarns. Rubbers are elastomers because they regain their initial shape after being deformed. Plastics can be made to flow and assume a desired shape through the application of heat and pressure.

Interactions between the long, individual molecules play an important role in determining polymer properties. Take, for example, polyethylene which is processed into a variety of forms including film and sheets:

$$H_2C{=}CH_2 \xrightarrow[\text{heat}]{\text{catalyst}} -[CH_2{-}CH_2]_n- \, . \quad (1.52)$$

Binding between the polyethylene molecules is due solely to short-range intermolecular forces. The most efficient arrangement for interaction is one with molecules assuming a crystalline configuration where chains are stacked neatly next to one another. This resembles a molecular crystal, as shown in Figure 1.15. The result is strong intermolecular forces and a product referred to as high density polyethylene (*HDPE*, $T_m = 145°C$).

A low density version is also available (*LDPE*, $T_m = 110°C$) by altering the synthesis conditions. In this case, the small side chains or branches depicted in Figure 1.16 form on the polymer backbone. The branches inhibit crystallization of the polymer, preventing individual molecules from coming into close proximity with one another. Intermolecular forces, in this case, are much weaker, yielding a more flexible polymer.

Figure 1.15 Linear arrangement of molecules within a crystalline polymer structure.

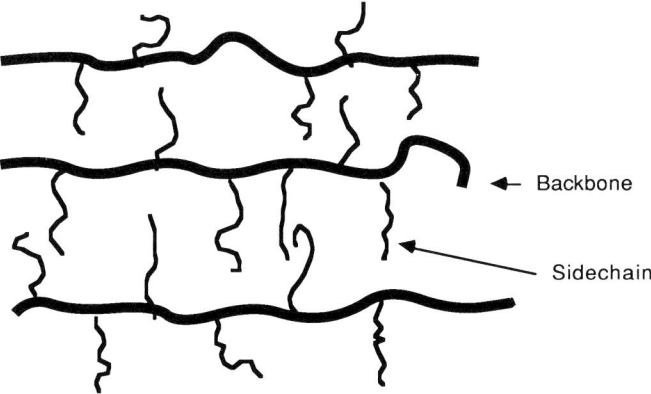

Figure 1.16 Polymer with side chains extending from backbone chains.

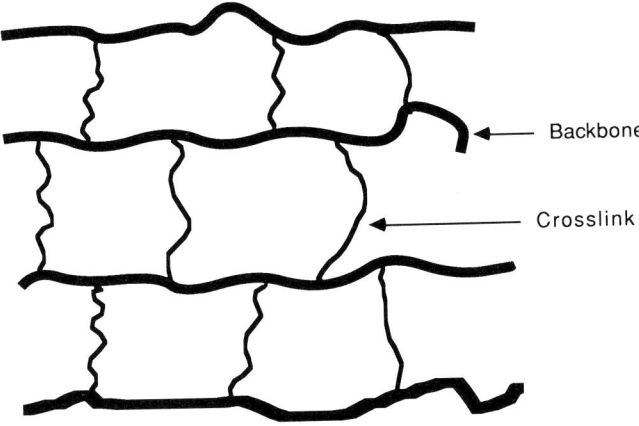

Figure 1.17 Crosslinking between polymer backbone chains.

An extension of branching is when the side chains actually link themselves to another polymer molecule, as shown in Figure 1.17. This phenomenon, referred to as crosslinking, adds greater rigidity to the polymer by preventing the chains from sliding past one another upon deformation.

Greater strength and rigidity can also be achieved by strengthening the polymer backbone. Backbones with saturated hydrocarbons (i.e., no multiple bonds) are able to bend and rotate about the single carbon-carbon bonds (e.g., polyethylene). Incorporation of rigid groups, like a benzene ring, significantly increases chain stiffness, as is the case with the polycarbonate repeating unit shown below:

$$-[O-C_6H_4-\underset{\underset{CH_3}{|}}{\overset{\overset{CH_3}{|}}{C}}-C_6H_4-O-\overset{\overset{O}{\|}}{C}]_n-$$

Depending upon the method of polymerization, two types of plastics can be formed: thermoplastics and thermosets. Thermoplastics consist of straight-chain molecules with no crosslinking (Figure 1.18). These materials can be thermally cycled several times and still maintain their molecular integrity upon cooling.

Thermosets, on the other hand, are crosslinked to form the amorphous, three dimensional network represented in

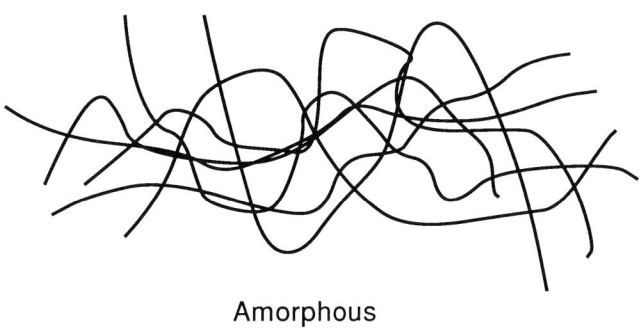

Amorphous

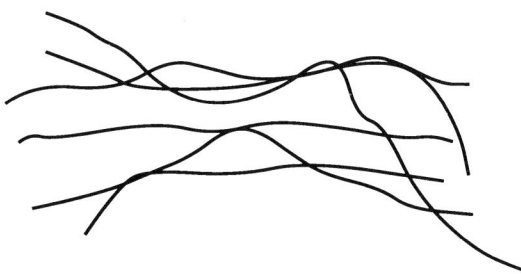

Partially Crystalline

Figure 1.18 The amorphous and partially crystalline molecular structures of the two general classes of thermoplastics.

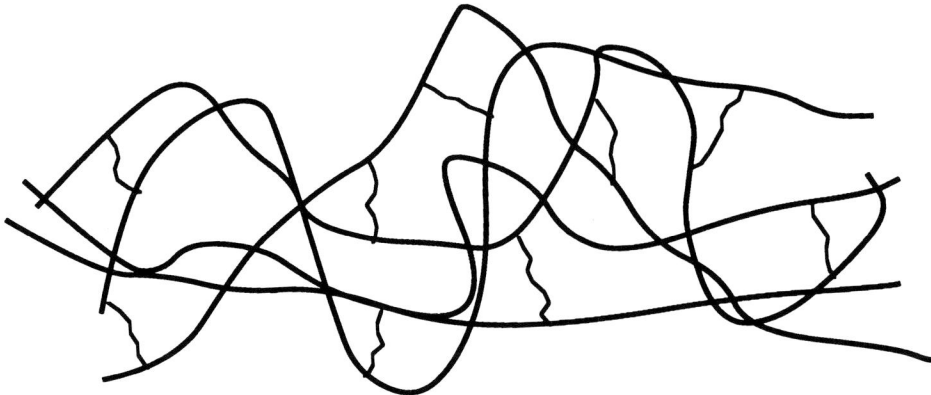

Figure 1.19 The highly crosslinked molecular structure characteristic of thermoset plastics.

Figure 1.19. Crosslinking gives them greater rigidity and chemical insolvency, but unfortunately makes them susceptible to breakdown at elevated temperatures. At high enough temperatures the crosslinks break and the polymer is irreversibly destroyed.

Electrochemistry

Electrochemistry deals with the interconversion between electrical and chemical energy. The supply of electrical current to cause a chemical reaction is called electrolysis. Hydrogen and oxygen produced from water is just one application of electrolysis. In the reverse process, a chemical reaction produces electrical energy as the system operates in the galvanic or voltaic mode. Lead-acid and nickel-cadmium are two common batteries used to produce power from chemical reactions.

The foundation for understanding electrochemistry is chemical thermodynamics. Recall the spontaneity of a reaction is determined by the value of Gibbs's free energy change (ΔG). Spontaneous reactions have negative values of ΔG while nonspontaneous reactions possess positive ΔG values. Reversal of the reaction's direction, as shown below, causes a switch in the sign of ΔG.

$$A + B \leftrightarrows C + D \quad (1.53)$$

The sign of ΔG determines whether or not a battery will operate under a given set of conditions.

The value for ΔG, however, is never printed on any of the cells or batteries, but instead voltage is displayed. Voltage is the pushing force that moves electrons through conductors. The larger the difference in voltage between two points the greater the pushing force will be and the more readily current will flow.

A key point to remember is that voltage is a measure of the difference in potential between two points. Voltage, like Gibbs's free energy, has no true absolute value and must be measured against a reference point. Elements in their pure, naturally occurring form at standard conditions are the zero points for ΔG^o. Since electrochemistry deals with oxidation-reduction reactions, scientists selected the reduction of hydrogen as the standard reference point:

$$2H^+ + 2e^- \rightarrow H_2 \quad E^o = 0.00 \text{ V}. \quad (1.54)$$

Standard conditions exist when solids are pure, solutions are 1 M in concentration, and gases have a partial pressure of one atmosphere. Measurements of the potential of many reduction reactions can now be made by using a H_2 reference electrode and the Nernst equation:

$$E = E^o - (RT/nF)\ln Q \quad (1.55)$$

where

E = cell potential
E^o = standard cell potential
F = Faraday's constant (96,487 coulomb/mole)
n = number of moles of electrons transferred
Q = reaction quotient

Chemists, by measuring values for E and Q under nonstandard conditions, are able to determine values for E^o. For those elements where measuring E^o directly is difficult, such as Na and Li, thermodynamic data is used to indirectly determine E^o using the equation below:

$$\Delta G = -nFE^o. \quad (1.56)$$

Table 1.7 shows electrode potentials compared to H_2 for various reactions under standard conditions. Notice the potentials are given for the reduction half-reaction and require a sign change to obtain the correct value for the reverse, oxidation half-reaction. These are called half-reactions because two electrode reactions are needed to have a complete cell and H_2 is one-half of the cell. The electromotive series table is one of the most valuable tools for constructing a cell.

Return now to the previous discussion of the importance of ΔG and reaction spontaneity. The battery must supply power when a load is connected to it. In other words, the

Review of General Chemistry

Table 1.7 Half-cell Reduction Potentials versus a Normal Hydrogen Electrode [1]

Reaction	E^o_{red} (V)
$Li^+ + e^- \to Li(s)$	−3.05
$K^+ + e^- \to K(s)$	−2.93
$Ca^{2+} + 2e^- \to Ca(s)$	−2.87
$Na^+ + e^- \to Na(s)$	−2.71
$Mg^{2+} + 2e^- \to Mg(s)$	−2.36
$Al^{3+} + 3e^- \to Al(s)$	−1.66
$2H_2O + 2e^- \to H_2(g)$	−0.83
$Zn^{2+} + 2e^- \to Zn(s)$	−0.76
$Cr^{3+} + 3e^- \to Cr(s)$	−0.74
$Fe^{2+} + 2e^- \to Fe(s)$	−0.44
$Cr^{3+} + e^- \to Cr^{2+}$	−0.41
$Ni^{2+} + 2e^- \to Ni(s)$	−0.25
$Sn^{2+} + 2e^- \to Sn$	−0.16
$Pb^{2+} + 2e^- \to Pb(s)$	−0.13
$2H^+ + 2e^- \to H_2(g)$	−0.00
$AgBr(s) + e^- \to Ag(s) + Br^-$	+0.10
$S(s) + 2H + 2e^- \to H_2S(aq)$	+0.14
$Cu^{2+} + e^- \to Cu^+$	+0.15
$AgCl(s) + e^- \to Ag(s) + Cl^-$	+0.22
$Cu^{2+} + 2e^- \to Cu(s)$	+0.34
$Cu^+ + e^- \to Cu(s)$	+0.52
$I_2(s) + 2e^- \to 2I^-$	+0.54
$O_2(g) + 2H^+ + 2e^- \to H_2O_2(aq)$	+0.68
$Fe^{3+} + e^- \to Fe^{2+}$	+0.77
$Ag^+ + e^- \to Ag(s)$	+0.80
$2Hg^{2+} + 2e^- \to Hg_2^{2+}$	+0.92
$NO_3^- + 4H^+ + 3e^- \to NO(g) + 2H_2O$	+0.97
$Br_2 + 2e^- \to 2Br^-$	+1.09
$O_2(g) + 4H^+ + 4e^- \to 2H_2O$	+1.23
$Cr_2O_7^{2-} + 14H^+ + 6e^- \to 2Cr^{3+} + 7H_2O$	+1.33
$Cl_2(g) + 2e^- \to 2Cl^-$	+1.36
$MnO_4^- + 8H^+ + 5e^- \to Mn^{2+} + 4H_2O$	+1.49
$Au^{3+} + 3e^- \to Au(s)$	+1.50
$MnO_2 + 4H^+ + 2e^- \to Mn^{2+} + 4H_2O$	+1.61
$H_2O_2(aq) + 2H^+ + 2e^- \to 2H_2O$	+1.78
$Co^{3+} + e^- \to Co^{2+}$	+1.81
$F_2 + 2e^- \to 2F^-$	+2.87

battery has to spontaneously supply power and therefore have a positive potential. The tremendous usefulness of this finding is best illustrated by an example. A cell is constructed using H_2 and Cu electrodes:

$$2H^+ + 2e^- \to H_2 \quad E^o = 0.00 \text{ V}$$
$$Cu^{2+} + 2e^- \to Cu \quad E^o = +0.337 \text{ V} \quad (1.57)$$

For a cell or battery to operate, one electrode must be a source of electrons (anode/oxidation) and the other a sink (cathode/reduction). Therefore, one of the half-cell reduction reactions must be reversed while keeping in mind that the overall cell potential is positive. Making the copper electrode the cathode yields the following spontaneous, overall cell reaction:

Anode: $H_2 \to 2H^+ + 2e^-$; $E^o = 0.00$ V
Cathode: $Cu^{2+} + H_2 \to Cu$; $E^o = +0.337$ V

$$Cu^{2+} + H_2 \to 2H^+ + Cu; \quad E^o = +0.337 \text{ V} \quad (1.58)$$

Consider a second example using the reactions for a satellite battery and emphasizing the important role of the electrolyte.

Electrical current is the transfer of charge through a solid or liquid conducting medium. Metals, such as copper, transport electrons from atom to atom according to band theory. Electrolytic conduction is quite different because mobile ions carry the charge. Ions move under the force of electrostatic attraction to the oppositely charged electrode so that imbalances of charge don't occur. In effect the electrolyte "neutralizes" the charge of the oxidized/reduced species which allows electrons to continuously flow from the anode to the cathode.

The role of the electrolyte becomes more apparent when looking at the nickel-cadmium cell. Again, selection of the anodic and cathodic reactions must yield a positive cell voltage:

$Cd + 2OH^- \to Cd(OH)_2 + 2e^-$; $E^o = 0.81$ V
$NiOOH + H_2O + e^- \to Ni(OH)_2 + OH^-$; $E^o = 0.49$ V
$Cd + 2NiOOH + 2H_2O \leftrightarrow Cd(OH)_2 + Ni(OH)_2$; $E^o = 1.30$ V
(1.59)

Potassium hydroxide, in the form of K^+ and OH^- ions, provides the hydroxide ions necessary for the electrochemical reactions to occur.

Not all cells, however, can be recharged once their energy supply is exhausted. Two broad classes of cells/batteries are the nonrechargeable primaries and rechargeable secondaries. Primaries are the disposable batteries used for flashlights, cameras, radios, and tape players. The chemistry of the discharge reaction in these cells cannot be reversed to reform the original reactants by supplying an external current. Secondary cells are more forgiving because they can be discharged and recharged (cycled) more than once. Look again at the reactions for the Ni-Cd cell only this time power is supplied to make it an electrolytic cell. The external power source must have a higher potential than the battery in order to overcome the 1.3 V pushing force and reverse the current's natural direction.

Summary of Key Equations

$E = h\nu = \dfrac{hc}{\lambda}$ (1.7) Planck's equation

$P_1V_1 = P_2V_2$ (1.9) Boyle's law

$V_1/T_1 = V_2/T_2$ (1.10) Charles's law

$PV = nRT$ (1.11) Ideal gas law

$P_{TOT} = P_A + P_B + P_C \ldots$ (1.12) Dalton's law

$K = \dfrac{\Pi[\text{Prod}]^y}{\Pi[\text{React}]^y}$ (1.31) Equilibrium constant

$\Delta H_{rxn} = \sum n \, \Delta H_f^{\text{Prod}} - \sum n \, \Delta H_f^{\text{react}}$ (1.33) Enthalpy of reaction

$\Delta S_{rxn} = \sum nS_{\text{prod}} - \sum nS_{\text{react}}$ (1.42) Entropy of reaction

$\Delta G = \Delta H - T\Delta S$ (1.44) Gibbs's free energy

$E = E^o - \dfrac{RT}{nF} \ln Q$ (1.55) Nernst equation

$\Delta G^o = -nFE^o$ (1.56) Relationship of Gibbs's free energy and voltage

References

1. Whitten, K., Gailey, K., and R. Davis, *General Chemistry*, 3rd ed., Saunders College Publishing, 1988.

Discussion Questions

1. Predict the type of bonding (ionic, covalent, or metallic) present in each of the following substances:
   a. Li_2O _____   b. CH_4 _____
   c. N_2H_4 _____   d. $ZnCl_2$ _____
   e. Al _____

2. Water has a specific heat about twice that of ethylene glycol.
   a. Explain which one would make the better cooling fluid.
   b. Why is a mixture of the two considered optimum for an automotive coolant system?

3. Match the element with the correct category.
   a. sulfur   _____ conductor
   b. germanium   _____ nonconductor
   c. beryllium   _____ semiconductor

   Using band theory, explain the difference in electrical conductivity of these three elements.

4. Discuss the differences between thermosetting and thermoplastic polymers.

5. There are two major types of batteries: nonrechargeable primaries and rechargeable secondaries.
   a. What are the two major differences between these?
   b. What are possible energy sources which can be used to recharge fuel cells on satellites?

6. Enthalpy, entropy, and Gibbs's free energy are all determining factors in the analysis of chemical reactions.
   a. Which parameter can be interpreted as a measure of the disorder in a system?
   b. Which parameter is a measure of a reaction's heat flow?
   c. Which parameter describes the balance of the influence of the other two on a chemical reaction?

Problems

1. Astronaut helmet visors incorporate small amounts of silver chloride (AgCl) which darkens upon exposure to light. When light of quantum energy E_q strikes the AgCl, the following reaction occurs:

   $$AgCl + E_q \rightarrow Ag + Cl$$

   The energy required to dissociate 1 mole of AgCl is 580.0 kJ.
   a. What is the maximum wavelength of light which would cause this reaction?
   b. What is the corresponding frequency of this energy?

2. Calculate the pressure inside a 5.00 ℓ vessel containing 1,000 g of O_2 at 10°C? Assume the gas behaves ideally.

3. A NASA scientist proposes that barium peroxide, BaO_2, be used on a space capsule to supply emergency oxygen. On heating, BaO_2 decomposes as follows:

   $$2BaO_2(s) \rightarrow 2BaO(s) + O_2(g) \ .$$

   a. What mass of BaO_2 would be needed to supply enough oxygen to fill a 10,000 liter capsule at 1.00 atm pressure and 25°C?
   b. What volume capsule could be filled if the temperature were kept at 10°C with the same mass of BaO_2 as calculated for part a?

4. Both hydrazine (N_2H_4) and ammonia (NH_3) are capable of being fuels in rocket engines. When combusted they produce liquid water and gaseous nitrogen.
   a. Write a balanced equation for both of these processes.
   b. Calculate ΔH^o for each of the above reactions given the following values:

   ΔH_f^o for $NH_3(g) = -46.25$ kJ/mole
   ΔH_f^o for $H_2O(\ell) = -285.5$ kJ/mole
   ΔH_f^o for $N_2H_4(\ell) = 50.42$ kJ/mole

   c. On the basis of mass (or weight) would hydrazine or ammonia be the better fuel? (Show calculations)

5. Write the rate law expression for the reaction $A + B \rightarrow C$ given the following rate data:

Experiment	Initial[A]	Initial[B]	Initial Rate of Formation of C
1	0.10 M	0.10 M	3.0×10^{-4} M/min
2	0.20 M	0.20 M	2.4×10^{-3} M/min
3	0.10 M	0.20 M	6.0×10^{-4} M/min

6. Calculate the potential for the following voltaic cell:

   $$Pb(s) + 2AgNO_3(aq) \rightarrow Pb(NO_3)_2(aq) + 2Ag$$
   $$(0.50M) (0.10M)$$

Chapter 2
Review of Astrodynamics

Space chemistry is an essential building block of successful aerospace activities. It drives the satellite system from launch through its entire mission lifetime and describes the interaction of the system with volatile environments. The path of a space structure in the gravitational field of a central body is described by astrodynamics. Knowledge of astrodynamics is necessary to our development of space chemistry in the near Earth and interplanetary environments.

The velocity, altitude, and orientation of a satellite with respect to its environment are driving factors in its design. This chapter will derive basic relationships describing the orbital nature of a satellite which circles one celestial body or is transferred from one planet to another. An operational understanding of terms used to describe the dynamic motion of man-made satellites is a prerequisite for ascertaining the key aspects of chemistry in support of space travel.

First, the theory of circular orbits is developed from the principle of conservation of energy. Elliptical orbits are then explained by an extension of the basic theory of circular orbits. Perturbations to the ideal Earth orbit due to the oblateness of the Earth (bulging at the equator) and the atmosphere are discussed. The orbital decay of a satellite is of special importance since this causes the space system to migrate through a wide range of environmental conditions. Methods are described to change between different types of planetary orbits and from one planet's orbit to another.

Circular Orbits

Law of Universal Gravitation

Two bodies attract each other by the law of universal gravitation:

$$\vec{F} = \frac{GMm}{r^2} \hat{r} \qquad (2.1)$$

where

$\vec{F}$ = force of attraction between M and m
G = universal gravitational constant, $6.6732 \times 10^{-11} N\text{-}m^2/kg^2$
M = mass of the central body
m = mass of orbiting body (assume of $m \ll M$)
r = distance between center of masses of m and M
$\hat{r}$ = unit vector in the radial direction between the center of masses of m and M

This expression applies for any two masses in space, but for our applications it is assumed that one mass is much larger than the other. This larger mass is the central body. The smaller mass m orbits the central body at constant altitude, as shown in Figure 2.1.

The force that is exerted on the satellite is the same in magnitude but opposite in direction to the force exerted on the central body. The satellite orbits the central body since the satellite's mass is much less than that of the central body. This relationship may be simplified by combining G and M to form the gravitational parameter, as

$$\mu = GM. \qquad (2.2)$$

The gravitational parameter values for all the planets of our solar system and the Sun are included in Table 2.1. The gravitational parameter is proportional to the mass of the central body and is an absolute measure of the central body's ability to influence an orbiting object.

By incorporating equation 2.2 into equation 2.1, equation 2.3 is formed. This new representation could easily drop the vector notation since for a two-body problem the locations of the center of mass of each body completely describes the direction of the gravitational force. Yet, the vector notation will be kept for completeness and because atmospheric drag analysis will require it in this form.

$$\vec{F} = \frac{-\mu m}{r^2} \hat{r}. \qquad (2.3)$$

Newton's second law of motion states that a force exerted on an object will cause it to accelerate proportionally. The constant of proportionality is the mass of the body being acted upon or, in this case, the orbiting object. Newton's second law can be expressed mathematically for the prescribed radial motion as

$$\vec{F} = m\vec{a} = m\ddot{\vec{r}}. \qquad (2.4)$$

Equating equations 2.3 and 2.4 yields the following equa-

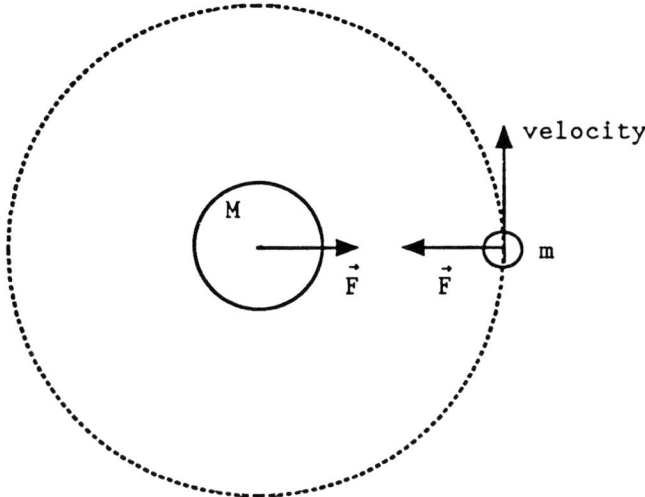

Figure 2.1 The gravitational force exerted on an orbiting satellite always points radially inward toward the central body and perpendicular to the velocity vector.

Table 2.1 Gravitational Parameters

	Mass (kg)	μ (km^3/s^2)
Sun	1.99×10^{30}	1.327×10^{11}
Mercury	3.35×10^{23}	2.232×10^{4}
Venus	4.88×10^{24}	3.257×10^{5}
Earth	5.97×10^{24}	3.986×10^{5}
Mars	6.45×10^{23}	4.305×10^{4}
Jupiter	1.90×10^{27}	1.268×10^{8}
Saturn	5.69×10^{26}	3.795×10^{7}
Uranus	8.72×10^{25}	5.820×10^{6}
Neptune	1.03×10^{26}	6.896×10^{6}
Pluto	5.38×10^{24}	3.6×10^{5}

tion of motion:

$$\frac{-\mu m}{r^2} \hat{r} = m \vec{\ddot{r}} . \quad (2.5)$$

The mass of the orbiting body may be eliminated in equation 2.5 and the terms rearranged to obtain

$$\vec{\ddot{r}} + \frac{\mu}{r^2} \hat{r} = 0 . \quad (2.6)$$

Equation 2.6 is the basic equation used to describe the motion of a small body (satellite) around a large central body.

Conservation of Energy

In the absence of an atmosphere, a satellite in a circular orbit will continue to circle the central body indefinitely. Since the force of gravity on the satellite is always acting radially inward toward the central body's center of mass it is constantly perpendicular to the velocity of the satellite. Thus no force is directed along the direction of the displacement of the satellite. From basic physics it is known that this equates to no work being done on the satellite. If there is no work done on the satellite its mechanical energy (kinetic and potential) remains constant. (Note that this simplistic development will also hold for noncircular orbits since a gravitational force is conservative.)

The observation that a satellite's energy is constant can be derived from equation 2.7. A short development is given in reference 1 yielding the so called vis viva equation

$$E = \frac{V^2}{2} - \frac{\mu}{r} \quad (2.7)$$

where

E = total specific mechanical energy
V = velocity of satellite
r = radius of satellite's orbit

The specific mechanical energy is simply the mechanical energy of a satellite divided by its mass. The first term in equation 2.7, $V^2/2$, is the specific kinetic energy of the satellite. This term is always positive. The second term, $-\mu/r$, is always negative and denotes the potential energy of the satellite per unit mass. This implies that as a satellite's distance from its central attracting body increases its potential energy attains smaller negative values. In the limit, the potential energy goes to zero as r approaches infinity. At this point the central body has no influence on the satellite. Yet, at all other points the satellite has a negative potential energy. This is an artifact of the derivation of this equation. As a satellite gets closer to its central body it attains a larger negative potential energy—a smaller value.

The balance between kinetic energy and potential energy determines the state of a satellite's orbit, as summarized in Table 2.2. If the potential energy is greater than the kinetic energy then the total specific mechanical energy E is negative and the satellite will not be able to escape the pull of its attracting body and thus will remain in orbit. (The various types of noncircular orbits possible will be described later.) Similarly, if the kinetic energy exceeds the potential energy, or $E > 0$, the satellite will escape the influence of its central body; it is on an escape trajectory. For example, when a spaceship travels to Mars it must be given sufficient energy to escape Earth's gravitational pull. As the spacecraft approaches Mars its kinetic energy must be small enough so that it can be captured by the gravitational attraction of Mars.

Elliptical Orbits

Characteristics of an Ellipse

Up to this point only satellites in circular orbits have been considered. This assumption forced the radius vector and the satellite's velocity vector to have a constant perpendicular orientation. Even though the circular orbit is a very specific subset of elliptical orbits it is also the most common orbit encountered. Fortunately, all the material covered thus far is still applicable to elliptical orbits.

Table 2.2 Orbital Energies

$E < 0$, remains in orbit
$E = 0$, minimum escape trajectory
$E > 0$, escape trajectory

Table 2.3 Eccentricity Values

$e = 0$ circular orbit
$0 < e < 1$ elliptical orbit
$e = 1$ parabolic orbit
$e > 1$ hyperbolic orbit

Before investigating attributes of elliptical orbits a complete description of an ellipse is necessary. The upper portion of Figure 2.2 shows an ellipse while the lower portion overlays an elliptical orbit in a similar format. The definition of an ellipse follows.

A conic section is a locus of points in the plane of a fixed focal point F *and a fixed line* d *(the directrix) such that the ratio of the distance from any point on the curve from* F *to its perpendicular distance from* d *is a constant* e *(eccentricity). If* e < 1, *the conic is an ellipse* [2].

Referring to Figure 2.2, the measurement *a* is called the semimajor axis and is half of the largest dimension of the ellipse. The quantity $2a$ is called the major axis. The semiminor axis *b* is half of the shortest dimension of the ellipse. The foci are located the distance "ae" from the center of the ellipse.

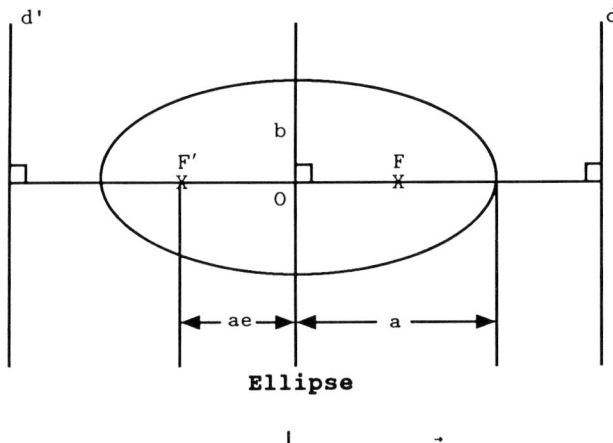

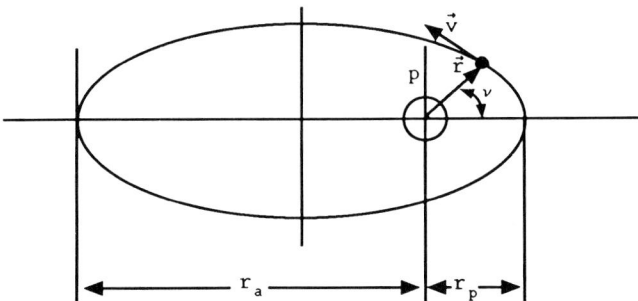

Figure 2.2 The central body is located at one of the foci for an elliptical orbit.

Orbital Parameters *a* and *e*

Many of these terms are used in describing the orbit of a satellite. The central attracting body is located at one of the two foci. The orbiting body's closest approach is called its *perigee*. The radius of perigee r_p is shown in Figure 2.2. For an Earth satellite, an r_p value of 8,350 km equates to an altitude of about 1,972 km since the mean radius of the Earth is 6,378 km. Although orbital dynamics usually use "radius" values, most satellite orbits are described in terms of altitudes.

The *apogee*, shown as r_a, is the point where the satellite attains its highest altitude. For a circular orbit the apogee and perigee are identical since the satellite maintains constant altitude. *The terms* apogee *and* perigee *are used for Earth orbits while* apoapsis *and* periapsis *are more general terms.*

From Figure 2.2 it can be seen that

$$2a = r_a + r_p \qquad (2.8)$$

where $2a$ = major axis.

From this relationship the importance of the semimajor axis *a* is evident. It represents the average distance between the center of mass of the central body and the satellite. This is a key parameter used in describing satellite orbits.

The equation for specific mechanical energy, equation 2.7, for an orbiting satellite can be rewritten more generally as

$$E = -\frac{\mu}{2a}. \qquad (2.9)$$

The time it takes for a satellite to circle a central body (e.g., the Earth) is called the orbital period and is represented by

$$P = \frac{2\pi}{\sqrt{\mu}} a^{3/2}. \qquad (2.10)$$

The amount that a satellite's orbit is elliptical is measured by the eccentricity *e*. This quantity was used in describing the general ellipse in Figure 2.2. The definition of eccentricity using orbital parameters is

$$e = \frac{r_a - r_p}{r_a + r_p} = \frac{r_a - r_p}{2a}. \qquad (2.11)$$

A circular orbit has an eccentricity equal to zero. A satellite in a noncircular orbit about a central body has an eccentricity value between zero and one. An orbit with an eccentricity of one will have a parabolic trajectory, which is a minimum escape trajectory. Lastly, eccentricity values greater than one result in hyperbolic orbits which are escape trajectories. Table 2.3 summarizes the eccentricity values for each type of orbit.

Angular Momentum

The specific angular momentum of an orbiting satellite is represented by

$$\vec{h} = \vec{r} \times \vec{V} = rV \sin \theta \, \hat{k} \qquad (2.12)$$

where

$\vec{h}$ = angular momentum vector
$\vec{r}$ = radius vector
$\vec{V}$ = velocity vector
θ = angle between $\vec{r}$ and $\vec{V}$

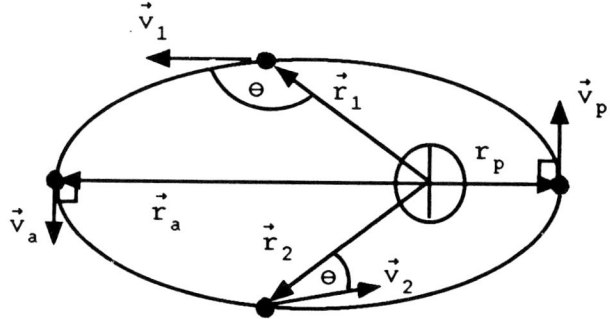

Figure 2.3 Angular momentum is constant about any orbit: as velocity increases, the radius decreases.

Figure 2.3 shows the radius vectors $\vec{r}_1$, and $\vec{r}_2$ drawn from the center of mass of the attracting body to the center of mass of the satellite. The velocity vector is always tangent to the satellite's orbit at apogee and perigee. The angular momentum is constant for a satellite as long as there are no external nonconservative forces acting on it, such as atmospheric drag. Conservation of angular momentum holds under the same conditions as conservation of energy.

Conservation of angular momentum helps to describe some basic interrelationships of orbital motion. In an elliptical orbit the angular momentum is constant therefore it is the same at apogee and perigee.

$$\text{Thus, } r_a V_a = r_p V_p \, . \qquad (2.13)$$

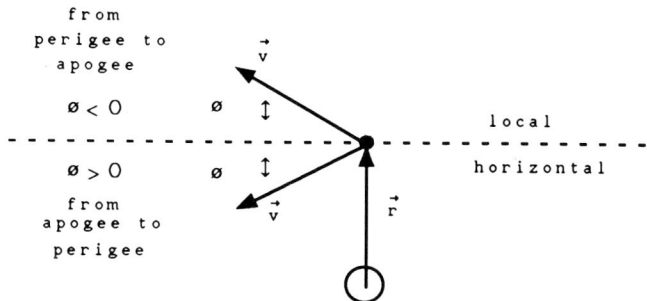

Figure 2.4 The flight path angle describes the satellite's deviation from a path along the local horizontal.

This implies that as the radius increases (from r_p to r_a) the orbital velocity will decrease (from V_p to V_a). Again, this same observation can be made from conservation of energy. As a satellite is going from perigee to apogee it is gaining altitude and the angle between the velocity vector and the radial position vector is greater than 90°. Conversely, as a satellite progresses from apogee to perigee the satellite is losing altitude and the angle is less than 90°.

A more convenient representation of the orientation between the radius vector and velocity vector is the flight path angle, ϕ. The point where ϕ is equal to 90° is defined as the local horizontal. For this scenario, the flight path angle is equal to zero. Figure 2.4 shows the flight path angle for situations in an elliptical orbit. The angle ϕ is positive ($\theta > 90°$ from Figure 2.3) as a satellite goes from perigee to apogee. The flight path angle may be equated to the aeronautical angle of attack which when negative means the craft is descending in altitude and vice versa. The magnitude of the specific angular momentum can now be defined as

$$h = rV \cos \phi \, . \qquad (2.14)$$

Other Orbital Parameters: p, v, and r

The semilatus rectum, p, is shown in Figure 2.2. This quantity can be defined in terms of specific angular momentum or semimajor axis and eccentricity; namely,

$$p = \frac{h^2}{\mu} = a(1 - e) \, . \qquad (2.15)$$

The semilatus rectum is a constant for a given satellite orbit like a, e, and h. Yet, for an elliptical orbit the position of the satellite is constantly changing with respect to the body it is orbiting. A representation for the satellite's position is a function of its location within its orbit. The true anomaly, v, as drawn in Figure 2.2, describes the location of a satellite in its orbit as the angular progression past perigee. True anomaly is 180° at the apogee point and 0° or 360° at perigee. The radius of a satellite's orbit is given by

$$r = \frac{p}{1 + e \cos v} \, . \qquad (2.16)$$

Note that for a circular orbit the true anomaly is undefined. This is not a concern since the radial position for a satellite in a circular orbit does not change over time.

By combining equation 2.16 with previous relationships a number of useful equations may be developed:

$$r_p = \frac{p}{1 + e} = a(1 - e) \qquad (2.17)$$

and

$$r_a = \frac{p}{1 - e} = a(1 + e) \, . \qquad (2.18)$$

Review of Astrodynamics

The orbital velocity of a satellite, V_o, may be described by

$$V_o = \sqrt{\frac{2\mu}{r} - \frac{\mu}{a}}. \quad (2.19)$$

For a circular orbit $r = a$ and equation 2.19 reduces to

$$V_c = \sqrt{\frac{\mu}{r}}. \quad (2.20)$$

Orientation of a Satellite's Orbit

Thus far the size and shape of a satellite's orbit have been defined by a and e. The position of the satellite within its orbit is prescribed by the true anomaly, ν. However, six orbital elements are needed to uniquely describe the location of a satellite in orbit about some central body. The other three elements determine the orientation of a satellite's orbit with respect to the central body.

The first parameter, inclination, is represented by i and is defined as the angle between the central body's equatorial plane and the satellite's orbital plane. It may alternatively be defined as the angle between the orbiting object's angular momentum vector, $\vec{h}$, and the central body's spin vector, $\vec{k}$. The argument of perigee, ω, represents the location of the perigee point with respect to the northward crossing of the equatorial plane. Argument of perigee is measured in degrees in the direction of satellite motion to perigee after the northward crossing of the equatorial plane. An ω of 180° means that the perigee of the orbit is located above the equator as the satellite crosses the equator on its southerly pass while an ω of 90° means that the perigee occurs a quarter of an orbit after it crosses the equatorial plane on a northerly pass. The longitude of ascending node, Ω, is the longitudinal location along the equator where the satellite crosses it on an ascending (northward) pass measured eastward from the vernal equinox. An Ω of zero degrees equates to a satellite crossing the equatorial plane in a northerly direction at the vernal equinox. Figure 2.5 shows the various orbital elements.

The inclination of Earth satellites may be classified in

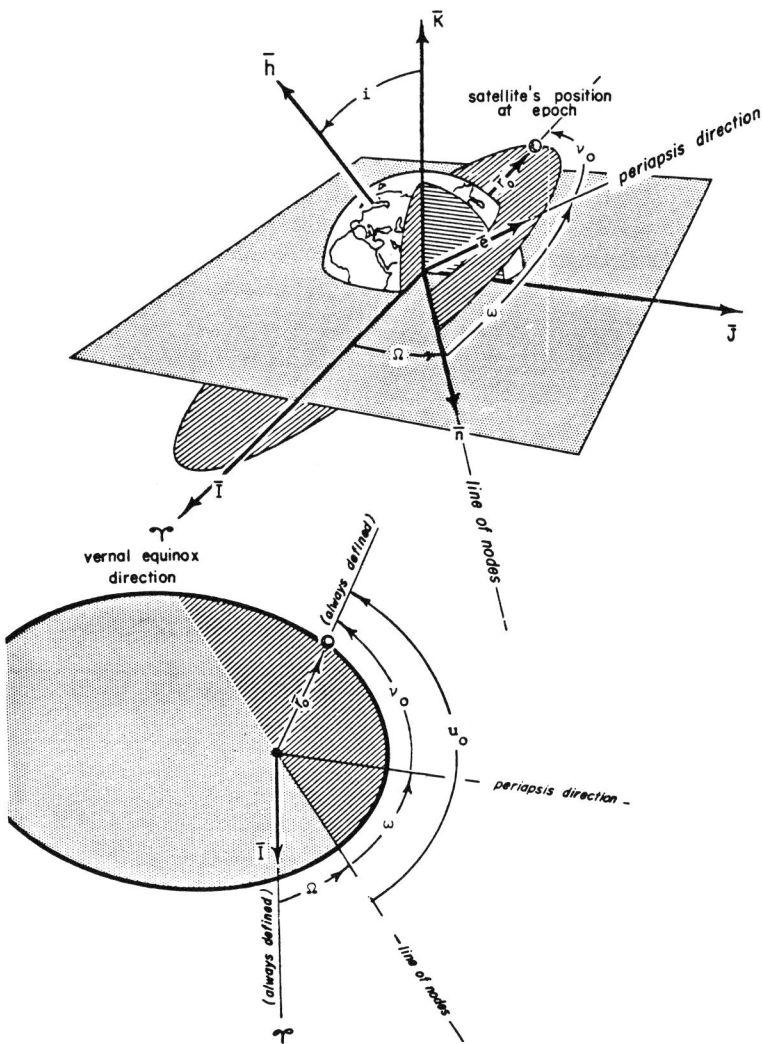

Figure 2.5 The orbital plane orientation is described by three orbital elements (i, ω, Ω) while the orbit's shape is described by e [1].

one of three groups and must be less than 180°:

$i < 90°$ direct orbit
$i = 90°$ polar orbit
$i > 90°$ retrograde orbit

Direct orbits rotate in the same general direction as the rotation of the Earth, west to east. A satellite in a direct orbit can only reach latitudes as high as its inclination. For example, a 28° inclination satellite will only traverse latitudes within 28° north and south of the equator. A low altitude satellite with such an inclination has limited usefulness since it encounters such a small amount of the Earth. On the other hand, if the satellite's altitude is high enough even a satellite with an inclination of 0° may be able to see much of the Earth. This is a major factor in the use of the geosynchronous altitude. This unique orbit will be covered later in this chapter.

Polar orbits have an inclination of 90° which allows them to cover the entire surface of the globe over time. Satellites with inclinations above 90° actually travel in a westward direction, opposite to the rotation of the Earth.

The longitude of ascending node for Earth orbits will vary over time. This perturbation is caused by a bulging at the equator of the Earth causing a slightly greater gravitational pull on a satellite as it approaches the equator than when it is at higher latitudes. If the Earth were a perfect sphere, then this effect, and other orbital perturbations, would be eliminated. The change in Ω over time can be represented by

$$\dot{\Omega} \cong -9.964 \left(\frac{R}{a}\right)^{3.5} (1 - e^2)^{-2} \cos i \frac{\deg}{\text{day}} \quad (2.21)$$

where R = Earth's equatorial radius, 6378.145 km.

Similarly, the value for the argument of perigee will also vary over the lifetime of a satellite as

$$\dot{\Omega} \cong 4.982 \left(\frac{R}{a}\right)^{3.5} (1 - e^2)^{-2}(5[\cos i]^2 - 1) \frac{\deg}{\text{day}}.$$

(2.22)

From equation 2.22 it can be seen that for an inclination of 63.4°, the argument of perigee will remain fixed. A number of highly elliptical satellite systems, mostly Soviet, use this characteristic of orbital motion to keep the satellite's orbital perigee in the southern hemisphere. This forces the apogee to remain in the northern hemisphere where the slow passage and wide coverage are useful for surveillance and communication missions. This type of orbit is referred to as a Molniya orbit.

Geostationary Orbit

A unique orbit warranting individual analysis is the geostationary Earth orbit (GEO). A satellite in GEO has a constant altitude of 35,787 km and thus an orbital period of 1436.1 minutes with an inclination of 0°. This orbit was first considered by Arthur C. Clarke years before the first man-made Earth satellite was launched. He noted that if a satellite were placed in a high enough orbit its angular velocity could be matched to that of the Earth. That is to say, the orbital period of the GEO satellite is 24 hours. An object in GEO appears to remain fixed above some point on the equator despite the fact that its orbital velocity is over 3 km/s. GEO satellites are assigned specific longitudinal locations and because of this the group of GEO satellites are said to make up the geostationary belt, depicted in Figure 2.6.

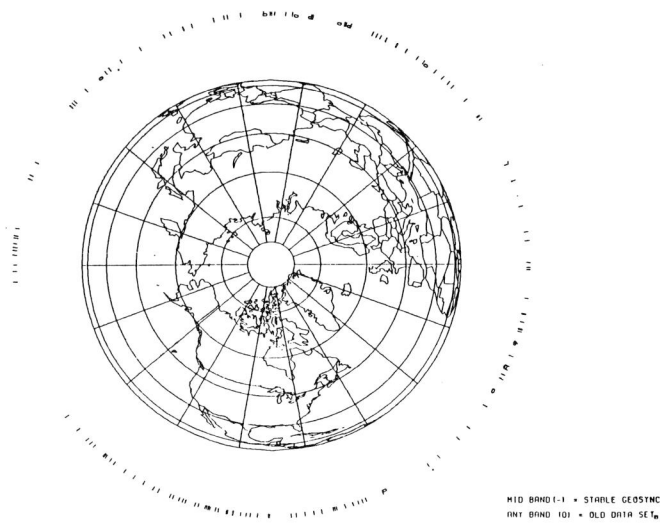

Figure 2.6 The distribution of satellites about the geostationary belt.

Geostationary satellites that have nonzero inclinations are called geosynchronous. They have identical orbital periods to geostationary satellites but their inclinations cause their ground trace to be a figure eight instead of a point.

Satellites at these altitudes will encounter a significantly different environment than low Earth orbit (LEO) satellites. Satellites with orbital periods less than 128 minutes (or average altitude less than 2,000 km) are considered LEO. The difference in environments will be covered in Chapter 3. Due to the unique characteristics of GEO the mission support requirements will also differ from the requirements for a LEO satellite.

Orbital Maneuvers

Satellites are not launched directly into GEO. They usually are placed in an inclined LEO parking orbit and then transferred into GEO. The move from LEO to GEO takes two independent orbital maneuvers. The Hohmann transfer

Review of Astrodynamics

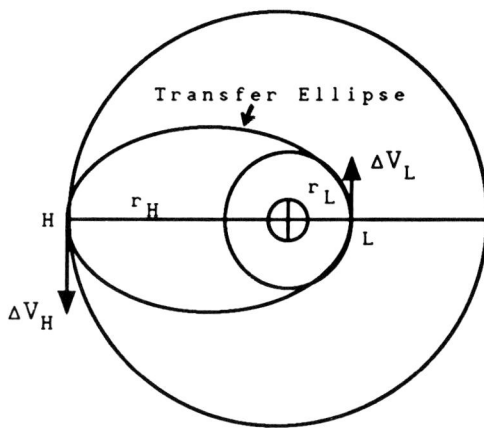

Figure 2.7 A Hohmann transfer consists of an elliptical trajectory between two circular orbits.

makes the satellite orbit larger while a plane change is required to change the inclination.

The Hohmann transfer, Figure 2.7, is the simplest and most energy efficient of all possible transfers between two circular orbits. It consists of the following three major steps:

1. Establish a circular parking orbit.
2. Provide an impulse or velocity increment (ΔV_L) to establish the Hohmann transfer ellipse.
3. Provide an impulse to produce a velocity increment (ΔV_H) to circularize the final orbit.

After establishing a circular parking orbit, an impulse (produced by a rocket burn) is imparted to the satellite so that the apogee of the newly formed transfer ellipse will equal the radius of the destination orbit. This first impulse, which gives ΔV_L, should be made tangential to the orbit. This thrust will cause this point to become the perigee of the transfer ellipse. The velocity needed at this point (perigee of transfer ellipse) is given by equation 2.19 modified to

$$V_p = \sqrt{\frac{2\mu}{r} - \frac{\mu}{a}} = \sqrt{\frac{2\mu}{r_L} - \frac{\mu}{(r_L + r_H)/2}}. \quad (2.23)$$

The velocity increment required at low point L is

$$\Delta V_L = V_p - \sqrt{\frac{\mu}{r_L}}. \quad (2.24)$$

This impulse is required to add enough velocity, ΔV_L, to the first circular orbit's velocity, $V_{CL} = \sqrt{\mu/r_L}$, to achieve the required V_p, as given by equation 2.23, to produce the required transfer ellipse. A similar impulse is then required at high point H to circularize the transfer ellipse. The final required velocity is therefore

$$V_{CH} = \sqrt{\frac{\mu}{r_H}}. \quad (2.25)$$

At point H (apogee of transfer ellipse) the spacecraft has a velocity

$$V_a = \sqrt{\frac{2\mu}{r_H} - \frac{\mu}{(r_L + r_H)/2}}. \quad (2.26)$$

Again, the impulse at H yields the difference between equations 2.25 and 2.26, namely

$$V_H = V_{CH} - V_a. \quad (2.27)$$

A Hohmann transfer may be made by itself or in conjunction with an orbital plane change. A plane change requires an impulse perpendicular to the orbital plane to change the orientation of the angular momentum vector and thus a satellite's inclination. Again, only the simplest plane change maneuver will be considered. This maneuver would occur as a satellite in a circular orbit crosses the equator. This scenario produces a required velocity increment

$$\Delta V = 2V_o \sin(\gamma/2) \quad (2.28)$$

where

ΔV = out-of-plane velocity increment, km/s
V_o = orbital velocity, km/s
γ = change in inclination

A plane change requires much more fuel than does a Hohmann transfer. From a physics point of view this can be explained by observing that it is much more difficult to change the direction of the angular momentum vector than it is to change its magnitude. Table 2.4 outlines some characteristic values for impulses.

The transfers require the same magnitude of delta velocity if performed in the reverse direction. Note that a change in inclination of 30° requires almost as much propulsive impulse as a transfer from LEO to GEO. At a 500 km altitude, orbital velocity is 7.61 km/s. A delta velocity impulse of this amount directed tangentially to the orbital plane will cause the satellite to escape the Earth's gravitational attraction. This same impulse applied normally to the orbital plane can only change the satellite's inclination by 60° if done from a 500 km circular orbit. From equation 2.28 it can be seen that a plane change maneuver will be more efficient if it occurs when the satellite's velocity is lower. For a Hohmann transfer, the lowest velocity occurs at the apogee of the transfer ellipse; hence, this is the location where most plane changes are made. As a matter of fact, the second burn of the Hohmann transfer and the plane change burn are often performed simultaneously.

Interplanetary Trajectories

Sending a space probe from Earth to another planet is quite similar to placing an Earth satellite into a prescribed orbit. One major difference is that an interplanetary trajectory will cause the probe to come under the influence of three bodies

Table 2.4 Orbital Maneuvers

Hohmann Transfer	Total ΔV(km/s)	Plane Change (alt-500 km)	ΔV(km/s)
100 km to 500 km	0.23	$i = 0°$ to $i = 15°$	1.99
100 km to 1000 km	0.49	$i = 0°$ to $i = 30°$	3.94
100 km to 2000 km	0.94	$i = 0°$ to $i = 45°$	5.83
500 km to GEO	4.54	$i = 0°$ to $i = 60°$	7.62

during its transit. A second difference is that a simple elliptical transfer (like a Hohmann transfer) will not usually work; a patched conic approach must be taken. The Hohmann transfer is the most energy efficient transfer but the slowest. For interplanetary missions this may not be feasible. For a Hohmann transfer it would require over 8 months to reach Mars and 45 years to arrive at Pluto.

The patched conic maneuver usually takes place in less than half of a transfer ellipse. The exact timing of launch and rendezvous with the target planet is a function of the orbits of the two planets. The patched conic starts with an impulse to escape the Earth's sphere of influence (SOI). The probe will then move toward the other planet under the influence of the Sun's gravitational field following a transfer ellipse similar to that used for moving a satellite from LEO to GEO. As the probe approaches the target planet and its SOI, the probe will slow. A final, precise thrust may be required to place the probe into its final mission profile.

The design of an interplanetary mission requires four steps [5]. First, the two planets concerned must be located. This requires the use of an ephemeris—a record of the position of the planets over time. Second, a transfer ellipse must be designed. Due to the differential in planetary orbital periods and phasing combinations, the possibility of a rendezvous launch window is usually restricted to a few minutes in certain days in a few select years. This fact puts significant constraints on the launching of interplanetary missions. Third, the departure hyperbola must be designed. The goal of this step is to provide the proper velocity vector for the start of the transfer ellipse to the target planet. Fourth, the arrival hyperbola must be developed to transition the probe from Sun-orbiting. The patched conic transfer for a Mars probe is shown in Figure 2.8.

Orbital Decay

A major topic in the study of man-made Earth satellites is orbital decay. Desmond King-Hele's work *Satellite Orbits in an Atmosphere* [3] represents the best coverage of this perplexing problem. Basically, a satellite's orbit is impeded by drag forces due to interaction with the atmosphere. The less massive the object the more effect atmospheric drag will have on it.

The aerodynamic drag force per unit mass is given by

$$\frac{D}{M} = \frac{1}{2}\rho V^2 \left(\frac{AC_D}{M}\right) \qquad (2.29)$$

where

D = drag force
M = satellite mass
ρ = atmospheric density
V = velocity of satellite
A = cross-sectional area of the satellite
C_D = coefficient of drag, typically 2.0–2.2

Often this equation is simplified to

$$\frac{D}{M} = \frac{1}{2}\rho V^2 \delta \qquad (2.30)$$

where

$$\delta = \frac{AC_D}{M}.$$

The parameter δ is nearly equivalent to the area-to-mass ratio since $C_D \approx 2.0$. So as δ increases, the drag increases since the area-to-mass ratio must rise proportionally. The δ term is often referred to as the ballistic coefficient.

The change in orbital parameters, a and e, as a function of time can be shown to be

$$\dot{a} = -\frac{a^2 \rho V^3 \delta}{\mu} \qquad (2.31)$$

and

$$\dot{e} = -\rho V \delta (e + \cos \nu). \qquad (2.32)$$

The general effect of atmospheric drag on an elliptical orbit is described by the last two equations and is shown in Figure 2.9. The negative signs denote that both a and e decrease over time due to atmospheric drag, and these changes are proportional to the ballistic coefficient. The drag effects at perigee will act to reduce apogee height while affecting perigee height very little. Thus, for satellites in elliptical orbits the rate of orbital decay is strongly dependent on perigee height. More precisely, orbit contraction is a function of the atmospheric density at perigee. Since the

Review of Astrodynamics

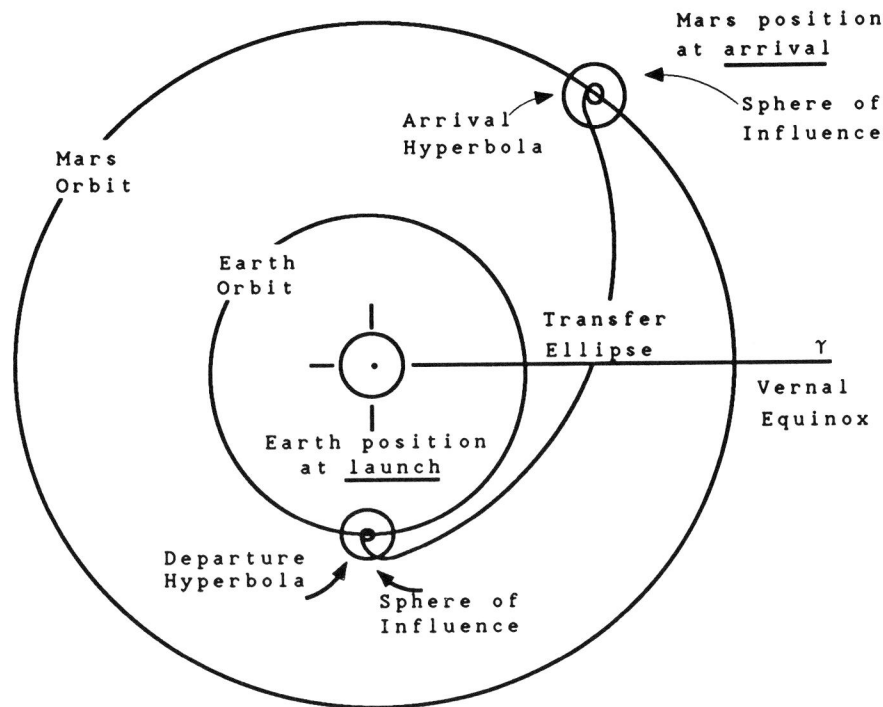

Figure 2.8 An interplanetary patched conic generally has three flight segments: departure hyperbola, transfer ellipse, and arrival hyperbola.

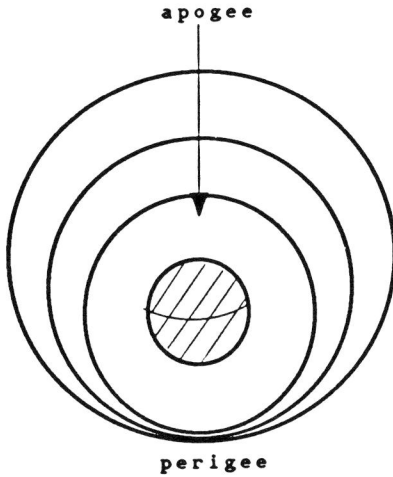

Figure 2.9 As a satellite experiences atmospheric drag, the primary effect is to change an elliptical orbit to a circular one by a lowering of the apogee altitude.

density varies exponentially, a small change in altitude will cause a great change in density and thus in orbital decay.

Once the orbit becomes circular, $e = 0$, the change in semimajor axis per orbit, Δa (in meters), is given by

$$\Delta a = -2\pi\delta a^2 \rho . \qquad (2.33)$$

where

- a = semimajor axis or radius of a circular orbit, meters
- ρ = atmospheric density at altitude, kg/m³

For a circular orbit, a large δ results in a quickly contracting orbit. The term δ is proportional to the satellite's area-to-mass ratio; thus, for long-lived orbits a small cross-sectional area and a large mass are best.

Orbital Lifetime

The characteristics that lead to a slowly contracting orbit (high perigee and low δ) also produce long orbital lifetimes. For a satellite in a circular orbit, the lifetime L_c is given as

$$L_c = \frac{H}{\sqrt{\mu a}\,\rho\delta} \qquad (2.34)$$

where

- ρ = atmospheric density, kg/m³
- H = density scale height, km
- μ = Earth's gravitational parameter, (km)³/s²
- L_c = lifetime for a circular orbit with semimajor axis a.

Note that as atmospheric density and ballistic coefficient (roughly area-to-mass ratio) values increase, the orbital lifetime decreases. Similarly, the lifetime is proportional to the density scale height which will be covered in more detail in Chapter 3.

For satellites in orbits between 300 and 900 km, solar activity has a major effect on lifetime calculations. Increased solar activity heats up the atmosphere; as it expands, larger density values are sensed at the same geometric altitudes. Even though chemistry principles predict that higher temperatures produce lower densities, the atmosphere is not a continuum. So when it expands, portions of

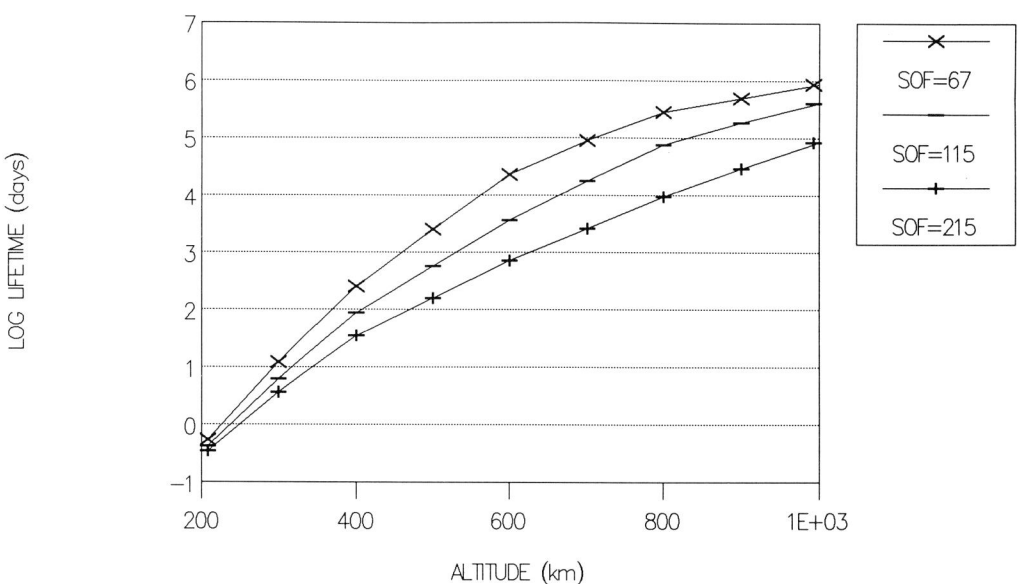

Figure 2.10 As solar activity increases, the orbital lifetime for a given satellite and altitude decreases.

the atmosphere move up in altitude which actually produces greater densities. Some altitudes will have lower densities but are too low to affect space operations. The periods of maximum solar activity in 1979–1980 and 1988–1990 caused a large number of satellites to decay more than during other years.

Figure 2.10 shows typical satellite lifetimes at a number of circular altitudes at varying solar activity levels. The $F_{10.7}$ cm solar flux (SOF) will be outlined in Chapter 3. The effect of solar activity on lifetime will also be discussed in more detail in Chapter 3. In general, increased solar activity causes the atmosphere to expand and results in higher atmospheric densities at orbital altitudes and therefore shorter lifetimes.

The lifetime of satellites in elliptical orbits can be found by using the previously derived circular orbit lifetime, L_c. By examining the decay of hundreds of satellites in elliptical orbits an empirical relationship was developed for the lifetime of satellites in elliptical orbits[4]. The lifetime for a satellite with an eccentricity, e, and a perigee altitude equal to the average altitude for a circular orbit with a lifetime L_c is given by

$$L_e = L_c \left\{ 1 + m(H_p) \tan^{3/2}\left(\frac{\pi e}{2}\right) \right\} \quad (2.35)$$

where

$$\log_{10} m(H_p) = 6.7836 - 2.8025 \log_{10} H_p + 0.3447(\log_{10} {}^2 H_p)$$

H_p = density scale height at perigee

The function $\log_{10} m(H_p)$ is simply an empirically derived relationship which depends greatly on the density scale height at perigee, H_p. Overall, the bracketed term in equation 2.35 is simply a factor of how much longer L_e is than L_c. As expected, as e goes to zero $L_e = L_c$. More sophisticated approaches have been used since, but this method has provided the most reliable results for the authors. Figure 2.11 shows lifetime values for a series of orbital conditions. For comparison, note that for this plot log 0 corresponds to 1 day, log 2 is close to 3 months, log 3 is about 1 year, log 4 is about 25 years, log 5 is just over 250 years, and log 6 is nearly 2,800 years.

The last major parameter which may affect the lifetime of a satellite is its area-to-mass ratio or A/M. The larger this ratio, the more drag perturbs the satellite's orbit thus reducing its lifetime. Figure 2.12 depicts the lifetime for a variety of A/M ratios with circular orbits under average solar activity. The A/M ratios have units of m^2/kg.

Review of Astrodynamics

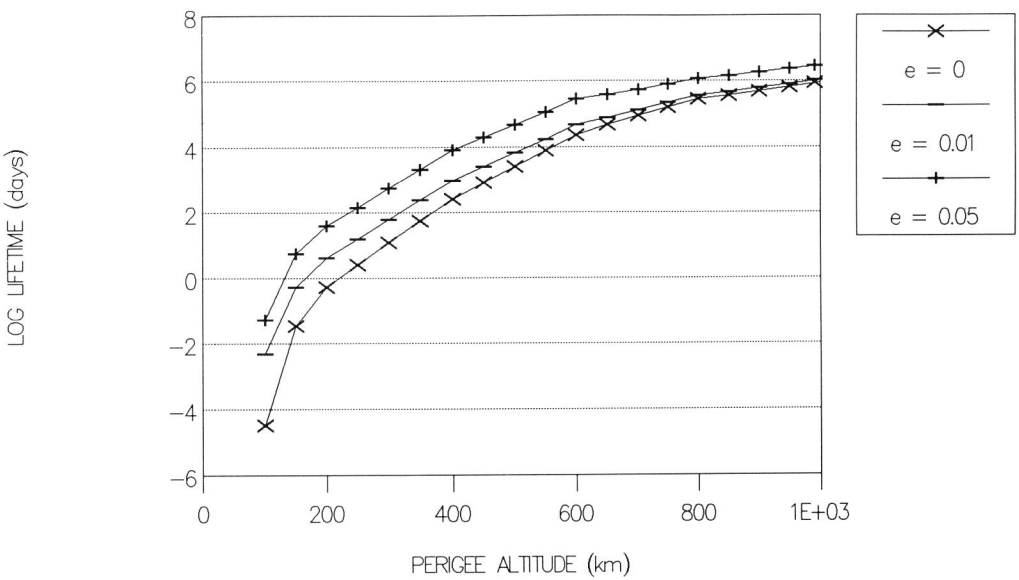

Figure 2.11 Increased orbital eccentricity with a common perigee results in greater orbital lifetimes.

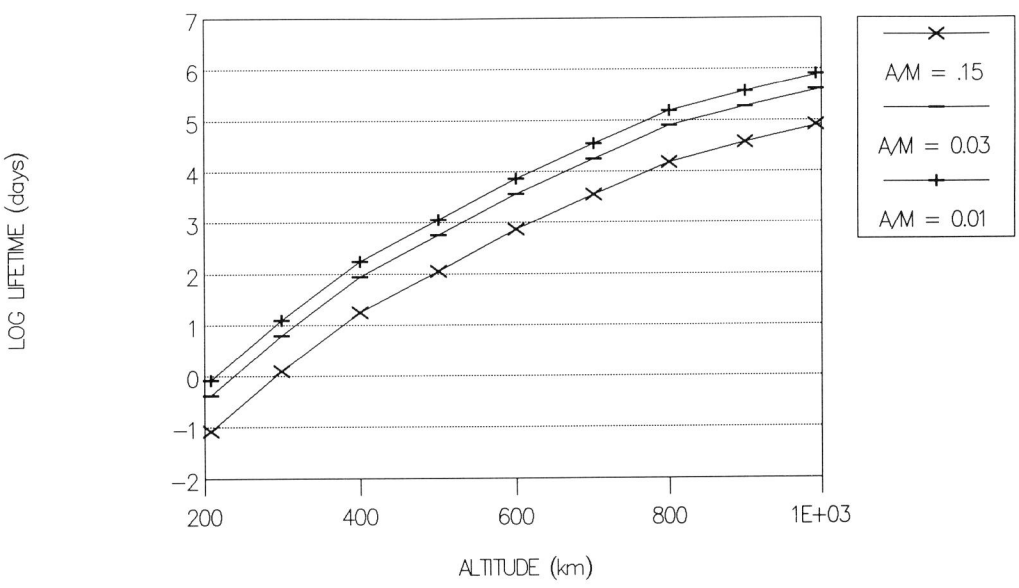

Figure 2.12 The orbital lifetimes of satellites with larger area-to-mass (*A/M*) ratios will be affected more by atmospheric drag.

Summary

When Kepler proposed his laws of planetary motion and Newton precisely outlined the physics of orbits, they could never have imagined the impact on future generations. There are now hundreds of operational satellites and thousands more abandoned pieces of hardware in orbit that crisscross at high speeds relative to each other. Being subjected to the harshest of environments, satellites orbiting the Earth are built to withstand the rigors of launch and cosmic radiation and particulate hazard during operation. The study of astrodynamics allows the planner to ascertain the location of the satellite in the dynamic space environment as a function of time. If the system is in the wrong orbit, the power of thermochemistry propels it to the correct one. Electrochemistry principles describe the lifeblood of the operating system of the satellite. The use of chemical reactions to produce electricity provides the satellite with the power necessary to maintain its proper orientation and perform its mission.

Many attributes of space chemistry concern the interaction of the satellite system with the environment. Many of these effects actually change the orbital energy of the satellite, thus causing it to change its orbital parameters. A change in altitude of only a few hundred kilometers may cause different environmental concerns (e.g., atomic oxygen vs. atmospheric drag) to dominate. Interaction with the Earth's magnetic field and its associated trapped energetic particles is much greater for high inclination Earth satellites. If this hazard is too great, then the user must apply astrodynamics techniques to maneuver away from threats posed by the environment or other systems. Similarly, successful completion of manned missions depends on the use of chemistry to provide life support systems.

Astrodynamics simply defines the rules of the road in space. These rules have been developed to provide a basic understanding of orbital trajectories as related to simple two-body motion. The introduction of an environment to the problem causes the orbital decay and eventual reentry of many satellites. The static composition and dynamic characteristics of the Earth's atmosphere will be covered in the next chapter.

Summary of Key Equations

$\vec{F} = \dfrac{GMm}{r^2}\hat{r}$ (2.1) Law of universal gravitation

$\ddot{\vec{r}} + \dfrac{\mu}{r^2}\hat{r} = 0$ (2.6) Orbital motion about a central attracting body

$E = \dfrac{V^2}{2} - \dfrac{\mu}{r}$ (2.7) Total specific mechanical energy

$T = \dfrac{2\pi}{\sqrt{\mu}} a^{3/2}$ (2.10) Orbital period

$e = \dfrac{r_a - r_p}{r_a + r_p}$ (2.11) Orbital eccentricity

$p = \dfrac{h^2}{\mu} = a(1 - e)$ (2.15) Semilatus rectum

$r = \dfrac{p}{1 + e \cos \nu}$ (2.16) Radial position

$r_p = \dfrac{p}{1 + e} = a(1 - e)$ (2.17) Perigee radius

$r_a = \dfrac{p}{1 - e} = a(1 + e)$ (2.18) Apogee radius

$V_o = \sqrt{\dfrac{2\mu}{r} - \dfrac{\mu}{a}}$ (2.19) Orbital velocity

$\Delta V = 2V_o \sin \dfrac{\gamma}{2}$ (2.28) Velocity increment for an impulsive plane change

$L_c = \dfrac{H}{\sqrt{(\mu a)}\rho\delta}$ (2.34) Circular orbit lifetime

$L_e = L_c \left\{ 1 + m(H_p) \tan^{3/2} \left(\dfrac{\pi e}{2} \right) \right\}$ (2.35) Elliptical orbit lifetime

References

1. Bate, R. R. et al. *Fundamentals of Astrodynamics.* Dover Publications, Inc., N.Y., 1971.
2. *CRC Standard Mathematical Tables.* Edited by William H. Beyer, CRC Press, 24th edition, 1976.
3. King-Hele, D. *Satellite Orbits in an Atmosphere.* Blackie and Son Ltd., Bishop Briggs, Glasgow, 1987.
4. Jacchia, L. G., and Slowey, J. "Formulae and Table for the computation of Lifetimes of Artificial Satellites." SIAO Research in Space Science Special Report No. 135, 16 Sept 1963.
5. Brown, C., *Aero 556 Spacecraft Design*, Professor Publishing, Boulder, CO, 1985.

Discussion Questions

1. When a satellite orbits the Earth the satellite exerts a force on the Earth of equal magnitude but opposite direction to the force the Earth exerts on the satellite. Why doesn't the Earth orbit the satellite?

2. A satellite's position and velocity can be uniquely defined with six components but astrodynamicists often describe a satellite's state with six orbital elements.
   a. List the six classical orbital elements.
   b. What two orbital elements describe the size and shape of an orbit, respectively?
   c. What three orbital elements describe the orientation of a satellite's orbit with respect to the Earth?

Review of Astrodynamics

d. What does true anomaly (ν) represent?
e. What eccentricity value is the most common for satellites in low Earth orbit?

3. Assuming that atmospheric drag is negligible, a satellite will orbit the Earth indefinitely.
   a. Why doesn't the angular momentum of a satellite, with respect to Earth, change as it orbits?
   b. Why would atmospheric drag affect a spacecraft's orbit?
   c. What orbital parameters are affected first by atmospheric drag?

4. Why do the Soviets place their Molniya satellites into 63.4° inclination orbits?

5. Is a Hohmann transfer the fastest way to move between two circular orbits? Explain the advantage of using a Hohmann transfer trajectory.

6. Why is it so much more fuel intensive to change a satellite's orbital inclination instead of its semimajor axis?

7. When an interplanetary probe is deployed, it progresses through three spheres of influence. Explain this process and describe each phase.

8. Why do elliptical orbits tend to circularize under the influence of atmospheric drag?

9. If you wanted a satellite launched into a 600 km circular orbit to remain in orbit as long as possible, what design parameters would you alter?

10. The gravitational force exerted on a satellite is a function of the Earth's mass, yet it is often called an inverse square force field. Explain.

11. A satellite that orbits the Earth has a negative total specific mechanical energy.
    a. Is its kinetic energy positive or negative? Explain.
    b. Is its potential energy positive or negative? Explain.

12. Explain the effects of varying solar activity and satellite area-to-mass ratio on orbital lifetime.

Problems

1. Assume that a new planet was discovered millions of kilometers beyond Pluto. It has a mass of 4.26×10^{24} kg. What is the gravitational parameter for our new planet?

2. A satellite about Mercury is in a circular orbit with a semimajor axis of 4,560 km.
   a. What is the total energy of the satellite?
   b. What is its velocity?

3. An Earth satellite has an eccentricity of 0.015 and a perigee altitude of 500 km.
   a. What is its apogee altitude?
   b. What is its semimajor axis?

4. The Hubble Space Telescope was launched into a circular orbit with an average altitude of 615 km and an inclination of 28°.
   a. What is its orbital velocity?
   b. Is this a direct or retrograde orbit? Explain.
   c. How quickly will its longitude of ascending node (Ω) regress? (Express answer in degrees per day.)

5. The development of a geostationary satellite normally requires three phases: 1) establish circular orbit, 2) provide impulse to establish transfer ellipse, and 3) provide impulse to circularize orbit.
   a. If the parking orbit for a GEO satellite is 300 km circular, what impulse is required to effect the transfer ellipse to GEO?
   b. What impulse is required at geosynchronous altitude to circularize the orbit?

6. The Space Shuttle is normally launched into a 28° circular orbit with an average altitude of 350 km.
   a. What ΔV is necessary for the Shuttle to move into a 57° orbit of the same altitude?
   b. Will it take more or less fuel than the Shuttle for the Hubble Space Telescope (615 km, $e = 0.0$, $i = 28°$) to be moved to a 57° inclination? Explain.

Chapter 3
Near Earth Space Environment

Millions of years ago the Earth's atmosphere consisted of toxic, poisonous gases. As living cells evolved, organisms began to grow and adapt to their surroundings. They eventually consumed the deadly compounds and produced, as by-products, oxygen, carbon dioxide, nitrogen, etc. The buildup of these chemicals as an atmosphere created a protective shield about the Earth that over time has produced a habitable planet. Our atmosphere's present state is only a split second snapshot in its million, million year lifetime. Luckily for us the Earth is now enveloped in a life-giving cloud of gases. This situation is unique to the solar system. The presence of living organisms on Earth is a direct result of our atmosphere and the presence of life produces chemicals which reinforce the composition and durability of this unique atmosphere. No other atmosphere in our solar system provides this function.

Our atmosphere is still changing slowly over time due to natural phenomena, but the industrialization of the world has caused greater and much faster changes to Earth's atmosphere. Low altitude air pollution from automobiles, wood burning, and industries has been a concern for decades. The last decade has seen the rapid increase in CO_2 and depletion of O_3 in the upper atmosphere which may greatly affect the future habitability of our planet.

Yet the basic constituents of our near Earth environment have changed very little over the past decades. The give and take between the terrestrial biosphere and the atmosphere is a special chemical equilibrium process which never quite balances. People on the Earth see the atmosphere as independent, fast moving weather fronts, slow moving fog, and heat waves. This dynamic behavior is the result of non-uniform heating at the Earth's surface by the Sun and the rotation of the Earth and is a direct function of the composition of the atmosphere.

These dynamic activities set the boundary conditions for the chemical cycles that sustain our planet's atmosphere. The constituents of the sea level atmosphere diffuse as the altitude increases with a corresponding lessening effect of gravity and an increasing influence of solar radiation. These phenomena create a complicated system governed by chemical reactions which can be described only through a vast number of in situ measurements.

This chapter presents the structure of Earth's atmosphere by molecular composition, temperature, density, and pressure. These parameters will determine a satellite's capability to perform its mission for its operational lifetime. Emphasis is placed on altitudes above 100 km (upper atmosphere) because below this height an orbiting satellite will remain aloft for less than one orbit.

General Description: Up to 100 km

Earth's atmosphere may be grouped in layers which describe each region. Table 3.1 outlines the names and Greek or Latin derivatives for each.

These terms describe the general characteristics of each atmospheric layer. The first two spheres are not parts of our atmosphere in the strict sense of the word, but both are vital to the overall well-being of the near Earth environment. The lithosphere is the stone and the soil of Earth which holds many gaseous substances essential to the healthy state of our atmosphere. Limestone ($CaCO_3$) contains tens of thousands of times more CO_2 than does our atmosphere. When limestone is formed it consumes the CO_2 but it is released back into the atmosphere when limestone is decomposed. This reaction can be written as

$$CaCO_3 + H_2O + CO_2 \rightleftarrows Ca^{2+} + 2HCO_3^- . \quad (3.1)$$

A similar reaction which involves the transforming of silicate rocks (e.g., diopside $CaMgSi_2O_6$) into limestone can be written as [2]

$$CaMgSi_2O_6 + CO_2 \rightleftarrows MgSiO_3 + CaCO_3 + SiO_2 . \quad (3.2)$$

The hydrosphere consists of the condensed water residing on the Earth's surface. This includes oceans (97%), snow/polar sheets (2.4%) and freshwater lakes/rivers (0.6%). Over three-quarters of the Earth is covered with water which exchanges compounds with the atmosphere. As a result, the Earth's atmosphere has the largest relative abundance of water vapor of any in the solar system [2]. The oceans of

Table 3.1 Atmospheric Layers

Atmosphere	= 'atmos'	= vapour,	'sphaira'	= sphere
Lithosphere	= 'lithos'	= stone,	'sphaira'	= sphere
Hydrosphere	= 'hydro'	= water,	'sphaira'	= sphere
Troposphere	= 'tropos'	= turning,	'sphaira'	= sphere
Stratosphere	= 'stratus'	= layered,	'sphaira'	= sphere
Mesosphere	= 'meso'	= middle,	'sphaira'	= sphere
Thermosphere	= 'thermo'	= heated,	'sphaira'	= sphere
Exosphere	= 'exo'	= outer,	'sphaira'	= sphere

our planet actually contain about 60 times as much carbon dioxide as does the lithosphere. The CO_2 contained in water reacts to form carbonic acid which is the reason for the natural acidity of rain water (pH = 5.6). This series of reactions may be written as

$$CO_2 + H_2O \rightleftarrows H_2CO_3 \rightleftarrows H^+ + HCO_3^- \qquad (3.3)$$

Constituents

The true atmosphere begins at the surface of the Earth and extends up to 85–100 km, as shown in Figure 3.1.

The first layer, the troposphere, is characterized by turbulence and mixing due to differential heating and extends to altitudes as high as 15–25 km. Particles may rise vertically tens of kilometers in minutes to days [3]. The lower edge of the troposphere occurs at sea level where the average temperature is 290 K (63°F or 17°C). The composition of the atmosphere at sea level is given in Table 3.2. Note that hydrogen is not a key element at low altitudes, but becomes important around an altitude of 150 km [4].

As one progresses in altitude through the troposphere, the temperature decreases almost linearly to 220 K (−53°C or −63.4°F) up to 10 km [5]. The tropopause is the theoretical transition from the troposphere to the stratosphere. The actual transition from the troposphere to the stratosphere varies by season and latitude. The troposphere is the thinnest of all layers but it contains about 75% of the total mass of Earth's atmosphere which is 5×10^{18} kg [2].

The stratosphere rises from the tropopause (10–15 km) to an altitude of 50 km. The temperature is fairly constant from the troposphere up to 25 km then steadily increases to about 280 K (45°F or 7°C) at the upper edge of the stratosphere. This layer of the atmosphere is characterized by so little vertical movement that years are required for molecules to transit the entire stratosphere. The generation and depletion of ozone occurs in the stratosphere. This vital molecular constituent will be discussed in more detail later in this chapter.

The mesosphere extends from 50 to 85 km in altitude. The temperature decreases evenly down to 180 K (−117°F or −83°C) at the mesopause (top of the mesosphere) [5]. The first three atmospheric layers—troposphere, stratosphere, and mesosphere—show very little chemical activity since most of the high energy solar radiation never penetrates to these altitudes. High energy radiation has shorter wavelengths and as such is absorbed by molecules at higher altitudes [2]. The changes in temperature are very slight up to 85 km (the mesopause).

The thermosphere marks the beginning of increased ultraviolet (UV) radiation absorption and a rapid increase in temperature [5]. In the thermosphere we transition from the lower atmosphere to the upper atmosphere. The lower atmosphere shows small erratic changes in physical parameters such as temperature and atmospheric density (Figure 3.2). The upper atmosphere is characterized by roughly an exponentially varying model. In the thermosphere the temperature increases with altitude.

The temperature in the exosphere (exospheric temperature, T_e) is determined by solar activity. The effects of solar activity on our atmosphere will be discussed more fully in the section covering the upper atmosphere. Very high temperatures, from 600 K to 2,000 K may be experienced in the exosphere. Actually, at altitudes above 500 km the term temperature is no longer appropriate. The mean free path of atoms is so large that they do not continually interact with each other. Some of the particles escape Earth's gravitational pull, others collide, while others fall back toward Earth [5]. Under these conditions atmospheric heat is not conducted to or from a satellite very efficiently. Heat is only absorbed via radiations, i.e., direct sunlight. This physical phenomenon becomes very important in designing thermal and attitude control systems for satellites.

Ozone

Before continuing on to a detailed analysis of the upper atmosphere model, a short discussion of the ozone depletion/growth cycle will be covered. Ozone is trioxygen, represented by O_3. Most of the ozone resides in a region from 15–45 km altitudes, centered around 25 km. The concentrations of O_3 may vary greatly but are usually a few tenths of a part per million (ppm) and may achieve a maximum of 10 ppm (10^{-5}) [2].

Ozone is important because it absorbs biologically lethal radiation in the 200–300 nm wavelength range [1]. Ultraviolet radiation with a wavelength of 290 nm or less will kill living cells, while N_2 and O_2 will only absorb and filter out radiant energy below 230 nm [2]. Thus, without ozone in the atmosphere, life, as we know it, would not exist on Earth. The UV radiation, between 290–320 nm, which is not completely absorbed by ozone has been linked to the occurrence of skin cancer [1]. This is only a small taste of what we could expect if ozone is significantly depleted from our atmosphere. The absorption of UV and visible radiation by ozone in the atmosphere also drives the circulation in the stratosphere and mesosphere [1]. Yet, while ozone protects our life on Earth it is dependent on atmospheric oxygen for its creation.

UV radiation possesses sufficient energy to break the

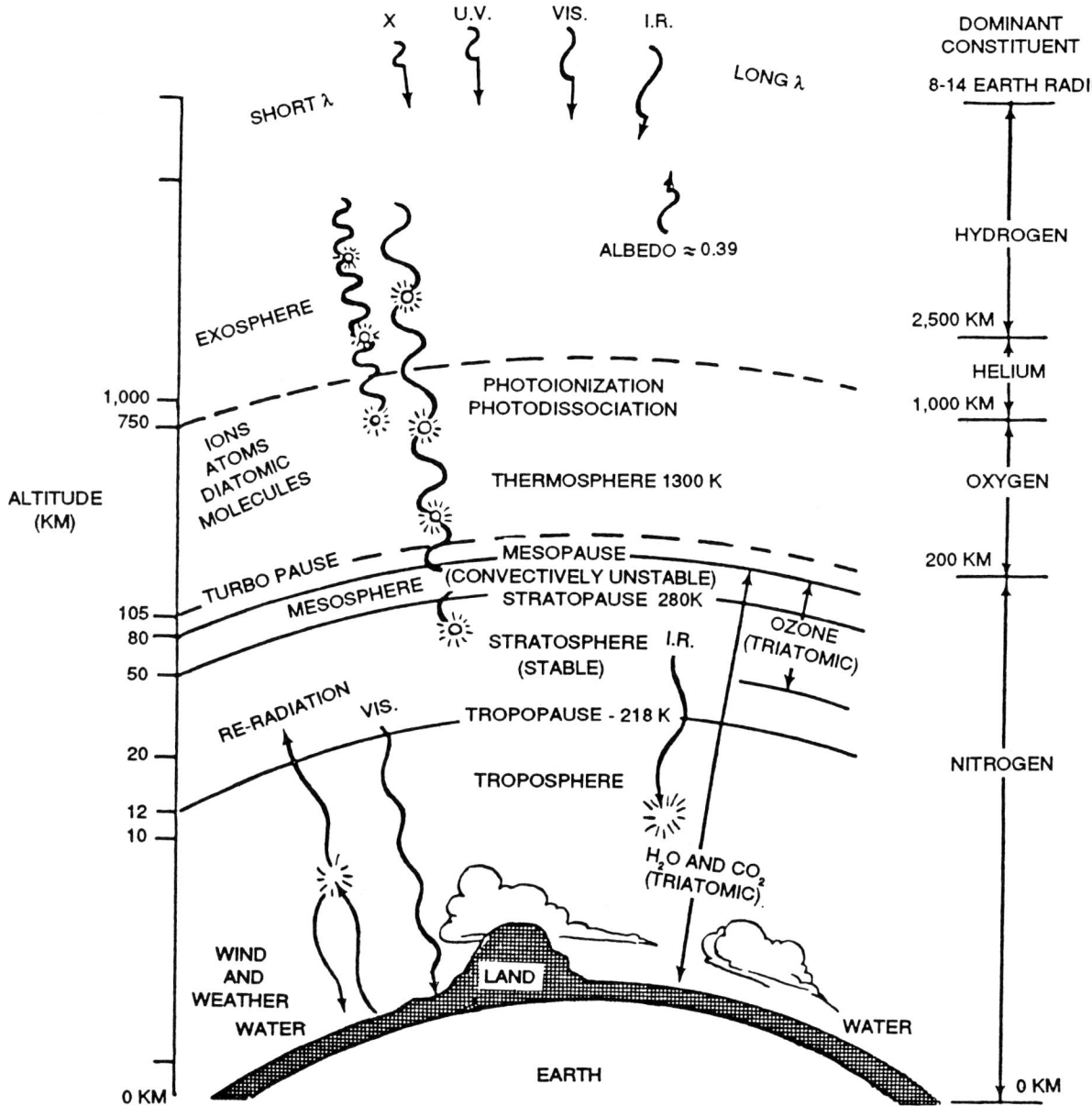

Figure 3.1 Earth's atmosphere is distinctly layered. It is partitioned mainly by the dominant physical reactions and major elemental constituents [3].

Table 3.2 Composition of Earth's Atmosphere at Sea Level [4]

Molecule	Volume Mixing Ratio*	Molecular Weight (amu)
N_2	0.78110	28.0134
O_2	0.20955	31.9988
Ar	0.009343	39.9480
He	0.000038	4.0026
		28.960 (average)

*Volume mixing ratio is equivalent to mole fraction.

covalent, double bond of molecular oxygen and create highly reactive atomic oxygen as follows:

$$O_2 + h\nu \rightarrow O + O . \quad (3.4)$$

The O=O bond strength is 498 kJ/mole, and therefore, the maximum wavelength of UV light capable of breaking the bond is determined by Planck's equation

$$\begin{aligned}\lambda &= hc/E \\ &= \frac{(6.63 \times 10^{-34} \text{ J} \cdot \text{s})(3 \times 10^8 \text{ m/s})}{(498 \times 10^3 \text{ J/mole})(1 \text{ mole}/6.02 \times 10^{23})} \\ &= 240 \text{ nm}\end{aligned}$$

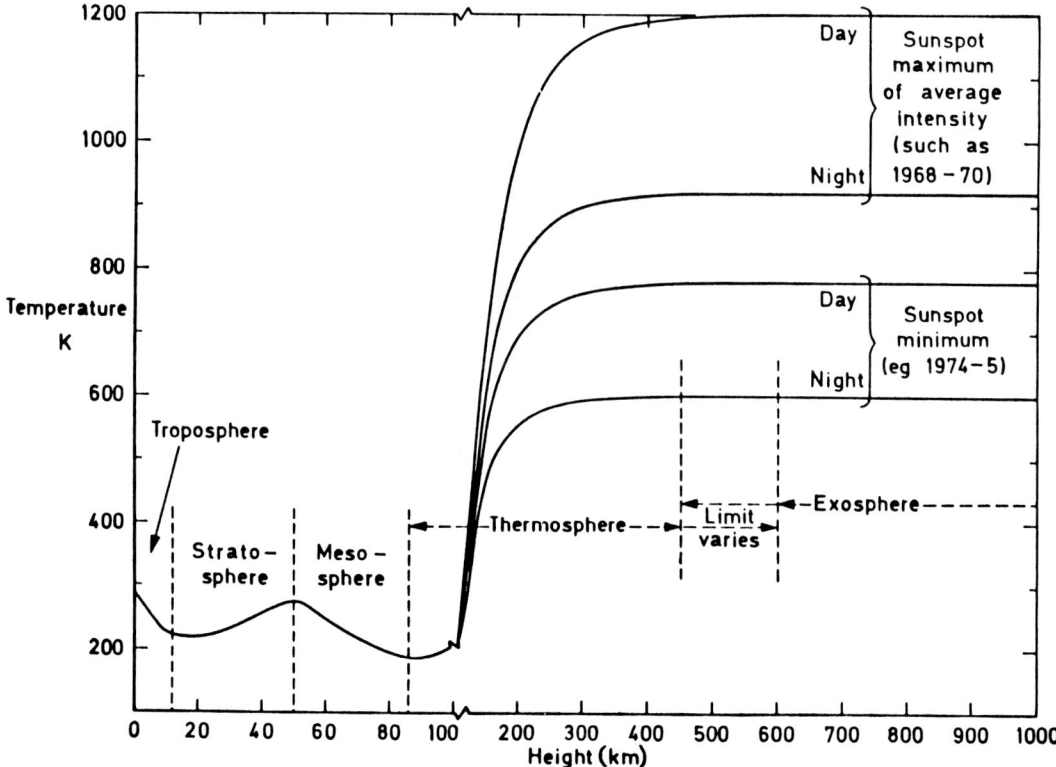

Figure 3.2 Average atmospheric temperature from sea level to 1000 km, showing a nearly sinusoidal variation as a function of altitude until the thermosphere is reached [5].

Another trace molecule, M, may participate in a reaction which forms ozone by the following catalytic process:

$$O + O_2 + M \rightarrow O_3 + M . \qquad (3.5)$$

The reaction shown in equation 3.5 is very important because it highlights a key aspect of ozone formation; namely, that it may be triggered by a "third party" molecule.

Yet ozone does not linger for long during daytime hours since it is destroyed by the absorption of photons of radiant energy ($h\nu$) according to

$$O_3 + h\nu \rightarrow O_2 + O . \qquad (3.6)$$

The reaction given in equation 3.6 does not occur below 90 km so ozone is usually only destroyed above 90 km by this reaction. Below 60 km nearly all of the products of the reaction in equation 3.6 recombine via the reaction in equation 3.5 to recreate more ozone. It has been found that the reaction described in equation 3.6 actually only accounts for about one-fifth of ozone depletion.

Combination of ozone and atomic oxygen with nitric oxide (NO) and nitrogen dioxide (NO_2) serves to deplete the ozone supply by [1]

$$NO + O_3 \rightarrow NO_2 + O_2 \qquad (3.7)$$
$$NO_2 + O \rightarrow NO + O_2 . \qquad (3.8)$$

The reaction given in equation 3.8 is important because it removes the source of ozone, atomic oxygen (O). This series of reactions, shown in equations 3.7 and 3.8, accounts for about 70% of the removal of ozone. This process is not as devastating as one might expect since during the day NO_2 is broken down, yielding atomic oxygen when it absorbs radiation as follows:

$$NO_2 + h\nu \rightarrow NO + O . \qquad (3.9)$$

The reaction described in equation 3.9 produces more atomic oxygen, the building block of ozone. The wavelengths of the radiant energy ($h\nu$) which drive the reactions of equations 3.4, 3.6, and 3.9 are different but similar in magnitude.

Recently the greatest concern about the depletion of ozone has been the effect of chlorofluorocarbons (CFC) on

Figure 3.3 The Upper Atmosphere Research Satellite (UARS) will measure the dynamics of ozone creation and depletion in the atmosphere. Compliments of NASA.

the atmosphere. A major source of CFCs in America is leaking automobile air conditioners. The major compounds being discussed are $CFCl_3$ (Freon-11) and CF_2Cl_2 (Freon-12) [1]. These two substances produce chlorine in the atmosphere by

$$CFCl_3 + h\nu \rightarrow CFCl_2 + Cl \qquad (3.10)$$
$$CF_2Cl_2 + h\nu \rightarrow CF_2Cl + Cl . \qquad (3.11)$$

These chlorine atoms will destroy the ozone supply by combining with O_3 to produce $ClO + O_2$

$$Cl + O_3 \rightarrow ClO + O_2 . \qquad (3.12)$$

It has been reported that the effects of ozone depletion have already started but more detailed data acquired over a long period of time is required before any significant conclusions may be made. Special attention has been given to the atmosphere above the poles where there is little global circulation to cleanse the region of ozone-depleting molecules. Concerns have been raised for years about the onset of this global environmental problem, yet data acquired by satellites and aircraft is incomplete. As a result, the Upper Atmospheric Research Satellite (UARS), shown in Figure 3.3, was designed to fill this data gap. General Electric built the UARS which was launched by the Space Shuttle in 1991 [6]. This NASA-funded and designed spacecraft will reside in a 57° inclination circular orbit at an altitude of 600 km. These orbital parameters permit the satellite to have global coverage every 36 days. The mission duration of 15 months is limited only by the warming and loss of onboard cryogenic liquids.

The mission is being timed to include two northern hemisphere winters. This allows the analysis of the seasonal warming of the stratosphere which is of special interest to atmospheric researchers. Ongoing activities to quantify the extent of ozone depletion in our atmosphere and its ramifications on our global environment are very important. The use of chemistry to describe this dynamic phenomenon of worldwide proportions highlights the need for space chemistry as a separate field of study.

Upper Atmosphere

Temperature Profile

The upper atmosphere begins at the thermosphere where there is a rapid increase in temperature. Up to this point, 85 to 100 km altitude, there were very small changes in the atmospheric temperature from 290 K maximum to 190 K minimum. Over the next 100 km in altitude, up to 200 km, the temperature may rise to as high as 2,000 K [4]. For this section, the Jacchia '77 upper atmospheric model [4] and the U.S. Standard Atmosphere 1976 [7] will be used as the major sources.

The standard temperature profile begins with $T_o = 188$ K ($-121°F$ or $-85°C$) at the altitude $Z_o = 90$ km with a gradient $G_o = (dT/dZ)_{z_o} = 0$. This model is applicable to a height of $Z_x = 125$ km where the temperature increases asymptotically to T_e. The term T_e is the constant temperature experienced in the exosphere. The flattening of the temperature curve actually marks the beginning of this outer layer of our atmosphere, the exosphere.

At the inflection point at 125 km, the temperature and temperature gradient are dependent on T_e and can be expressed as

$$T_x - T_o = 110.5 \sinh^{-1} 0.0045(T_e - T_o) \quad (3.13)$$

and

$$G_x = 1.9\left(\frac{T_x - T_o}{Z_x - Z_o}\right), \quad (3.14)$$

where

T_x = temperature at inflection point

G_x = temperature gradient at inflection point

$$= \left[\frac{dT}{dZ}\right]_{Z = Z_x}$$

The actual temperature profiles are represented by

For $Z < Z_x$:

$$T = T_x + \frac{T_x - T_o}{\pi/2} \tan^{-1}\left\{\frac{G_x}{(T_x - T_o)/(\pi/2)} \times (Z - Z_x)\left[1 + 1.7\left(\frac{Z - Z_x}{Z - Z_o}\right)^2\right]\right\}. \quad (3.15)$$

For $Z > Z_x$:

$$T = T_x + \frac{T_e - T_x}{\pi/2} \tan^{-1}\left\{\frac{G_x}{(T_e - T_x)/(\pi/2)} \times (Z - Z_x)\left[1 + 5.5 \times 10^{-5}(Z - Z_x)^2\right]\right\}. \quad (3.16)$$

The exospheric temperature T_e is a function of solar activity; the greater the activity, the greater the exospheric temperature. The effect of solar activity on the dynamics of the atmosphere will be covered later in this chapter. A family of four temperature profiles for Earth's atmosphere are shown in Figure 3.4 [4].

Composition

A five component atmosphere is assumed at sea level consisting of nitrogen, oxygen, argon, helium, and hydrogen. The atmosphere is thoroughly mixed to altitudes of 10 km with molecular diffusion driving the structure of the atmosphere above 100 km. Hydrogen is not included in the Jacchia '77 model until 150 km where it becomes an important contributor to the environment. The volume mixing ratios (or mole fractions) of the other four components at sea level were listed previously in Table 3.2.

Above 100 km the molecular diffusion process is defined by the ideal gas law which can be expressed as

$$P = \frac{NRT}{A} \quad (3.17)$$

where

A = Avogadro's constant
  = 6.022×10^{23} mole^{-1}

N = total number density, particles/m^3

R = 8.314 J/(mole · K)

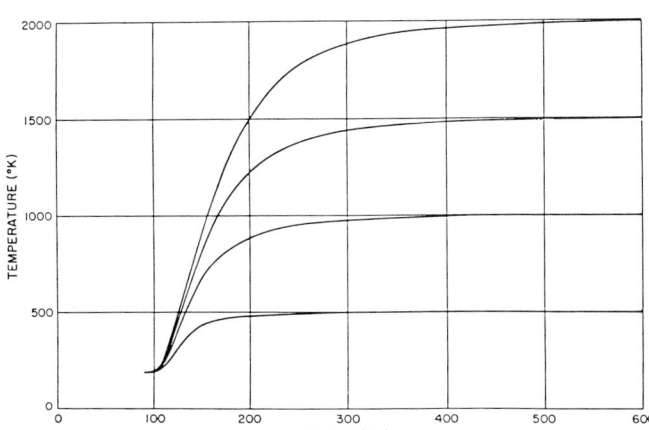

Figure 3.4 The temperature profile of the low Earth atmosphere depends greatly on solar activity; the greater the solar activity, the greater the exospheric temperature [4].

Dalton's law of partial pressures may be used to find the total pressure by summing the partial pressures of a series of gases

$$P_{TOT} = \sum P_i = \sum n_i kT \qquad (3.18)$$

where

- n_i = number density of ith gas
- k = Boltzmann's constant
  - = 1.380622×10^{-23} J/K
- P_i = partial pressure due to the ith gas.

At lower altitudes where mixing occurs and little vertical movement is experienced, the hydrostatic equation for pressure may be used [5]:

$$\frac{dp}{dy} = -pg \qquad (3.19)$$

where

- p = atmospheric density, kg/m^3
- g = acceleration due to gravity, m/s^2
- y = altitude, m .

Equation 3.19 shows that as altitude increases, the pressure will decrease as a function of the density and acceleration due to gravity. The pressure and temperature profiles will combine to give the number density of each diffusing element in the upper atmosphere. Figure 3.5 plots the number density for a variety of elements. Note that atomic oxygen is included on this figure—it will be of great interest later in the text when degradation of materials in LEO is discussed.

Similarly, the density as a function of altitude may also be found from the ideal gas law. Figure 3.6 plots atmospheric density and pressure versus altitude for average solar activity.

Approximations

King-Hele of the Royal Aircraft Establishment is widely known as the expert in the field of satellite orbits in an atmosphere [5]. During his many years of work he has produced numerous useful approximations of the structure of the upper atmosphere and its effect on a satellite's orbit. The development in reference 5 begins with

$$\frac{dp}{p} = -\frac{Mg}{RT} dy \qquad (3.20)$$

where

$h = \dfrac{RT}{Mg}$ is the pressure scale height in kilometers.

By integrating this last equation the so-called barometric

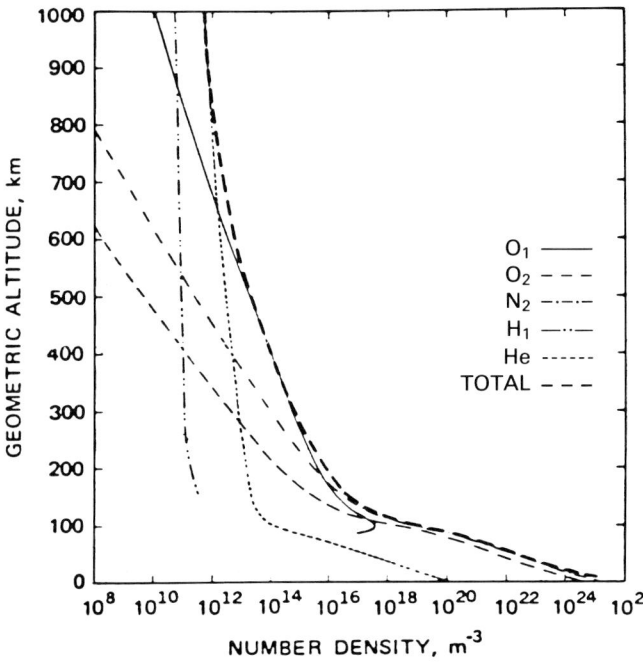

Figure 3.5 Number density of terrestrial atmospheric constituents versus altitude [7].

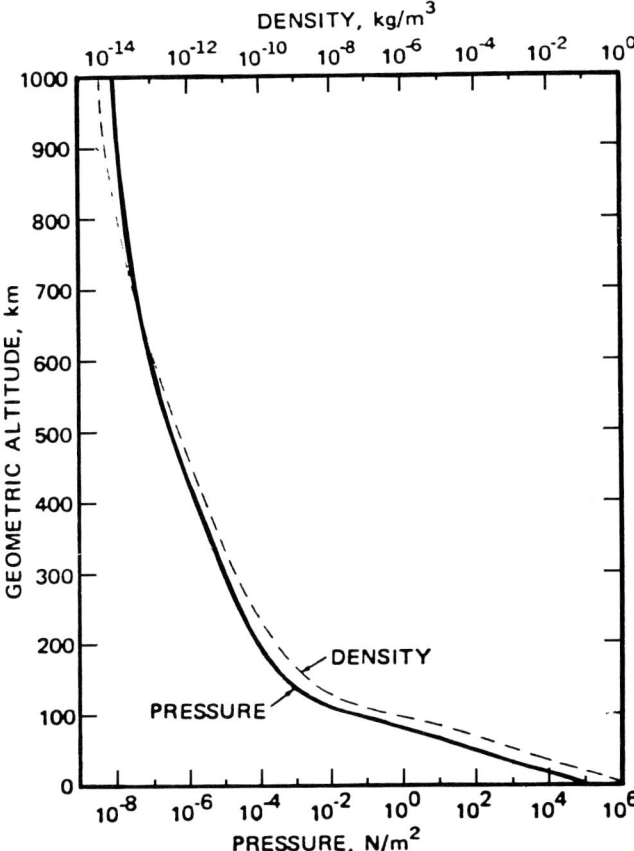

Figure 3.6 Atmospheric density and pressure versus altitude [7].

equation is obtained, giving the pressure as a function of altitude; namely,

$$p = p_o \exp[-(y - y_o)/h] \quad (3.21)$$

where
- p = pressure at altitude y, Pa
- y = altitude, km
- p_o = reference pressure at altitude y_o, Pa
- h = pressure scale height, km.

According to equation 3.21, the pressure varies exponentially in the upper atmosphere. As discussed earlier in the model from Jacchia '77, the density will vary similarly to the pressure. The resulting density profile is

$$\rho = \rho_o \exp - [(y - y_o)/H] \quad (3.22)$$

where
- ρ = atmospheric density at altitude y, kg/m^3
- ρ_o = atmospheric density at reference altitude y_o, kg/m^3
- H = density scale height, km

The density scale height, H, is a constant and is nearly identical to the pressure scale height, h. They are related as

$$\frac{1}{H} = \frac{1}{h} - \frac{Z}{r_o} \quad (3.23)$$

Here Z is a dimensionless parameter and r_o is the reference radius—the radius of Earth, r_E, plus the reference altitude, y_o. That is,

$$r_o = r_E + y_o. \quad (3.24)$$

Figure 3.7 shows the pressure and density scale heights as a function of altitude. The values of H and h are small in comparison to r_o so the difference between H and h is at most a few percent.

For this reason these two terms are used interchangeably as the scale height.

The exponential density profile provides a simple method to predict densities up to 500 km. A limitation of these relationships is that one must select y close enough to the reference altitude y_o so that the scale height can be assumed constant. Differences of less than 50 km are acceptable.

From the exponential model one can see that if the altitude is increased by one scale height, then

$$\frac{\rho}{\rho_o} = \frac{1}{e}, \quad (3.25)$$

the density will decrease by a factor of $e = 2.718$ [5]. A change in altitude of three scale heights will cause a decrease in density of 20. If $y_o = 250$ km and $H = 40$ km, then at $y = 370$ km the density is $= \rho_o/20$ or 5% of its value at the surface.

Above 500 km the scale height changes quickly due to the diffusive nature of the atmosphere and the lower mo-

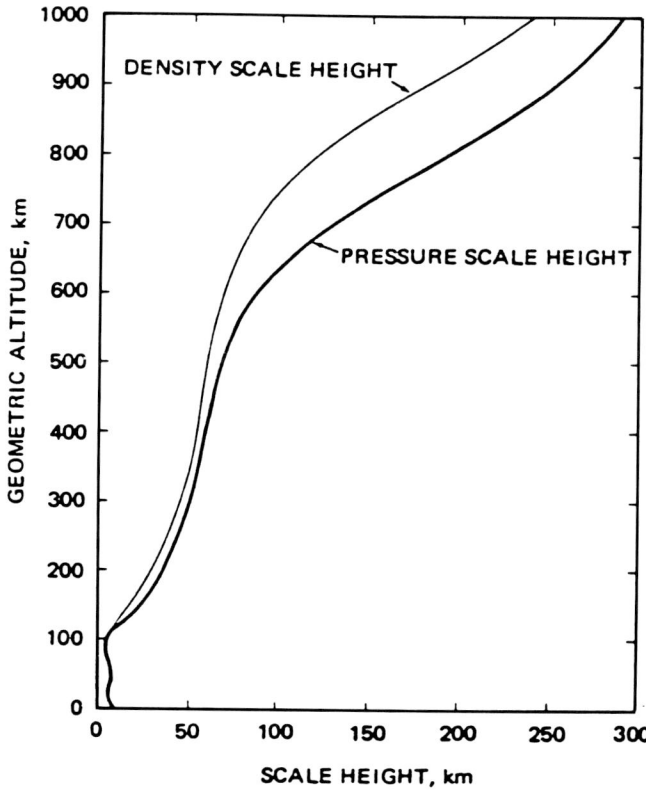

Figure 3.7 Pressure and density scale height versus altitude [7].

lecular weight species of gases will begin to dominate. In the Earth's atmosphere this causes an increase in the percentage by number of hydrogen atoms at higher altitudes.

The exponential character of Earth's atmosphere is directly responsible for the exponentially varying lifetimes shown in the astrodynamics chapter. Doubling the altitude of a satellite from 300 km to 600 km circular orbit will much more than double its lifetime. Variations in lifetime are even more drastic at higher altitudes. The lifetime for a 1,200 km circular orbit is a factor of 300 times greater than the lifetime of a satellite in a 600 km circular orbit.

Variations in Upper Atmosphere

Above 200 km the upper atmosphere's structure is determined by solar activity [5]. There are five different measurable effects which make the atmosphere vary from the static model [4]:

1. Variation with the solar cycle
2. Daily, or diurnal, variation
3. Variation with geomagnetic activity
4. Seasonal-latitudinal variations
5. Rapid density fluctuations, probably connected with gravity waves

Except for gravity waves, the variations occur on a regular

basis and may be roughly predicted using observed atmospheric/cosmic data. Some of these variations will occur over very short time intervals, thus making it more difficult to quantify their effects. Variations 1 (solar cycle) and 2 (daily) have the greatest influence on the state of the atmosphere. Each of these five variations will now be outlined with special emphasis on the two prominent ones (variations 1 and 2).

1. *Variation with the solar cycle*: The Sun's activity level varies roughly on 11-year cycles. As the activity level increases, i.e., more sunspots, the atmosphere is heated and expands. The extreme ultraviolet (EUV) solar radiation causes this heating. A convenient and accurate measurable parameter which quantifies the EUV is the $F_{10.7}$ flux. $F_{10.7}$ is the radiant energy measured at a wavelength of 10.7 cm by the National Research Council in Ottawa, Canada [2]. Units of this term are 10^{-22} W m^{-2} Hz^{-1}, but the $F_{10.7}$ radiation is normally referred to by its numerical value alone. For example, during low solar activity the $F_{10.7}$ EUV is around 70×10^{-22} W m^{-2} Hz^{-1} but is normally stated as 70 with the units implied.

The heating of the atmosphere by solar EUV determines the exospheric temperature by [4]

$$T_e = 5.48 \, (\overline{F}_{10.7})^{0.8} + 101.8(F_{10.7})^{0.4} \quad (3.26)$$

where

$\overline{F}_{10.7}$ = average 10.7 cm solar flux over last six solar cycles
$F_{10.7}$ = 10.7 cm solar flux
T_e = average exospheric temperature, K

This equation was formulated from over 40,000 satellite drag measurements taken from 1958 to 1975 and has proven reliable in the many years of its use [4]. Figure 3.8 shows the average exospheric temperature profile as a function of $F_{10.7}$.

As noted earlier, increased T_e results in higher densities at the same geometric altitude. This variation is clearly shown in Figure 3.9.

Historically the solar cycles have been about 11 years in length with variable shapes and magnitudes. Since the lifetime, maintenance needs, and effectiveness of low Earth satellites depend greatly on the solar activity, much work has gone into describing previous solar cycles and predicting future ones. Table 3.3 lists equations used for these purposes [6,8,9].

Table 3.3 represents the best fit curves but in actuality the solar activity may vary greatly about these mean values. Table 3.4 lists the curves which depict the one sigma range of activity levels possible.

Tables 3.3 and 3.4 provide the best approximation of the global atmospheric effects due to solar activity. This in turn

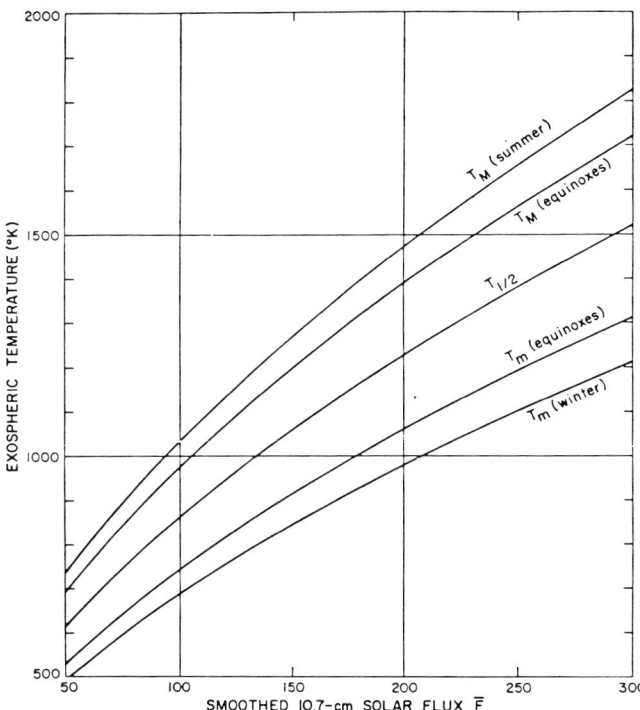

Figure 3.8 Exospheric temperature as a function of $F_{10.7}$ solar flux for various seasons [4].

allows us to quantify the atmospheric density in the upper atmosphere where it affects satellite activities the most.

The effect of increased solar activity is felt most by objects with large area-to-mass ratios orbiting at low altitudes. The vast quantity of space debris, remnants of space operations, in Earth orbit provides an especially interesting application of this phenomenon. The number of large objects (greater than 10 cm in diameter) cataloged by the U.S. Space Command's Space Surveillance Network exceeds 7,100 with 95% of this being debris. Over the years the growth of this population has been closely monitored since it could eventually pose a great hazard to operational space systems. Figure 3.10 shows the number of objects added to the catalog each year. Note that the only decreases in population occurred in periods of maximum solar activity: 1979–1980 and 1988–1989. The $F_{10.7}$ values in the cycle peak from 1988–1990, Cycle 22, have exceeded 200 for a long period. This is being described as the most energetic period of solar activity since measurements have been made. The solar maximum actually ocurred in July 1989. For this reason the large drop in 1988–89 is totally understandable [10]. In late 1990, despite the continued high solar activity, the cataloged on-orbit population began to increase again. The apparent reason for this reversal was that increased atmospheric drag had already cleansed most of the debris that could be affected. In 1991, the dropoff in solar activity was significant and a proportional increase in the catalog population resulted. This phenomenon high-

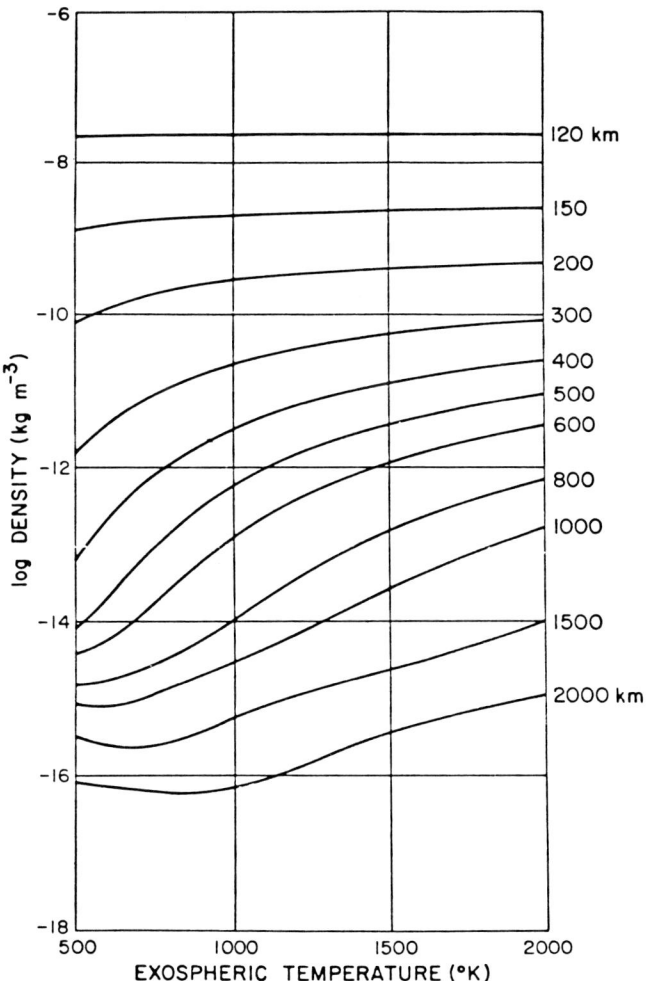

Figure 3.9 Total atmospheric density as a function of exospheric temperature for various altitudes [4].

lights the fact that nature can correct for only a portion of man's pollution, as on the Earth. Mankind must be responsible for devising methods to control the remainder.

Solar activity also affects the amount of atomic oxygen present in low Earth orbits between 200-1,000 km. UV radiation below 240 nm wavelengths possesses sufficient energy to break the double bond of O_2 and create O. Figure 3.11 shows the atomic oxygen flux for a typical Space Shuttle orbit during various levels of sunspot activity. The fluence or total number of oxygen atoms impacting the surface can be determined by multiplying the flux times the cross-sectional area and time in orbit. The effect of this fluence will be examined later in Chapter 5.

2. *Daily, or diurnal, variation*: The upper atmosphere has a higher temperature during the afternoon than in the early morning, much as we experience at the surface of the Earth. The daytime maximum of the exospheric temperature may be up to 400 K higher than the nighttime minimum. These extremes occur at around 3 p.m. and 3 a.m., respectively, depending on latitude and season [5]. The average exospheric temperatures outlined in the last section actually are obtained by taking the mean of the temperature values throughout the day. During the two equinoxes (Earth intersects solar orbital plane twice a year) these peak temperatures and thus peak densities will occur along the Earth's equatorial plane. Figure 3.12 depicts this variation for an average exospheric temperature of 1,000 K [4].

During the June solstice, the Earth is below the Sun's orbital plane and as a result the northern hemisphere experiences summer. The upper atmosphere is warmed in a similar way with the maximum exospheric temperature occurring over the northern hemisphere. As seen in Figure 3.13, the temperature profile during June solstice shows two large peaks.

Table 3.3 $F_{10.7}$ Best Estimate Curves

Time Interval	Duration (months)	Values of t (months)
Jan 82–Nov 86	59	1–59
$F_{10.7}(t) = -3.33831 \times 10^{-5}(t)^3 + 0.040812(t)^2 - 4.542(t) + 193.92$		
Dec 86–Dec 87	13	60–72
$F_{10.7}(t) = F_{10.7}(59)$		
Jan 88–Apr 91	40	73–112
$F_{10.7}(t) = -0.0023(t-72)^3 + 0.13812(t-72)^2 + 66.337$		
May 91–Dec 92	20	113–132
$F_{10.7}(t) = F_{10.7}(112)$		
Jan 93–Apr 99	75	133–208
$F_{10.7}(t) = 0.0034(t-132)^3 - 00.03895(t-132)^2 + 140.0$		
May 99–Aug 2002	40	209–248
$F_{10.7}(t) = -0.00338(t-208)^3 + 0.2025(t-205)^2 + 65.0$		

Table 3.4 $F_{10.7}$ One Sigma Curves

Time Interval	Duration (months)	Values of t (months)
Jan 82–May 87	65	1–65

$F_{10.7}(\min) = -0.000204(t)^3 + 0.053302(t)^2 - 4.34(t) + 172.0$
$F_{10.7}(\max) = -0.000029(t)^3 + 0.03626(t)^2 - 4.34(t) + 208.0$

June 87–Apr 91	47	66–112

$F_{10.7}(\min) = -0.00206(t\text{-}65)^3 + 0.12375(t\text{-}65)^2 + 59.0$
$F_{10.7}(\max) = -0.00162(t\text{-}65)^3 + 0.11408(t\text{-}65)^2 + 71.0$

May 91–Dec 92	20	113–132

$F_{10.7}(\min) = F_{10.7}(112; \min)$
$F_{10.7}(\max) = F_{10.7}(112; \max)$

June 93–Apr 97	76	133–206

$F_{10.7}(\min) = 0.0003(t\text{-}132)^3 - 0.03428(t\text{-}132)^2 + 125.0$
$F_{10.7}(\max) = 0.00039(t\text{-}132)^3 - 0.04415(t\text{-}132)^2 + 155.0$

May 99–Aug 2002	40	209–248

$F_{10.7}(\min) = -0.00275(t\text{-}208)^3 + 0.165(t\text{-}208)^2 + 59.0$
$F_{10.7}(\max) = -0.00408(t\text{-}208)^3 + 0.24375(t\text{-}208)^2 + 70.0$

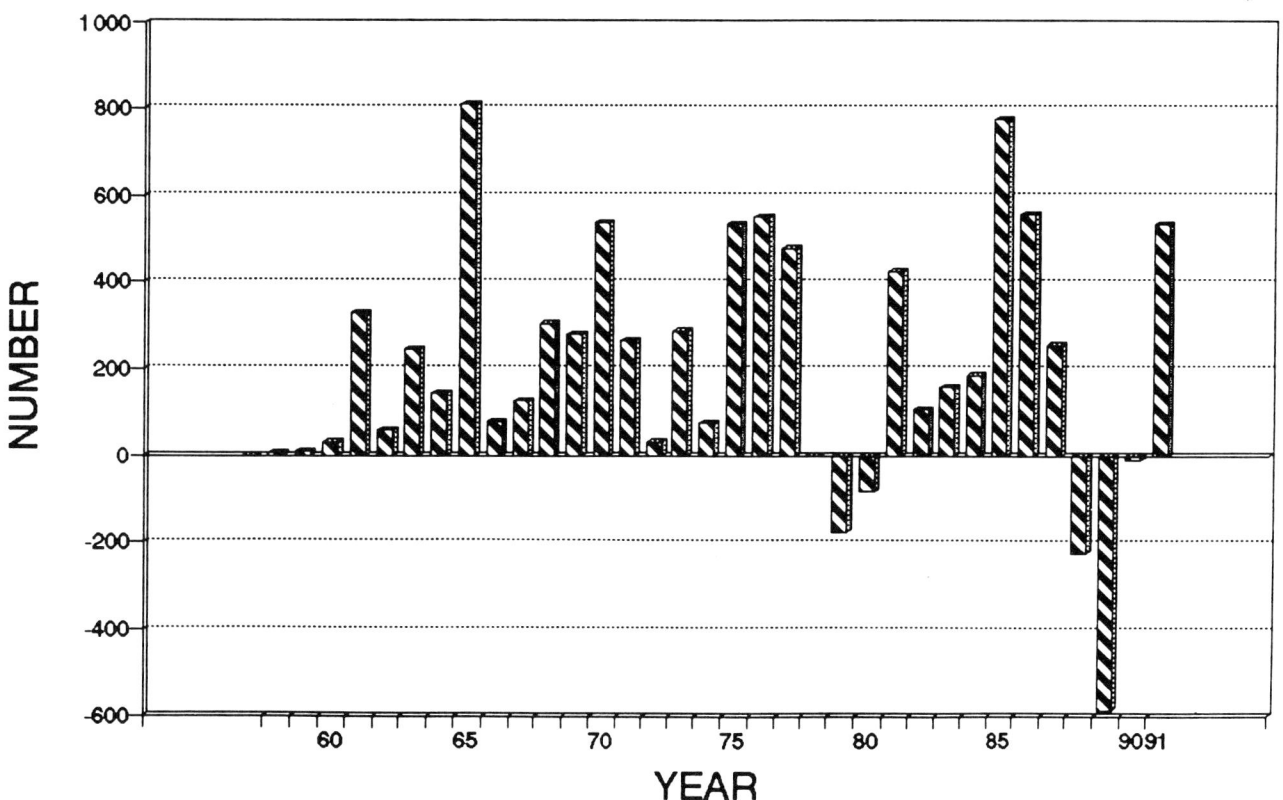

Figure 3.10 The annual change in the on-orbit population of satellites is greatly affected by solar activity.

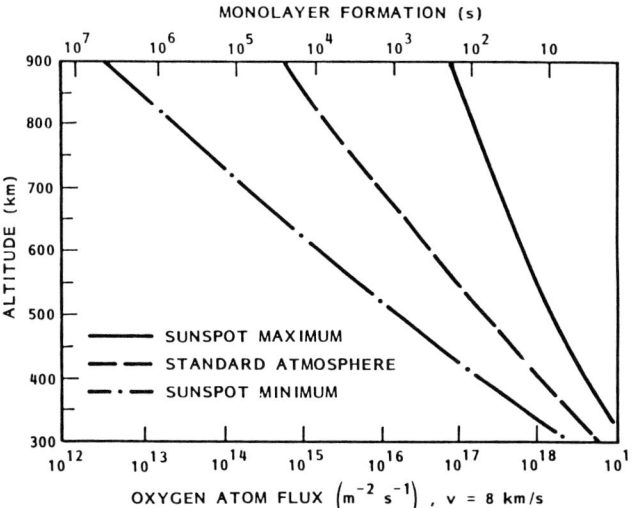

Figure 3.11 Atomic oxygen flux at three levels of solar activity. Flux is along the horizontal axis and is expressed in atoms per square meter per second. Shuttle velocity relative to the atomic oxygen is assumed to be 8 km/s [11].

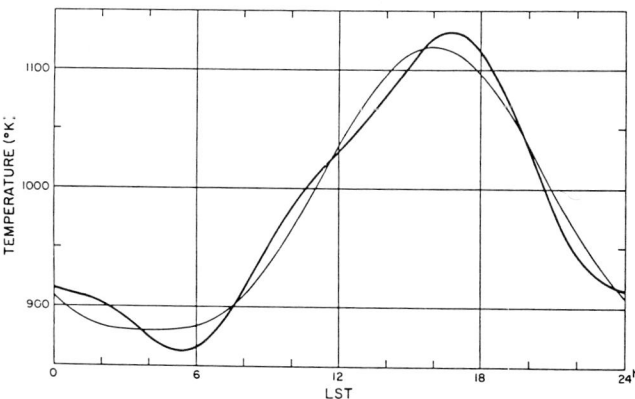

Figure 3.12 The diurnal variation of exospheric temperature at the equator may be an important factor for some satellite operations [4].

3. *Variation with geomagnetic activity*: While describing the solar activity up to this point, it has been assumed that the geomagnetic activity of the Sun has been very low, $K_p = 0$. When $K_p \neq 0$, there is an increase in exospheric temperature, scale height, and density depending on magnetic latitude. The increase in temperature is denoted by ΔT_G [4]. As the geomagnetic activity increases, more heating takes place at higher magnetic latitudes. Figure 3.14 details the effect on the global temperature profile as K_p goes from 0 (quiet) to 9 (very active). These plots reinforce our appreciation of the importance of the time of year on the structure of the upper atmosphere. We can begin to appreciate how difficult it is to predict the lifetime of orbiting satellites due to the drastic variation possible in the heating of the atmosphere by a wide variety of effects.

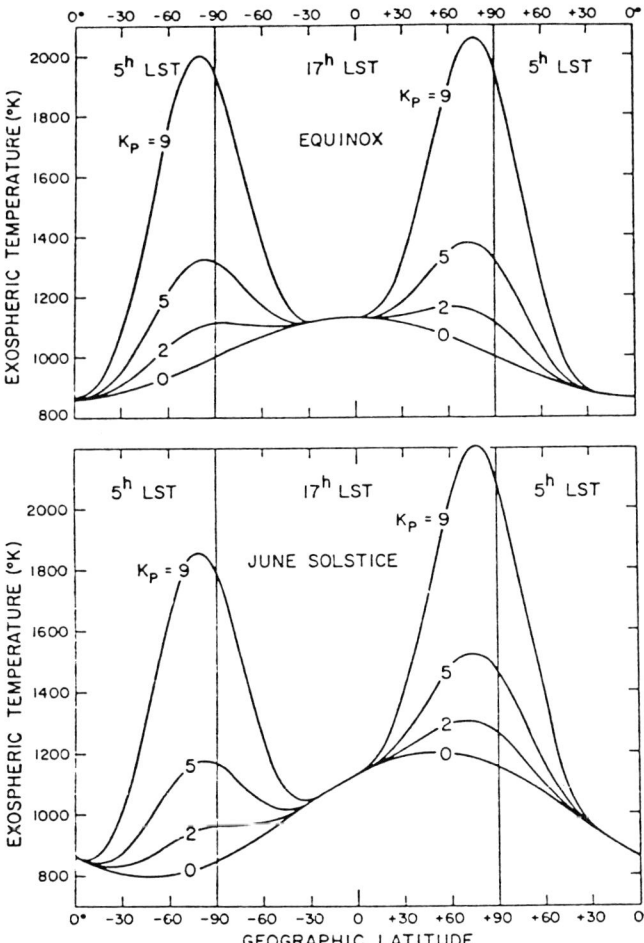

Figure 3.13 The profile of the exospheric temperature at June solstice as a function of geographic latitude [4].

4. *Seasonal-latitudinal variations*: There are two types of seasonal-latitudinal variations in the atmosphere. The first is an alteration in temperatures in the thermosphere. The second is a phenomenon confined to the stratosphere and mesosphere (below 150 km). The second low altitude effect will not be considered here, although it should be noted that the region from 66 km to 100 km actually has a reverse temperature profile than what we might expect. During the summer the minimum temperature in the range is encountered while the maximum temperature occurs in the winter [4].

Two components of the thermospheric temperature variation are dependent on the season and latitude. The first effect is the basic background variation due to changes in the time of year and location on the Earth. This annual alteration of the atmosphere is independent of atmospheric height. The actual change in the number density is best represented as a function of latitude on the Earth and the Sun's declination. (The Sun's declination is an angular measurement of how much the Earth is above or below the Sun's orbital plane, similar to latitude.)

Near Earth Space Environment

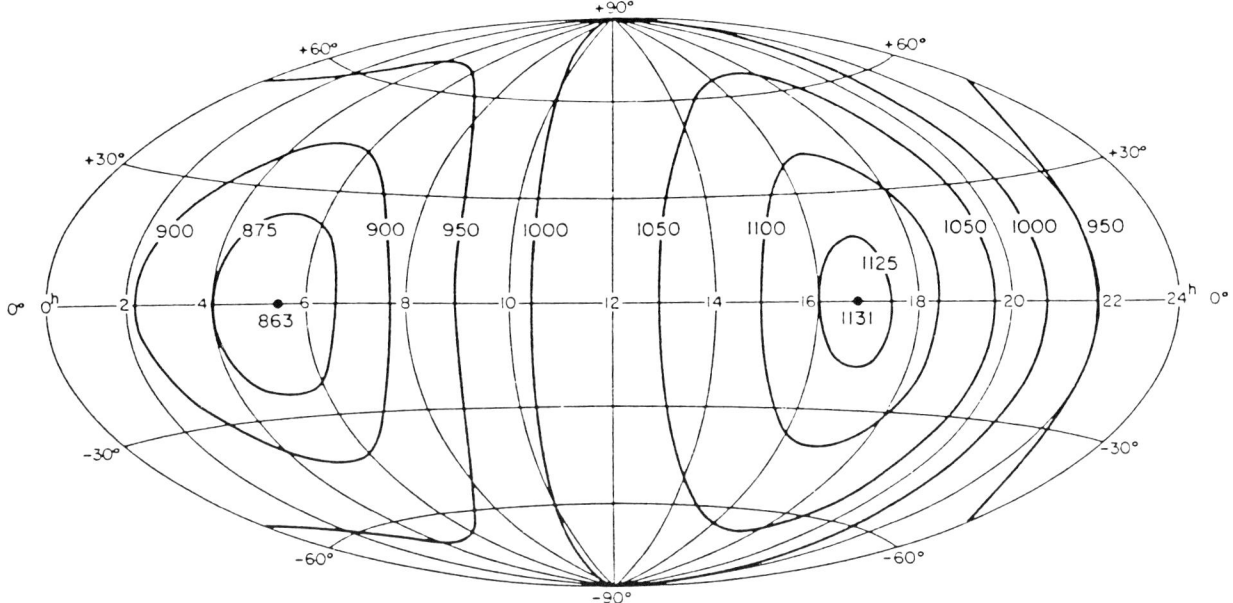

a) Equinoxes.

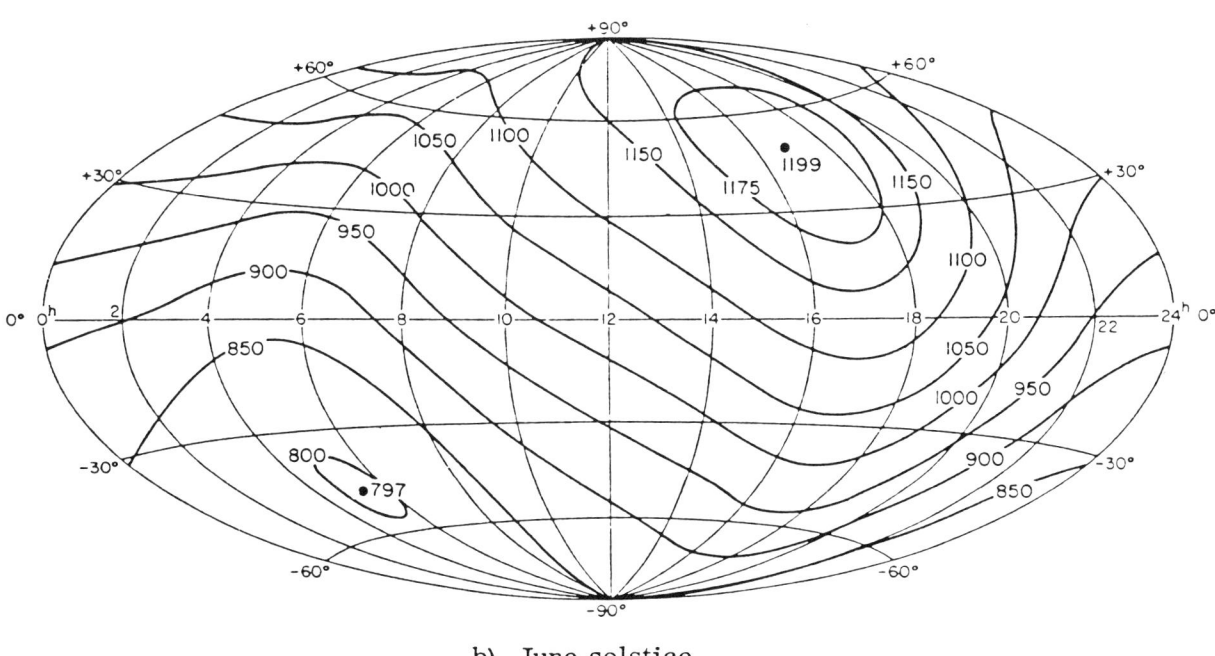

b) June solstice.

Figure 3.14 The influence of geomagnetic activity on exospheric temperature is very pronounced but smaller in magnitude than the diurnal effect [4].

A type of variation which is very difficult to resolve from the seasonal-latitudinal is the semiannual variation. The only way to differentiate between the two is by using measurements from satellites in different orbits over a long period of time [4]. For the practical applications of this book the effects are quite similar: a change in upper atmospheric temperature occurs due to increased EUV deposited on the atmosphere.

5. *Rapid density fluctuations*: Data acquired by the Explorer 32 spacecraft has suggested that density waves may exist in the height range from 286 km to 510 km [4]. The

density waves may extend past these altitudes but the perigee/apogee heights of the satellites used to determine the range of 286 km to 510 km did not allow a proper verification. These waves seem to occur most often at higher latitudes in late evening and early morning hours [4]. These waves also produce a change in the composition of the atmosphere where they exist. Similar to the geomagnetic variation there is a decrease in helium in conjunction with an increase in the nitrogen and argon density [4].

Summary of Variations

The variety of phenomenon that may cause the atmosphere to change in structure and composition provides a significant amount of uncertainty for the space researcher. Long-term analyses involving spacecraft charging, material integrity, and satellite lifetimes require knowledge of these dynamics. Luckily, the variations in the solar cycle and diurnal differences are the ones of most importance. They are physically easy to understand while also fairly simple to model. The other variations provide levels of uncertainty that are usually masked in our inability to accurately depict the temperature, atmospheric density, and number density in the upper atmosphere. Nearly all the models discussed were developed through the use of empirical data obtained from satellites. This data has provided a clear insight into global atmospheric conditions but may be lacking in its ability to precisely depict the effects of all possible variations. For these reasons it is reasonable to account for the first two variations to yield acceptable answers. As a matter of fact, most researchers merely account for the variations in the solar cycle when representing the upper atmosphere with the understanding that they are using a global average with error bars dependent on the other variations.

Atmospheric Density Profiles

For reference purposes Tables 3.5, 3.6, and 3.7 provide atmospheric parameters for exospheric temperatures of 500 K, 1100 K, and 1800 K, respectively [4]. These correspond to low, average, and high solar activity levels.

Trapped Radiation Profiles

Charged particles or ions from the universe and the Sun continually bombard the Earth and become trapped in Earth's geomagnetic field. These radiation belts, named after their discoverer James Van Allen, occupy the volume surrounding the planet which is bounded by the region of interaction between the solar plasma and magnetosphere [12]. Van Allen radiation is of critical importance to all near Earth orbits due to the energy and reactivity of these particles. Although many different types of ions exist in space, only protons and electrons exist in sufficient quantity and possess enough energy to be of consideration.

When high energy protons strike a satellite they may disrupt operations. Solar arrays are actually designed to account for a percentage loss of their capability due to proton damage. The impacting of protons may also temporarily "blind" sensitive optical sensors used to keep a satellite oriented properly. A permanent failure of this type could render a satellite useless.

The proton belt is divided into two zones on the basis of energy. Figure 3.15 shows the isoflux contours for selected energies. The outer proton zone, which extends outward from four earth radii, is defined by particle energies of less than 4 MeV [12]. Flux and energy variations do occur in this region due to the changing levels of solar activity. Stronger magnetic fields, however, prevent such large flux variations from occurring in the inner proton zone (greater than 4 MeV).

Similar to the proton zones surrounding the Earth are the two electron zones that are divided by particle energy at approximately 2.5 to 3 Earth radii (Figure 3.16). Again, much larger flux and energy variations take place in the outer zone than in the inner zone. Inner zone densities also

Table 3.5 Atmospheric Parameters: $T_e = 500$ K

Height (km)	Temp (K)	Log Number Density (#/m³)					Mean Molec. Wt.	Density Scale Ht (km)	Log Density (kg/m³)
		N_2	O_2	O	He	H			
100	191.2	—	18.326	17.668	13.801	—	28.36	5.67	−6.243
200	479.3	18.974	13.378	15.292	13.002	12.378	18.56	21.26	−10.112
300	495.6	14.712	10.265	13.728	12.605	12.239	14.96	28.48	−11.825
400	498.6	11.986	7.302	12.245	12.233	12.143	7.50	36.16	−13.218
500	499.4	9.391	—	10.812	11.874	12.052	2.67	85.55	−14.066
600	499.7	6.882	—	9.422	11.526	11.964	1.84	167.80	−14.415
700	499.8	—	—	8.072	11.189	11.789	1.52	220.49	−14.639
800	499.9	—	—	6.760	10.860	11.796	1.32	274.89	−14.816
900	499.9	—	—	—	10.541	11.716	1.20	337.33	−14.958
1,000	499.9	—	—	—	10.231	11.637	1.12	402.18	−15.076
1,250	500.0	—	—	—	9.490	11.451	1.04	535.72	−15.307
1,500	500.0	—	—	—	8.797	11.276	1.02	618.44	−15.494
2,000	500.0	—	—	—	7.535	10.958	1.01	723.59	−15.817
2,500	500.0	—	—	—	6.415	10.677	1.01	815.85	−15.100

Table 3.6 Atmospheric Parameters: $T_e = 1{,}100$ K

Height (km)	Temp (K)	Log Number Density (#/m³)					Mean Molec. Wt.	Density Scale Ht (km)	Log Density (kg/m³)
		N_2	O_2	O	He	H			
100	194.0	18.971	18.323	17.665	13.798	—	28.36	5.62	−6.247
200	956.4	15.562	14.391	15.650	12.980	11.280	21.70	33.43	−9.520
300	1,068.9	14.222	12.867	14.864	12.765	11.077	18.25	49.79	−10.559
400	1,089.7	13.022	11.497	14.174	12.590	11.006	16.51	60.03	−11.347
500	1,095.5	11.875	10.187	13.518	12.425	10.958	15.34	66.44	−12.032
600	1,097.7	10.765	8.920	12.884	12.266	10.917	13.65	71.99	−12.660
700	1,098.7	9.690	7.691	12.270	12.112	10.877	10.87	80.54	−13.234
800	1,099.2	8.645	6.498	11.673	11.963	10.840	7.74	98.30	−13.727
900	1,099.4	7.629	—	11.093	11.818	10.803	5.54	134.2	−14.110
1,000	1,099.6	6.641	—	10.528	11.676	10.767	4.41	191.31	−14.382
1,250	1,099.8	—	—	9.183	11.340	10.682	3.53	321.08	−14.803
1,500	1,099.9	—	—	7.923	11.025	10.603	3.19	377.15	−15.112
2,000	1,100.0	—	—	—	10.458	10.458	2.49	475.15	−15.627
2,500	1,100.0	—	—	—	9.942	10.330	1.88	632.59	−16.027

Table 3.7 Atmospheric Parameters: $T_e = 1{,}800$ K

Height (km)	Temp (K)	Log Number Density (#/m³)					Mean Molec. Wt.	Density Scale Ht (km)	Log Density (kg/m³)
		N_2	O_2	O	He	H			
100	195.6	18.969	18.321	17.663	13.796	—	28.36	5.59	−6.249
200	1,391.2	15.782	14.664	15.706	12.934	10.756	22.90	40.89	−9.355
300	1,707.4	14.856	13.620	15.136	12.759	10.509	20.31	68.60	−10.141
400	1,769.2	14.102	12.760	14.702	12.644	10.422	18.44	84.23	−10.708
500	7.786.7	13.394	11.953	14.296	12.541	10.377	17.19	96.25	−11.189
600	1,793.1	12.714	11.176	13.906	12.443	10.346	16.35	105.85	−11.618
700	1,796.0	12.055	10.423	13.530	12.348	10.319	15.65	113.63	−12.014
800	1,797.5	11.416	9.693	13.165	12.257	10.294	14.88	120.58	−12.385
900	1,798.3	10.794	8.983	12.810	12.168	10.271	13.85	127.69	−12.735
1,000	1,798.8	10.190	8.293	12.465	12.082	10.249	12.50	136.32	−13.064
1,250	1,799.5	8.750	6.649	11.642	11.876	10.197	8.34	176.03	−13.777
1,500	1,799.7	7.403	—	10.872	11.683	10.148	5.50	272.46	−14.283
2,000	1,799.9	—	—	9.471	11.333	10.060	4.01	575.57	−14.816
2,500	1,799.9	—	—	8.228	11.022	9.981	3.77	738.63	−14.143

include contributions from previous high altitude nuclear testing [12].

Charged particle fluxes become an important concern again once a satellite is placed well above the Van Allen radiation belts. Most notably, geosynchronous satellites experience fluxes of energetic particles and cosmic rays which in general exceed the hazard from the trapped radiation belts. Galactic cosmic rays may exceed 1 GeV which may cause single event upsets by damaging electronic components [13].

Summary

The study of Earth's atmosphere has shown it to be a complicated dynamic system of gases powered by solar radiation. The constant furor of chemical reactions about the Earth never reaches an equilibrium because of additions from Earth's biosphere and solar EUV. Many of the interactions with terrestrial sources provide essential components to our atmosphere, such as the CO_2 and O_2 exchange through our vegetation. Yet, more damaging deposits placed in our atmosphere have actually reduced the effectiveness of its lifegiving mechanisms. A pressing example of this type of phenomenon is ozone depletion via chlorine atoms released from aerosol chlorofluorocarbons. The growing levels of carbon dioxide in our atmosphere that may eventually lead to global warming are also of great interest.

The lower atmosphere, below 100 km, is marked by an irregular profile of temperature, pressure, and density while above 100 km the atmospheric structure is nearly exponential in nature. Approximations made by King-Hele in reference 5 have been used extensively to describe the upper atmosphere and its effects on satellite orbits. The variations of the atmosphere due to solar activity and day-night differences are highlighted while a variety of less important phenomenon, such as geomagnetic activity, seasonal-latitudinal variations, and gravity waves, are also discussed

Chemical Principles Applied to Spacecraft Operations

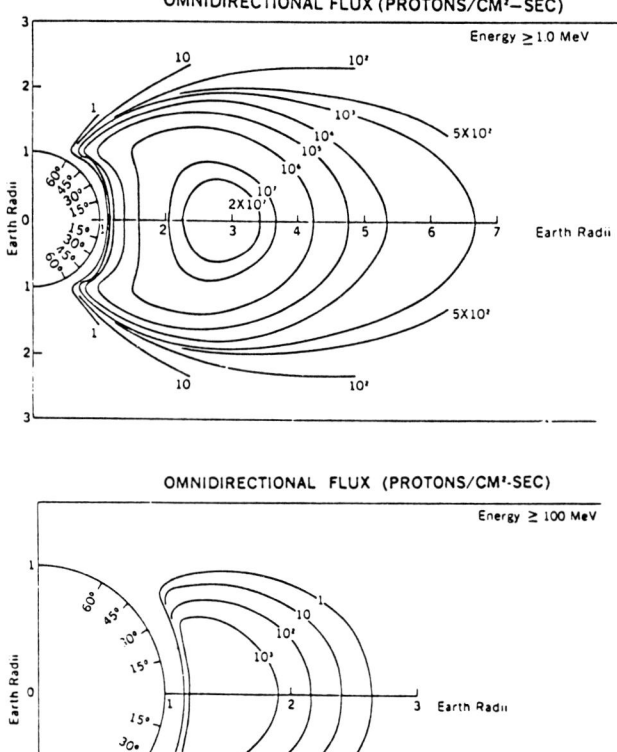

Figure 3.15 A. Distribution of trapped protons with energies greater than 1 MeV. B. Distribution of trapped protons with energies greater than 100 MeV [3].

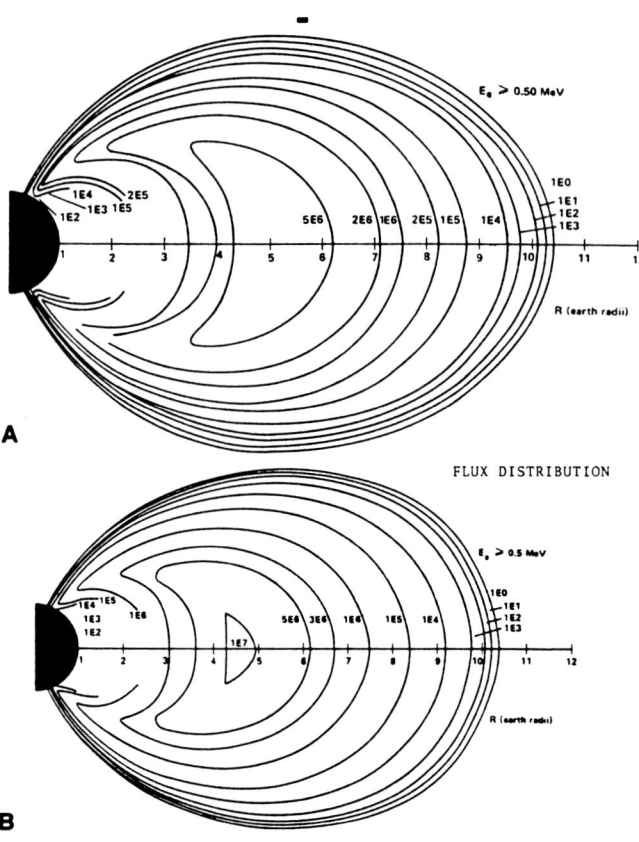

Figure 3.16 A. Distribution of trapped electrons during solar minimum with energies greater than 0.5 MeV. B. Distribution of electrons with same energy during solar maximum (cm^{-2} sec^{-1}) [3].

briefly. Data tables are provided with critical atmospheric parameters for low, average, and high solar activity levels.

The development of Earth's atmospheric structure emphasizes measurable factors which have a quantifiable effect on satellite systems. These attributes will be referred to in later chapters. Mission support system needs are driven by the orbital environment and the corresponding protection measures required. Of more direct concern is the use of this chapter when analyzing the interaction of the satellite with the environment. Atomic oxygen will be shown later to have a significant effect on many materials and its concentration is a function of altitude and solar activity level. The temperature profile at low orbital altitudes will have an effect on the atmospheric drag exerted on orbiting spacecraft. Satellites and their subsystems must also be protected from the radiation environment. Often measures taken to counter one environmental concern adversely affect another. This complex situation is one of the major reasons why this text is necessary.

Before discussing specific concerns with space chemistry in the Earth's atmosphere, the planetary atmospheres are described. This is essential so that concepts outlined later in this text may be applied to planetary and interplanetary missions.

Summary of Key Equations

$O_3 + h\nu \rightarrow O_2 + O$	(3.6)	Ozone depletion via radiation
$NO + O_3 \rightarrow NO_2 + O_2$	(3.7)	Ozone depletion via reaction with nitrogen oxide
$Cl + O_3 \rightarrow ClO + O_2$	(3.12)	Ozone depletion via reaction with chlorine
$P = \dfrac{NRT}{A}$	(3.17)	Ideal gas law
$\dfrac{dp}{dy} = -\rho g$	(3.19)	Hydrostatic equation
$p = p_o \exp[-(y - y_o)/h)]$	(3.21)	Exponential approximation of pressure profile
$\rho = \rho_o \exp\left(\dfrac{y - y_o}{H}\right)$	(3.22)	Exponential approximation of density profile
$T_e = 5.48(\overline{F}_{10.7})^{0.8} + 101.8(F_{10.7})^{0.4}$	(3.26)	Average exospheric temperature

References

1. Lewis, L. S. and Prinn, R. G., *Planets and Their Atmospheres*. Academic Press, Inc., Orlando, FL, 1984.
2. Wayne, R. P., *Chemistry of Atmospheres*. Claredon Press, Oxford, 1985.
3. Tascione, T., *Introduction to the Space Environment*, Orbit Book Co. (Malabar, FL), 1988.
4. Jacchia, L. G., "Thermospheric Temperature, Density and Composition: New Models," Research in Space Science, SAO Special Report No. 375, 15 March 1977.
5. King-Hele, D., *Satellite Orbits in an Atmosphere*. Blackie and Son Ltd., Bishopbriggs, Glasgow, 1987.
6. Marsh, G. "Investigating the Ozone Layer," Space, February, 1989.
7. "U.S. Standard Atmosphere, 1976," National Oceanic and Atmospheric Administration, National Aeronautics and Space Administration, U.S. Air Force, 1976.
8. Tobiska, W. K., "Predictive Model of the Orbit Decay of the Solar Mesosphere Explorer," Master Thesis, University of Colorado, 1985.
9. Euler, H. and Holland, R., "Solar Activity Statistical Estimation Techniques," Published Proceedings of a Workshop on Satellite Drag, March 18–19, 1982, Boulder, Colorado.
10. Sargen, H. H., "An Early Forecast for Sunspot Cycle 22," Seminar Notes, HAO Colloquium Series, 1984.
11. "Space Station Atomic Oxygen Effects," Presentation by Lockheed Missiles and Space Company to NASA-LeRC, July 11, 1986.
12. *Space Materials Handbook*, 3rd ed., ed. by J. Rittenhouse and J. Singletary, NASA SP-3051, 1969.
13. Williamson, M., "A Tough Place to Work," Space, June 1988, pp. 20–23.

Discussion Questions

1. Global warming has been speculated by a number of scientists.
    a. What *natural* phenomena could be contributing to this apparent increase in global surface temperatures?
    b. What manmade contributions to our environment could be adding to this possible upward trend in global temperatures?
2. As altitude increases why is there not a continuous increase in temperature or pressure?
3. The use of chlorofluorocarbons has been identified as a leading culprit of ozone depletion in Earth's atmosphere.
    a. What are the most common reactions that create ozone?
    b. What are the most common reactions that deplete ozone?
    c. What is the major effect of ozone depletion on our environment?
4. At higher altitudes lower molecular weight elements are more prevalent than higher molecular weight elements.
    a. Why does this occur?
    b. Taking this phenomenon to the limit, what would be the most common element at higher altitudes?
5. How does increased solar activity affect the atmospheric drag experienced by a low Earth satellite? How has this affected the population of orbiting satellites over the years?
6. What are the five major influences on the variation of the upper atmosphere in order of importance?
7. Solar radiation fuels many of the chemical reactions which occur in our atmosphere but it also may have an adverse effect on space systems.
    a. What is the worst Earth orbit for encountering adverse radiation effects?
    b. What is the best Earth orbit for avoiding radiation effects?
8. Explain the significance of having an exponentially varying atmospheric density profile on satellite operations?
9. Exospheric temperature (T_e) is dependent on solar activity in a rather complex function.
    a. Why isn't there a linear relationship between these quantities?
    b. What is the equation relating solar activity and T_e?

Problems

1. UV radiation is the trigger for many chemical reactions in the Earth's atmosphere. What are the frequency and wavelength of radiation which will trigger each of the following processes:
    a. $O_2 + h\nu \rightarrow O + O$
    b. $O_3 + h\nu \rightarrow O_2 + O$
    c. $NO_2 + h\nu \rightarrow NO + O_2$
2. Figure 3.6 shows the similarity of the density and pressure profiles in the near Earth environment.
    a. Using this figure, what are the values for pressure and atmospheric density at 350 km?
    b. Using Figure 3.7, what are the pressure and density scale heights at 350 km?
    c. An exponential atmospheric model implies that a change in altitude of one scale height will decrease the density (or pressure) by $1/e$. Using Figures 3.6 and 3.7, confirm this approximation.
3. The exospheric temperature determines the atmospheric density in low Earth orbit.
    a. If the 10.7 cm solar flux (F) averaged 120 over the last six solar cycles and it is presently 150, what is the exospheric temperature?
    b. Is this situation considered low, medium, or high solar activity?
4. Tables 3.3 and 3.4 list models for predicting the solar activity levels through the year 2002.
    a. What is the best estimate for July 2000?

b. What are the minimum and maximum values for July 2000?

5. Atomic oxygen flux is greater around 300–500 km than at 800–1,000 km. High solar activity, however, reduces this difference.
    a. Why does this occur?
    b. At 900 km, what is the variance in atomic oxygen flux between sunspot maximum and sunspot minimum?
    c. Using Tables 3.5 through 3.7, what is the number density for atomic oxygen for low, medium, and high solar activity levels?

6. For low solar activity (Table 3.5), the log of the atmospheric density is -15.076 for a 1,000 km altitude with a density scale height of 402.18 km
    a. Assuming an exponentially varying atmosphere, what is the density at 1,125 km?
    b. What is the atmospheric density at 460 km during periods of average solar activity (Table 3.6)?
    c. What is the atmospheric density at 460 km during periods at high solar activity (Table 3.7)?

Chapter 4
Planetary Environments

In our solar system planetary atmospheres can be neatly grouped into two distinct types; namely, the atmospheres of the inner planets where light gas escape has been the determining factor in their evolution and the atmospheres of the outer planets with substantial gaseous envelopes similar to their original compositions. Since the atmospheres of the outer planets are thought to have undergone little change since they were first formed 4 to 5 billion years ago, they will provide scientists with vital information concerning the creation of our solar system [1]. The atmospheres of the inner planets have evolved over time with some of the original light weight elements escaping and being replaced by heavier components outgassed from each planet. These constituents have developed as the result of solar activity and, additionally for the Earth, the activity of living organisms. This unique situation has resulted in Earth's oxygen rich atmosphere.

The inner planets are Mercury, Venus, Earth, and Mars while Jupiter, Saturn, Uranus, and Neptune constitute the Jovian planets. Note that even though Pluto is the outermost planet it is definitely not a Jovian planet, but neither is it classified as a terrestrial planet.

The clear distinction between the inner, or terrestrial, and the outer, or Jovian, planets can be summarized by examining a few key parameters. The mass of the inner planets is hundreds of times less than the outer planets, yet the terrestrial planets are composed of much denser material. The inner planets are made primarily of silicates and rocks while the outer planets largely consist of hydrogen and helium. With the exception of Venus, the terrestrial planets have very thin to nonexistent atmospheres. It is also interesting to note that the inner planets have very few natural satellites while the outer planets have many [3]. Table 4.6 at the end of this chapter summarizes key parameters of all the planets.

Basically, two methods are used to evaluate planetary atmospheres. First, observations from Earth have been used for centuries to determine the characteristics of other celestial bodies. Second, satellite systems have been designed and flown to provide in situ data of these far-off worlds. Table 4.1 lists the programs developed to explore our sister planets. These encounters have provided us with the vast majority of information available on the evolution and nature of the other main components of our solar system.

The shortened mission of Phobos 2 and the launching of the Venus Magellan probe ushered in a renewed interest in interplanetary exploration at the outset of the 1990s. Even though the Phobos mission experienced technical problems and was terminated prematurely, it has provided significant information about the Martian atmosphere. The University of Colorado is jointly analyzing this data with the Soviets [4].

The Magellan space probe was launched by the U.S. Space Shuttle on 4 May 1989, beginning its 15-month journey to Venus. This $550 million satellite went into orbit about Venus in August 1990 after its 800 million mile trip. The Magellan mission has provided the most accurate information ever on another planet's atmosphere, but the data obtained will take many years to analyze completely.

The 1990s have marked a resurgence in interplanetary missions with encounters with four planets planned. Five launches over eight years (1989–1997) will supply copious information about our solar system. When Magellan was propelled toward Venus, it had been nearly 11 years since the last interplanetary launch. This new era comes as the first family of probes leaves the solar system. Pioneer 11, in March 1990, became the fourth spacecraft to exit the solar system. Pioneer 10 and Voyagers 1 and 2 were the first three probes to venture into interstellar space. By scientific "luck of the draw" the four probes are leaving our solar system in different directions.

Mercury

Due to the proximity of Mercury to the sun, it is nearly impossible for any gaseous elements to stay in contact with the planet. For our purposes, Mercury has no substantial atmosphere. During the 1960s, scientists made measurements of the Mercurian atmosphere but found no concrete evidence of CO_2 around the planet. Figures 4.1 and 4.2 show the Mariner 10 in preparation and on the launch pad. The fly-by of the Mariner 10 probe to within 80 km of the surface in the mid-1970s provided a wealth of information

Table 4.1 Interplanetary Space Programs

Mercury—Mariner
Venus—Venera, Mariner, Zond, Pioneer, VEGA, Magellan
Mars—Mariner, Viking, Mars, Phobos
Jupiter—Pioneer, Voyager, Galileo
Saturn—Pioneer, Voyager, Galileo, Cassini
Uranus—Voyager
Neptune—Voyager
Pluto—None

Figure 4.1 The Mariner spacecraft being prepared for integration. *Compliments of NASA.*

Figure 4.2 The Mariner spacecraft prepared for launch. *Compliments of NASA.*

[5]. Using the ultraviolet spectrometer (UVS) experiment on board, helium was found to be the most abundant element, two orders of magnitude more prevalent than hydrogen. Hydrogen and helium may both be deposited by the interaction of the solar wind with Mercury's magnetosphere, thus are not really considered its atmosphere. The effects of atmospheric drag would be negligible for a space probe in comparison to solar effects on a satellite orbiting Mercury.

From Mariner 10 observations, it was found that the surface of Mercury was heavily cratered with many expansive, large cliffs as high as 2 km and as long as 1,500 km. The planet has a crust of light silicate rock and when its planetary core cooled, the cliffs formed as this crust was compressed and buckled [5]. The mostly metallic iron-nickel alloy planet exists without a shield of gases like the ones that protect other planets in our solar system [1]. Due to this condition the surface temperature of Mercury varies from $-210°C$ on its dark side to $510°C$ on its sunlit side ($-346°F$ to $950°F$). Its slow rotation rate causes its "hot day" to also be very long—59 Earth days [6].

Though this celestial body still appears to be in its infancy, the probability of it developing and sustaining an atmosphere in the future is quite low due to its geological makeup and close proximity to the Sun. A two satellite mission is planned for launch to Mercury in 1997 with arrival in 2002. The Soviet Institute for Space Research (IKI) is the prospective developer for this dual probe. It is expected that robotic landers and penetrators will perform a photographic survey and geochemical analyses [1].

Venus

Venus has been under human observation for centuries. As early as 1761 the atmosphere of this planet was examined closely, allowing scientists to identify gaseous constituents by measuring how the atmosphere affected incident sunlight. Optical studies of this planet at visible wavelengths

Planetary Environments

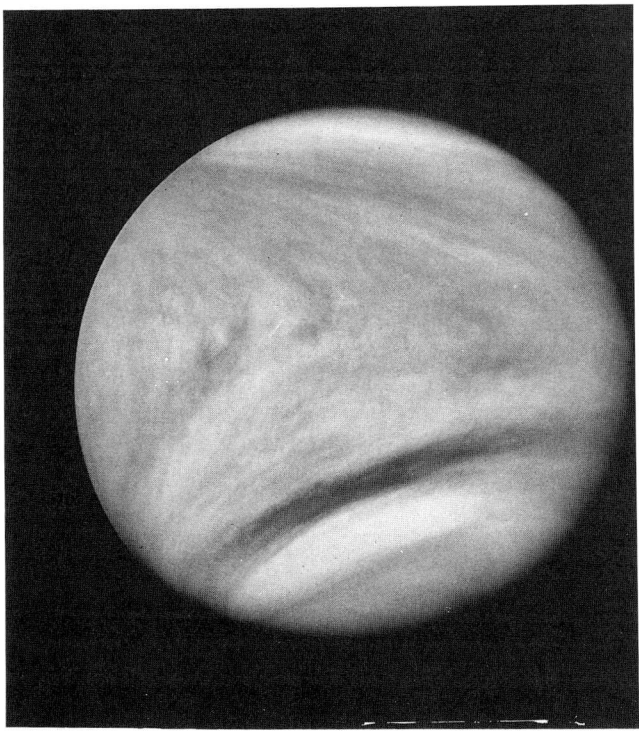

Figure 4.3 This ultraviolet photograph of Venus by a Mariner spacecraft clearly shows structure in the Venusian atmosphere. *Compliments of NASA.*

reveal an encircling of dense, pale lemon-yellow clouds. At first thought to contain vast quantities of H_2O and O_2, it is now known that CO_2 is the most abundant gaseous compound about Venus accounting for about 96.5% of its atmosphere with N_2 amounting to about 3.5% [7].

Mariner spacecraft first scanned the surface of Venus in 1962, marking the first observation of another planet by a man-made space probe. A view of the planet is shown in Figure 4.3. Though very similar in physical attributes to the Earth, Venus has distinctly different weather patterns and atmospheric composition. The existence of low altitude, high speed (hundreds of km/hr) clouds containing high levels of sulfuric acid is just one of these phenomena [5].

With all the current concern about the greenhouse effect the reader might surmise that an abundance of CO_2 would cause a very high Venusian surface temperature. This is indeed correct but additionally the solar power per unit area at Venus is 2,600 W/m^2 which is twice that reaching Earth [7]. As early as the 1930s, the temperature on Venus was predicted to be on the order of 400–500 K due to these two factors. One major source (or sink) of CO_2 for the Venusian, atmosphere is the abundance of CO_2 in carbonate rocks like those found on Earth. The major governing reaction for the exchange of CO_2 is given by

$$CO_2 + CaSiO_3 \leftrightarrows CaCO_3 + SiO_2 . \quad (4.1)$$

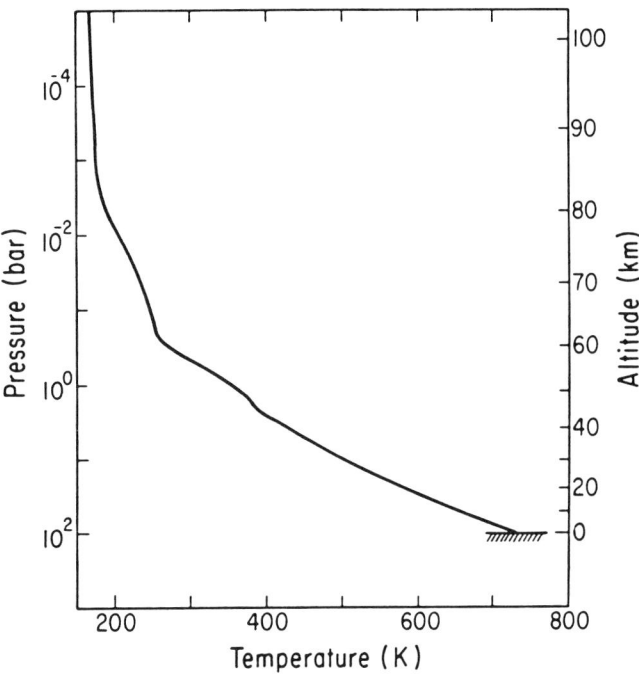

Figure 4.4 Pressure-temperature profile for the atmosphere of Venus. *Reprinted by permission of Academic Press [7].*

The reversible process shown in equation 4.1 is known as the Urey reaction (after H. C. Urey) and is a buffer reaction. A buffer reaction is one in which the gas pressure is a function of temperature alone [1]. Another theory about the evolution of the atmosphere of Venus proposes that liquid water and liquid hydrocarbons coexisted long ago on Venus. The H_2O dissociated into free hydrogen and free oxygen, with the oxygen subsequently combining with the hydrocarbons to form CO_2.

In the late 1950s the surface temperature of Venus was estimated to be 600 K. The cause of the high temperature has been attributed to many phenomena including frictional dissipation of wind energy, greater solar energy exposure, and the greenhouse effect [7]. Direct measurements from the surface by Soviet Venera spacecraft and the U.S. Pioneer-Venus probes place the planet's surface temperature in the 700–775 K range. The pressure and temperature distribution about Venus is plotted in Figure 4.4.

The graph in Figure 4.4 plots the atmospheric temperature and pressure of Venus, showing that it is essentially negligible above 80 km altitude. The clouds originally detected above the surface reside at altitudes between 50–70 km and provide a clear demarcation between two important layers in the atmosphere; namely, the troposphere, which is below the clouds, and the stratosphere, which is above the clouds. The major constituents of the atmosphere in these two regions are listed in Table 4.2.

The major difference between the two layers is the factor of 100 increase in H_2O below the clouds (troposphere).

Table 4.2 Venus Atmospheric Constituents by Volume Mixing Ratio [7]

Molecule	Troposphere	Stratosphere
CO_2	0.96	0.96
N_2	0.04	0.04
H_2O	10^{-4} to 10^{-3}	10^{-6} to 10^{-5}
CO	2×10^{-5} to 3×10^{-5}	5×10^{-5} to 10^{-3}
HCl	10^{-5}	10^{-6}

Additionally, sulfuric acid appears in significant quantities between 50–80 km altitude. The high surface temperature and composition of Venus produce a significant number of interaction cycles of chemical reactions between the atmosphere and minerals, best summarized by the following series of reactions [2]:

$$4CaCO_3 + 2O_2 + 4SO_2 \rightarrow 4CaSO_4 + 4CO_2 \quad (4.2)$$
$$FeS_2 + 2H_2O \rightarrow FeO + 2H_2S + 1/2\,O_2 \quad (4.3)$$
$$FeS_2 + 2CO_2 \rightarrow FeO + 2\,COS + 1/2\,O_2 \quad (4.4)$$
$$2H_2S + 3O_2 \rightarrow 2H_2O + 2SO_2 \quad (4.5)$$
$$2\,COS + 3O_2 \rightarrow 2CO_2 + 2SO_2 \quad (4.6)$$
$$4SO_2 + 2O_2 + 4CaCO_3 \rightarrow 4CaSO_4 + 4CO_2 \quad (4.7)$$
$$4CaSO_4 + 2FeO + 4CO_2 \rightarrow 2FeS_2 + 4CaCO_3 + 7O_2 \quad (4.8)$$

The net result of these activities leaves the planet with $CaSO_4$, CO_2, and ferrous compounds while the atmosphere contains the substances listed in Table 4.2 plus traces of sulfuric compounds (SO_3, H_2SO_4, S_8, S_4, H_2S, SO_2, etc.).

Adding to the chemical and physical dynamics of the planet are several regions of volcanic activity. Venus is only one of four bodies in our solar system that are known to be volcanically active. Jupiter's moon Io, plus Neptune's Triton (ice lava), and Earth are the others [6]. The giant sized volcanos of Mars do not appear to be active at the present time.

In the upper atmosphere (above the clouds) the major chemical activities involve the dissociation and recombination of CO_2. The ultraviolet radiation from the Sun causes CO_2 to break down into CO and O but CO_2 is usually quickly reformed. The continual formation and destruction of CO_2 is a major topic in the analysis of planetary atmospheres. While many texts have been written on this aspect alone of atmospheric evolution, here we will strive to emphasize the implications of each atmosphere's composition on space travel rather than dwell on a detailed analysis of its origin.

The Venus-Magellan probe, launched by the United States in 1989, has provided the most accurate information and data ever obtained on this planet. It encountered Venus in 1990 and began returning pertinent data immediately. It

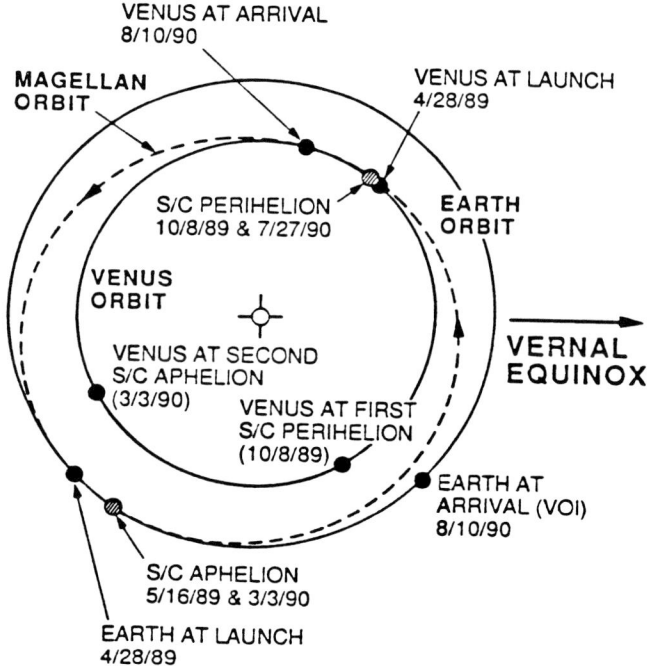

Figure 4.5 The patched conic path of the Magellan probe to Venus took over a year to complete [5].

will take many years to assimilate the vast quantity of new information being produced by this probe.

Figure 4.6 is an image of Venus in far ultraviolet light taken by the University of Colorado's ultraviolet spectrometer experiment on the Pioneer-Venus orbiter. The crescent Venus is seen in light emitted by atomic oxygen in the planet's upper atmosphere. The hydrogen is seen to surround the planet in a cloud extending many thousands of kilometers into space.

A practical applications-oriented description of the atmosphere of Venus and Mars has been developed by King-Hele [8]. This source will be the primary reference over the next two sections.

It is often useful to describe the atmosphere of other planets by direct comparison with the Earth's. The pressure at the surface of Venus is nearly 100 times greater than Earth while its temperature (750 K) is a factor of 2.5 greater. At lower altitudes (below 150 km) the atmospheric density above Venus is greater than at the surface of the Earth, but drops off more quickly as a function of altitude so that by the height of 150 km and above, the Venusian atmospheric density is less than for the comparable altitude above Earth. Just as Earth's atmosphere has day-night variations in atmospheric density, due to heating in the Venusian atmosphere, similar changes occur there with wider spreads down to lower altitudes than for Earth. The *Venus International Reference Atmosphere* [3] provides the most complete description of the atmosphere of Venus. Figure 4.7

Planetary Environments

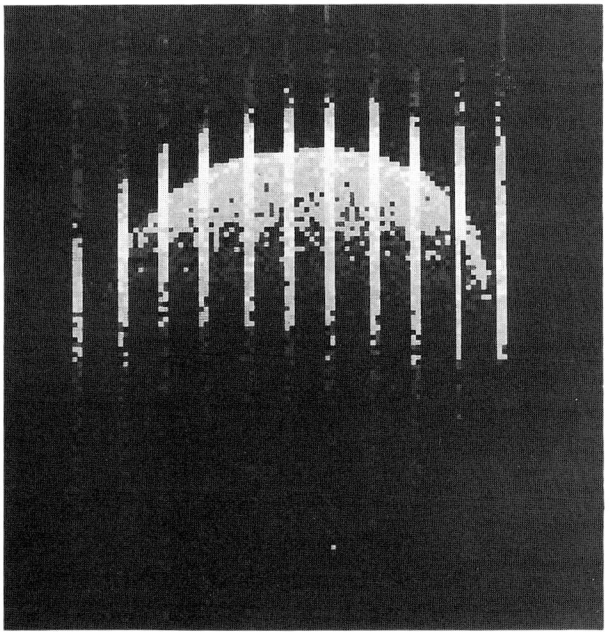

Figure 4.6 The imagery from the Pioneer Venus Orbiter depicts atomic oxygen concentration in Venus' atmosphere. The stripes result from periodically measuring the light emitted by atomic hydrogen during acquisition of the image. *Compliments of NASA.*

plots the atmospheric density of Venus as a function of altitude [8].

The lifetime of satellites in orbit about Venus may be directly linked to the lifetime of Earth satellites in similar orbits by accounting for differences in the atmospheres. The Venusian atmosphere has an abrupt daytime bulge which greatly affects the lifetimes of satellites in highly elliptical orbits. If the perigee of such an orbit coincides with the high density region, the satellite's orbital lifetime will be greatly shortened. The opposite is true if the perigee is located at the nighttime minimum. King-Hele [8] determined that the eccentricity cutoff between these two effects is 0.004. For a satellite in a nearly circular orbit, the average density encountered is about half that of the maximum density. This density curve is the dashed line in Figure 4.7. An equation is derived in reference 8 for the lifetime of a satellite in a Venusian orbit, L_V, in comparison to the orbital lifetime about Earth, L_E. The ratio of these orbital lifetimes is

$$L_V/L_E = (H_V/H_E)(\mu_E a_E/\mu_V a_V)^{1/2} (\rho_E/\rho_V) \quad (4.9)$$

where H is the respective planet's density scale height in kilometers. The ratio L_V/L_E is 1.1 at 150 km altitude, in-

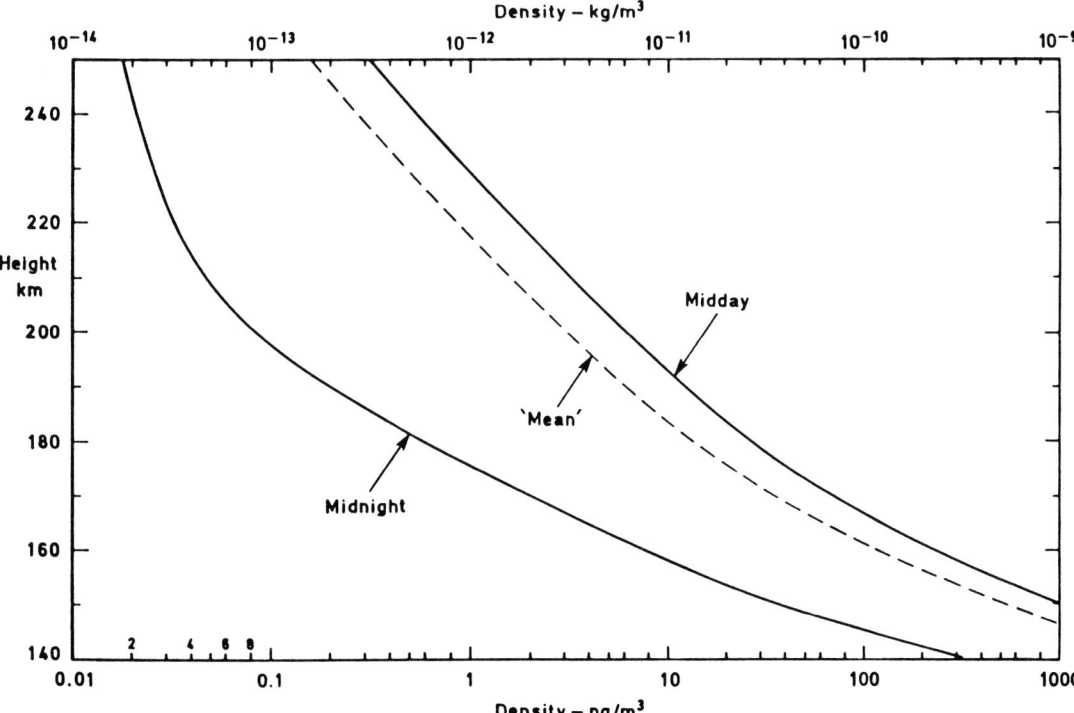

Figure 4.7 The average Venusian atmospheric density differs very little from the density encountered at midday [8].

creasing to 10 at 180 km, and 70 at 210 km [8]. As a result, the lifetimes of satellites in circular orbits about Venus at 180, 200, and 220 km are 9 days, 58 days, and 1 year; respectively. In summary, the orbital lifetimes are much greater about Venus as the altitude goes above 150 km compared to the lifetimes of satellites in similar Earth orbits.

The 3,500 kg Magellan space probe, employs a high resolution radar system to map the Venusian surface from its elliptical orbit (250/8,000 km) about Venus. This 3-hour nearly polar orbit will allow the planet's surface to be mapped in 243 days. The mission will provide a mapping of the Venusian surface better than we have of Earth, since most of Earth is covered by oceans [9]. Soviet scientists have proposed a 1998 encounter using six to eight 50 kg penetrators to derive in situ measurements of the planet's surface [1].

Mars

With the exception of Earth, no planet has received as much attention as Mars. It has been considered the prime candidate for extraterrestrial life [10] and its eery red appearance has piqued observers' imaginations for centuries. The Babylonians tracked the path of Mars in the night sky and gave it the name Nergal after its god of war. The planet's present name, Mars, was coined by the Romans to honor their own war god. Mars was originally considered to harbor life due to its changes in color on a regular basis similar to Earth's changing of the seasons. Yet, measurements made by the two U.S. Viking spacecraft have revealed no evidence of life on Mars at all. As a matter of fact, the temperature on the surface of the planet varies from $-124°C$ to $-21°C$ ($-19°F$ to $-6°F$), much too cold to support life as we know it [10].

The atmosphere of Mars, however, is quite similar to that of Venus, consisting primarily of carbon dioxide. But, the atmosphere of Mars changes significantly with season, time of day, latitude, longitude, and solar activity [2].

The total mass of the Martian atmosphere is less than 1% of the Earth's atmosphere. Its red appearance is not due to gaseous constituents but rather because of airborne ferrous oxide particulates. This dust is tossed up by strong surface winds that buffet the planet.

In general, an atmosphere's composition is greatly dependent on the makeup of the planet itself. Common elements found on Mars are oxygen (42%), silicon (21%), iron (13%), magnesium (5%), calcium (4%), aluminum (3%), sulfur (3%) with traces of chlorine, titanium, sodium, and hydrogen [11]. The large amount of oxygen available in the soil provides a source for many of the chemical reactions which control the state of the atmosphere. Similarly, the

Table 4.3 Atmospheric constituents of Mars by volume mixing ratios [2].

CO_2	0.953
N_2	2.7×10^{-2}
Ar	1.6×10^{-2}
O_2	1.3×10^{-3}
CO	7.0×10^{-4}
H_2O	3.0×10^{-4}

icy poles of Mars provide a large source of H_2O for atmospheric reactions.

The main reaction in the Martian atmosphere is carbon dioxide photolysis which occurs at wavelengths shorter than 204 nm as

$$CO_2 + h\nu \, (\lambda \leq 204 \text{ nm}) \rightarrow CO + O \, . \quad (4.10)$$

A whole series of reactions help to power this process. First, carbon monoxide must be able to recombine with oxygen to keep the CO_2 level in equilibrium:

$$CO + O \rightarrow CO_2 \, . \quad (4.11)$$

A similar pair of reactions for water dissociation and recombination are essential to the Martian atmosphere; namely,

$$H_2O + h\nu \rightarrow OH + H \quad (4.12)$$

and

$$OH + HO_2 \rightarrow H_2O + O_2 \, . \quad (4.13)$$

Yet, the H_2O concentration in the Martian atmosphere is kept low by the reaction

$$H + HO_2 \rightarrow H_2 + O_2 \, . \quad (4.14)$$

This prevents the HO_2 molecule from being available for H_2O production and also provides a source of H_2. The amount of H_2 present indirectly affects the amount of ozone and water in the atmosphere. The H_2 that does not escape may be dissociated into atomic hydrogen (H) which may decrease the abundance of ozone by the reaction [7]

$$H + O_3 \rightarrow OH + O_2 \, . \quad (4.15)$$

The hydroxide molecule (OH) provides a key building block for H_2O formation as shown in equation 4.13. The creation and absorption of water on Mars is an important area of discussion since it is imperative for sustaining life. A major source of H_2O, other than the previous chemical reactions, is crustal outgassing. At the same time, the H_2O atmospheric concentration is decreased by oxidation of iron and the formation of clay [7].

Recent measurements by the Soviet Phobos 2 probe may provide improved insight into the Martian atmosphere. Before the spacecraft failed on 27 March 1989, it acquired

Planetary Environments

Figure 4.8 The "canals" on the barren Martian surface resemble irrigation ditches and fueled early speculation that intelligent life existed on this planet. *Compliments of NASA.*

significant data on the CO_2, water, ice, and mineral dust present about Mars. The University of Colorado at Boulder will assist the Soviets in the analysis of this atmospheric data [12].

Much of the information available on the composition of Mars and its atmosphere came from the Viking program. One of its experiments measured the proportions of isotopes present on Mars. (Isotopes are atoms that have different atomic weights but are the same element (e.g., U-235 and U-238).) The ratio of isotopes on the planet and in its atmosphere can help determine the temperature of the rocks when they were formed. It is hypothesized that comparable isotope ratios for different planets indicate that they may have formed from the same part of the original solar nebula [11].

The isotope data acquired by Viking I and II showed a greater similarity between the Martian and Earth atmospheres than had been previously thought. Table 4.4 shows a comparison of several important isotope ratios for Mars and Earth.

Table 4.4 Isotope Ratios: Mars versus Earth [11]

	C_{13}/C_{12}	O_{18}/O_{16}	N_{15}/N_{14}
Earth	1/89	1/500	1/271
Mars	1/89	1/500	1/156

The lower ratio of nitrogen for Mars may be the result of its escape due to the planet's lower gravity [11]. Scientists have speculated that if nitrogen were abundant far in the past the ancient atmosphere of Mars could have been much more dense, possibly containing significant amounts of water. This abundance of water may have manifested itself as global Martian oceans, thus providing an explanation for the river-like geological characteristics of the planet's surface today [11].

A satellite orbiting Mars is exposed to distinctly different effects than a satellite orbiting Earth. Since the mass and therefore the gravity of Mars are smaller than Earth's, the orbital period is larger for a comparable circular altitude. For a satellite with h = 400 km, for example, the orbital

Figure 4.9 The Viking probe examines Mars, showing its rocky surface. *Compliments of NASA.*

period is 93 minutes about Earth and 117 minutes about Mars. Thus, one orbit about Mars exposes a satellite to the atmosphere longer. The density of the Martian atmosphere as a function of altitude is given in Figure 4.10.

The atmosphere of Mars, like that of Earth, is affected by the level of solar activity; namely, greater activity creates greater atmospheric density at a given altitude. Despite the different atmospheric constituents and planetary structures the orbital lifetimes of Martian satellites may be calculated and compared to Earth satellite lifetimes. For elliptical orbits having eccentricities greater than 0.01, the following equation of orbital lifetimes applies [8]:

$$L_M/L_E = (\rho_E/\rho_M)(\mu_E a_M/\mu_M a_E) \quad (4.16)$$

where the subscript M is for Mars. Yet, at altitudes from 300 to 500 km the Martian atmosphere has a density less than 1% of that at a comparable height above Earth. Thus, lifetimes of Martian satellites in nearly circular orbits are approximately 200 times longer than for satellites in comparable Earth orbits. During low solar activity this factor increases to 800. In summary, the lifetimes in Martian orbits are 300 to 800 times longer than for Earth satellites [8].

The Mars Observer is slated for launch in the 1990s. This probe will measure atmospheric constituents, surface topography, and mineral resources of the planet. Earlier observations by the Viking and Mariner missions showed that polar ice caps that change with the seasons actually consisted of carbon dioxide. The Mars Observer will orbit in a 350–400 km circular polar orbit by the mid-1990s. The Mars Rover, a sister mission to be deployed within a few years after the Mars Observer begins its mission, will have to overcome a number of technical problems due to the harsh environment at the planet's surface. First, due to the extremely cold temperatures in some regions ($-140°C$), special lubricants will have to be developed. Second, because of Mars' distance from the Sun, solar cells are not a particularly efficient way of providing power. Ways to combat both these difficulties will be addressed in following chapters, highlighting the contribution of applying chemistry principles to assure mission success.

Jupiter

Jupiter is the innermost of the massive, gaseous Jovian planets. It is nearly five times farther from the Sun than Earth and over 300 times more massive. As a matter of

Planetary Environments

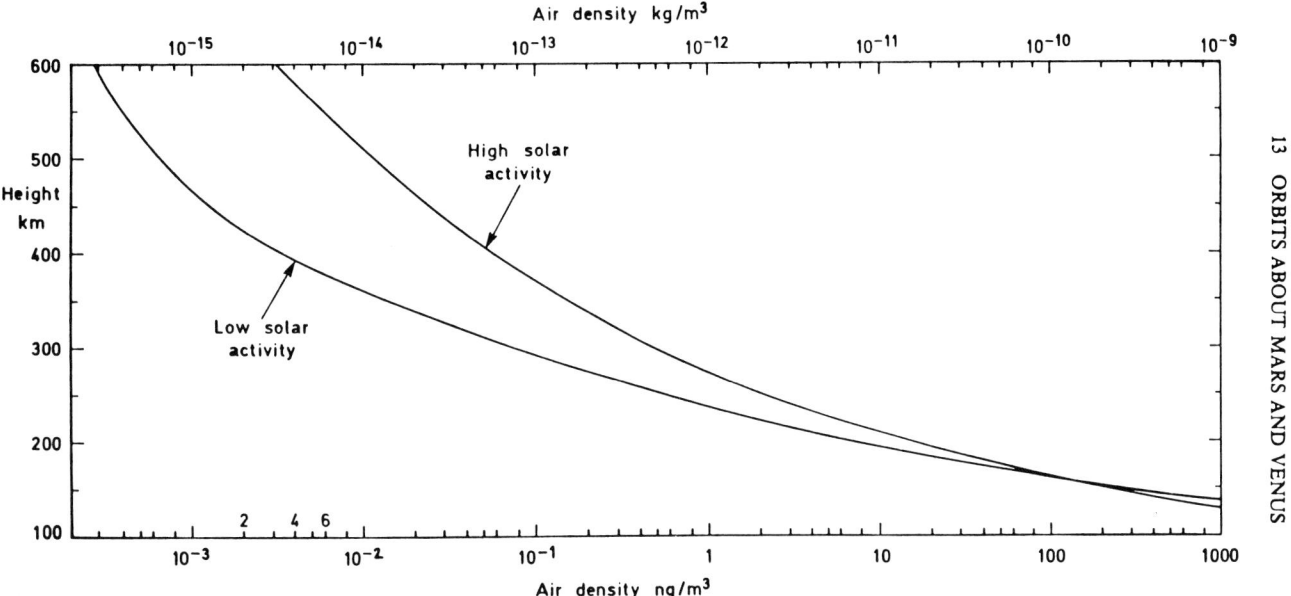

Figure 4.10 The atmospheric density about Mars increases with higher solar activity except at very low altitudes [8].

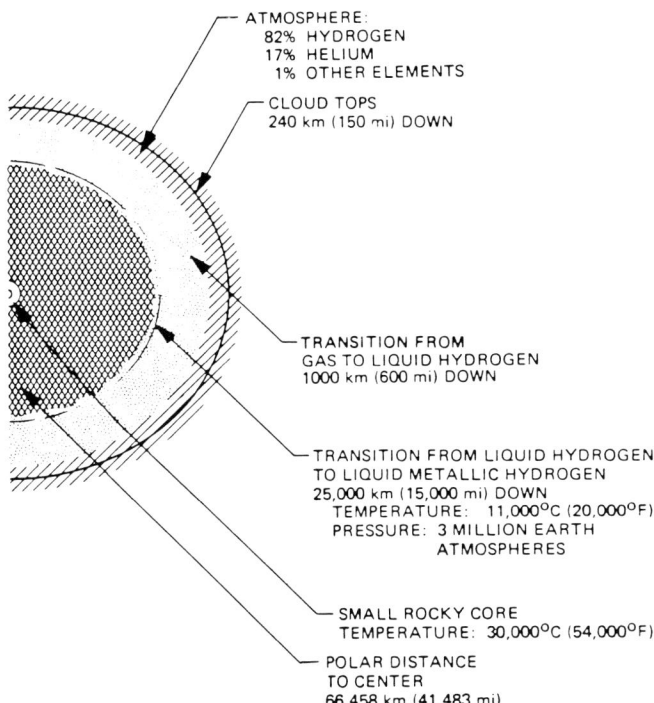

Figure 4.11 According to a commonly accepted model, Jupiter is mostly liquid hydrogen except for a very small core containing heavy elements [13].

Table 4.5 Jupiter's Atmospheric Constituents by Volume Mixing Ratio [2]

Hydrogen, H_2	0.89
Helium, He	0.11
Methane, CH_4	2.4×10^{-3}
Ammonia, NH_3	2.0×10^{-3}
Water, H_2O	5.0×10^{-6}
Carbon Monoxide, CO	2.0×10^{-9}
Phosphorous, Acetylene, Germanium Tetrahydride, Hydrogen Cyanide	Traces

This colossal planet essentially consists of a ball of liquid enveloped by a relatively thin layer of gases. Figure 4.11 shows the structure of Jupiter and its atmosphere. The only solid material that may exist in Jupiter is a small rocky core. At the center of the planet, temperatures may reach as high as 30,000°C (54,000°F). Around the core there is a large volume of hydrogen. Due to the large pressures encountered, the hydrogen which is usually a gas, exists as a liquid and a liquid metal. Liquid metallic hydrogen is so dense that it readily conducts heat and electricity [13]. Near the surface of the planet, the hydrogen returns to its gaseous state. This point roughly marks the beginning of Jupiter's atmosphere whose constituents are listed in Table 4.5.

Jupiter is rather hot and actually radiates more than twice as much heat as it absorbs from the sun. One current explanation is that the planet is still cooling down from its formation [13]. As massive as Jupiter is, it is far too small to have a sufficient gravity to compact its center to such a state that nuclear reactions (fusion) would occur to continue this heat production. Had it been large enough it might have

fact, the mass of Jupiter accounts for the majority of the planetary mass in our entire solar system [7]. Due to its huge mass, Jupiter's gravitational pull affects the orbits of other planets and may actually have prevented the formation of a planet at the location of the asteroid belt [13].

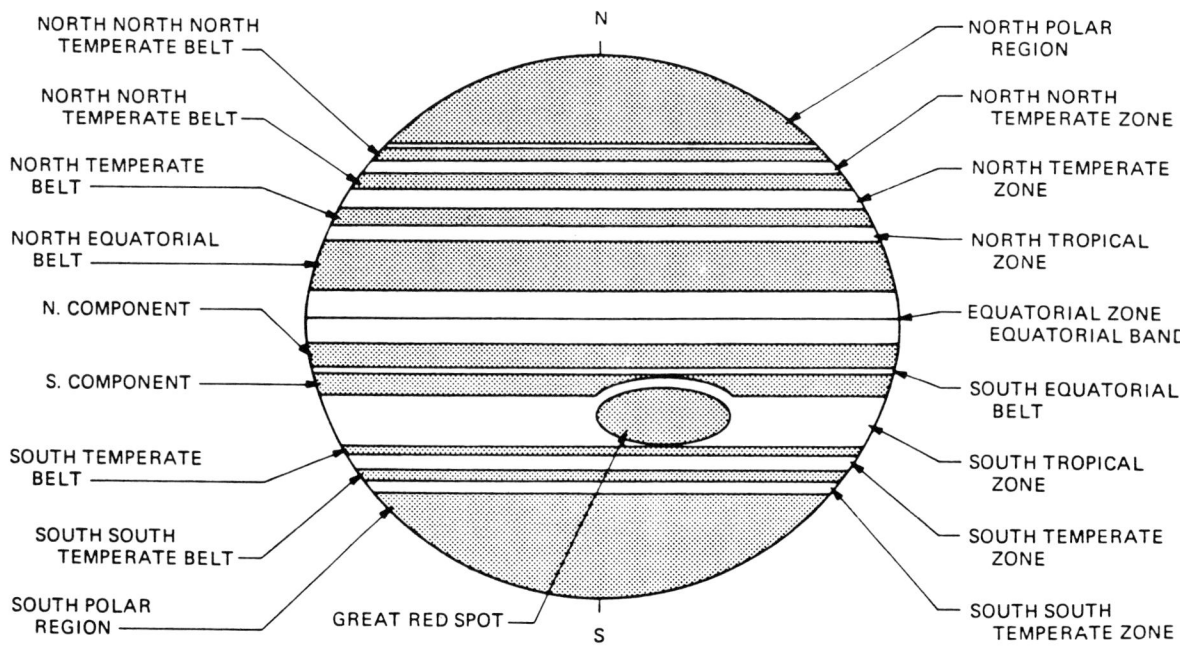

Figure 4.12 Jupiter's atmosphere exhibits a consistent banded appearance due to the planet's rapid rotation rate [13].

become another Sun, making our solar system a binary star system [13].

Many features of Jupiter can be seen from Earth. The view of this quickly rotating body is dominated by the Great Red Spot. This atmospheric storm spins counterclockwise like a hurricane would on Earth south of the equator. The spot has varied in intensity, color, and size over the years, and several times it has almost completely disappeared [13]. Its red color may be due to the presence of phosphorous or a complicated chemical chain reaction occurring below this cloud level. The other banded regions of Jupiter's atmosphere are the combined result of convective movements, (like thunder cloud formation) blurred into a band due to the planet's fast rotation rate: once every 10 hours. These dynamics result in surface winds on the order of 400 km/h [13].

The different colors of the clouds about Jupiter can be explained by their chemical composition. The highest altitude clouds (like the Great Red Spot) appear reddish because of phosphorous in them. Clouds at slightly lower altitudes appear white due to the ammonia crystals present. The lowest of the clouds are grey or brown. These contain a variety of ammonia, sulfur, and complex organic molecules [13]. Figure 4.13 depicts cloud regions by altitude, pressure, temperature, and chemical makeup. The layered nitrous-based clouds and steep temperature profile are distinguishing characteristics of Jupiter's atmosphere. Figure 4.14 is a photograph of Jupiter showing the banded nature of its atmosphere while anomalies, like the Great Red Spot and small storms, are readily apparent.

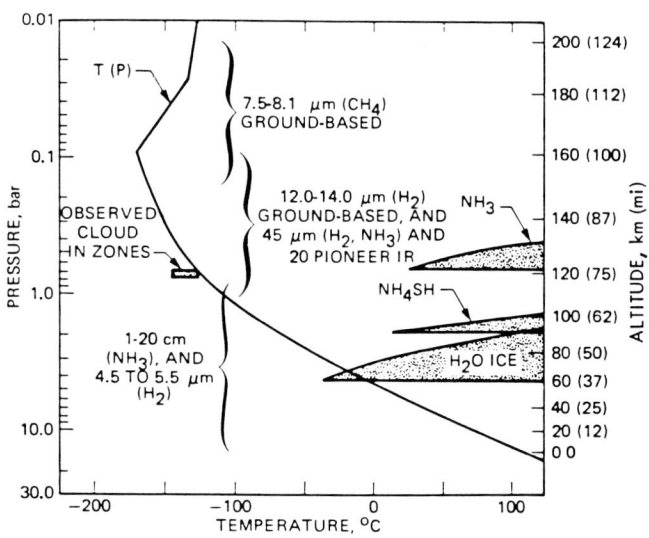

Figure 4.13 Jupiter's atmosphere is dominated by clouds between 60 and 130 km in altitude [13].

The dynamic behavior of the atmosphere about Jupiter can be described by a number of chemical reactions including the following [7]:

$$4CH_4 + h\nu \rightarrow C_2H_6 + C_2H_2 + 4H_2 \quad (4.17)$$
$$4NH_3 + h\nu \rightarrow N_2 + N_2H_4 + 4H_2 \quad (4.18)$$
$$4PH_3 + h\nu \rightarrow P_4 + 6H_2 \, . \quad (4.19)$$

These three reactions have a significant impact on the makeup of the atmosphere. Reactions 4.17 and 4.18 are the photodissociation of methane and ammonia, respectively.

Planetary Environments

Figure 4.14 The Great Red Spot, the major feature of Jupiter, is a large counterclockwise-rotating storm. *Compliments of NASA.*

These are two of the key molecules present in the vicinity of Jupiter. The last reaction is important because P_4 is the element that causes the red color in Jupiter's clouds. All three reactions produce hydrogen—the most abundant element present. However, these reactions are also reversible, allowing for the recombination of hydrogen.

In actuality the chemical reactions taking place in the atmosphere are much more complicated. Figure 4.15 shows the complete methane cycle which, in simplified form, is given by reaction 4.17.

Information on the Jovian planets has been greatly in-

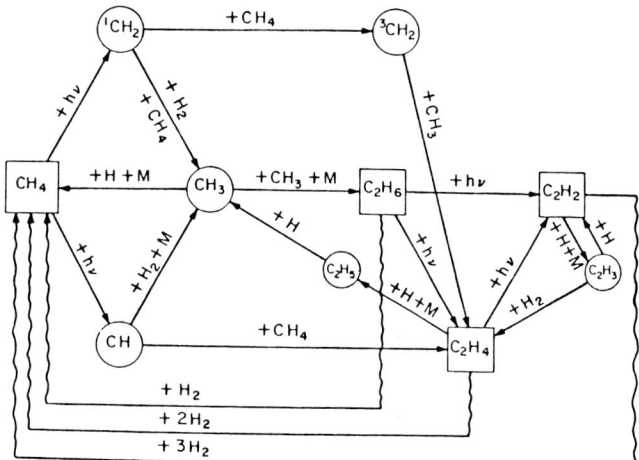

Figure 4.15 The methane cycle for Jupiter is driven by the abundance of gaseous hydrocarbons and their thermal and photochemical dissociation. *Reprinted by permission of Academic Press.*

creased by the flights of Voyager 1 and 2. In August 1989, Voyager 2, after 12 years of operation, provided the best pictures ever of Neptune. Figure 4.16 shows Voyager 2, possibly the most prolific scientific spacecraft ever built.

The 1989 launching of the Magellan probe to Venus marked a renewed emphasis on interplanetary exploration. This launch was quickly followed by the launch of the Galileo spacecraft bound for Jupiter. The Galileo mission consists of an atmospheric probe which will enter the low levels of Jupiter's atmosphere while another probe remains in orbit about the planet [6]. Much of the data gathering by the Galileo project will emphasize the Galilean moons Io, Europa, Ganymede, and Callisto. The probe will arrive at Jupiter in 1995 after a six year flight through the inner solar system where it will gain velocity for its eventual Jovian trajectory. The plutonium-powered radioisotopic thermoelectric generators (RTGs) used on the Galileo probe are necessary since there is so little solar energy at such great distances from the Sun. Later chapters will discuss space power concepts such as RTGs.

Saturn

Saturn is much like Jupiter in composition and size. It consists mainly of hydrogen and its atmosphere appears banded for similar reasons as Jupiter. The major atmospheric constituents are hydrogen (94%) and helium (6%) with traces of CH_4, C_2H_6 and PH_3, like Jupiter [2]. A major difference in the appearance of Saturn compared to Jupiter is its rings. They are composed of many low density particles of widely varying size, consisting mostly of ice and frosted rock [10].

The launch of the joint NASA/ESA Cassini mission to Saturn is planned for early 1996. It will take about 6.5 years to reach Saturn in the year 2003. A small atmospheric probe will be used to deploy eight parachuted instruments to mea-

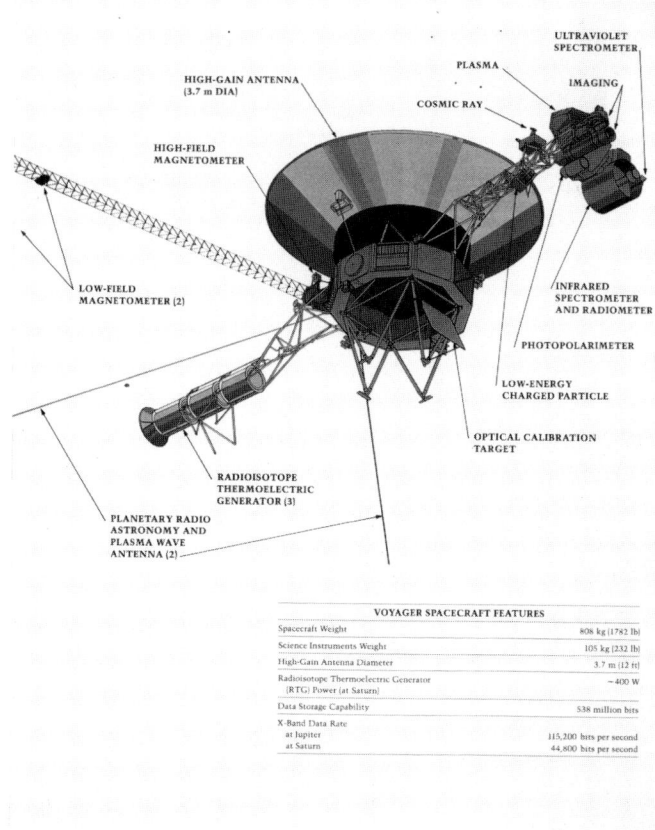

Figure 4.16 The spacecraft features of Voyager 2 which has been the workhorse of interplanetary probes. *Compliments of NASA.*

sure the components of the atmosphere. The Cassini probe will actively study the Saturnian system for five years examining the planet's rings and icy moons, including Titan [1]. Titan is unique in that it is the only moon in the solar system known to have an extensive atmosphere.

Uranus and Neptune

The last two Jovian planets are more like each other than are Jupiter and Saturn. Both Uranus and Neptune appear bluish green because of similar abundances of methane in their atmospheres. They are both colder and denser than Jupiter and Saturn, and are also thought to consist mainly of hydrogen in its liquid and gaseous states [10]. These two cooler planets may also contain ice on their surfaces and probably have a larger proportion of rocky material than either Jupiter or Saturn.

Since Uranus and Neptune are at such great distances from Earth, precise information about them and their atmospheres is scarce. It is known that there are significant amounts of hydrogen and helium in their atmospheres, but measurements have been too imprecise to determine an exact ratio. The accepted proportion of these two elements is 89% H_2 and 11% He—the ratio of these elements in our Sun [2]. Traces of other compounds have also been detected

Figure 4.17 Saturn's rings, consisting of icy and dirty particles, dominate its appearance. *Compliments of NASA.*

and models of the atmosphere/clouds are shown in Figures 4.18 and 4.19.

In Figures 4.18 and 4.19 note that the vertical axis for altitude is referenced to a temperature. This highlights the uncertainty with which these two planets (Neptune and Uranus) can be analyzed; it is difficult to actually ascertain where the planet's surface ends and the atmosphere begins. Despite similar cloud structures, Neptune's clouds obscure sensing of its surface much more than Jupiter's thin cloud formations. In addition, a set of rings was discovered photometrically about Uranus in 1977. Neptune also has at least two thin sets of rings, but none of these ring systems are visible from Earth.

Information about the outer planets is still being "discovered" a little more each year. In April 1989, the Voyager 2 space probe made an exciting find. Neptune was found to have a large dark spot similar to the Great Red Spot of Jupiter. Neptune's newly found spot appears to be a low altitude phenomenon, but further analysis should clarify this speculation. Four new moons and five rings were also discovered during the Voyager 2 flyby. Yet, the most exciting findings involved Triton, Neptune's largest moon, shown in Figure 4.20. It appears that there are active volcanoes on Triton that eject liquid nitrogen "lava" which create high-rising dark plumes. Scientists suspect that Triton is the object most like Pluto in our solar system. The study of planetary systems is truly an ongoing science which spacecraft will slowly unravel with the help of advanced technologies based on aerospace chemistry as the 21st century approaches.

Pluto

Pluto, though the outermost planet of our solar system, has more in common with the inner terrestrial planets than it has with the Jovian planets. It was discovered in 1930 and has been likened to a "celestial snowball" [6]. It is about

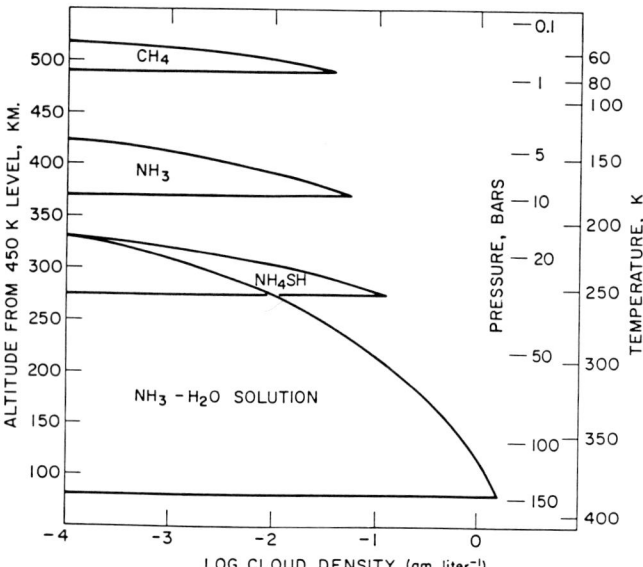

Figure 4.18 Uranus' atmosphere exhibits a layering of cloud structure similar to that of Jupiter's. *Reprinted by permission of Academic Press [7].*

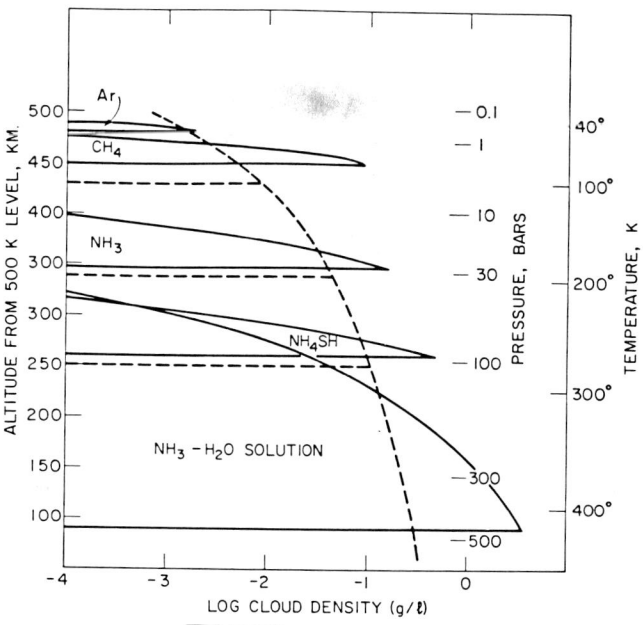

Figure 4.19 The outer Jovian planet, Neptune, exhibits the characteristic layered atmosphere of these colossal, gaseous planets. *Reprinted by permission of Academic Press [7].*

the size of our Moon and appears to be coated in a layer of methane ice. Yet, it has no measurable atmosphere. Pluto has a greatly inclined orbit, 17.2°, and a very elliptical orbit whereas most of the planets of our solar system have low inclination, nearly circular orbits. Of all the planets, Pluto is the only one that will not be visited by an Earth launched space probe this century [10]. Figure 4.21 shows an artist's conception of Pluto and its moon Charon. A number of missions have been suggested for visiting Pluto but there are presently no firm commitments to do so.

Summary

The nine planets of our solar system encompass a broad range of environments for spacecraft to encounter. The chemical reactions of photodissociation are a key to the state of many of the planetary environments. The effect of radiant energy on planetary environments differs due to the wide range of planet compositions, atmospheric constituents, and vicinity to the Sun. Recombination reactions help to maintain a semi-equilibrium state for most atmospheres. Our planet is unique in its atmospheric composition being rich in oxygen and nitrogen. This is the source and a byproduct of life on our planet.

The atmospheres of the terrestrial planets have evolved over time due to the major influence of the Sun energetically and gravitationally on them and the relatively small gravitational attraction of the inner planets. Thus, for the terrestrial bodies light gases can escape and the resulting atmospheres are thin but consist of fairly high molecular weight molecules. The atmospheres of the outer planets have changed very little with time and consist mostly of hydrogen and helium in proportions similar to our Sun's composition. The Jovian planets are also distinguished by their large size, low density, and numerous moons. The outer planets may provide essential information concerning the creation of our solar system. Table 4.7 summarizes key parameters of the planets of our solar system while Figure 4.22 highlights the difference between Earth and the Jovian planets.

References

1. Paine, T. O., "The Next 40 Years in Space," Spaceflight, Vol 32, pp. 14–17, Jan 90.
2. Wayne, R. P., *Chemistry of Atmospheres.* Clarendon Press, Oxford, UK, 1985.
3. Brandt, J. C., and Hodge, P. W., *Solar System Astrophysics.* McGraw-Hill Book Company, New York, 1964.
4. "Voyager, Mission to Outer Planets," NASA Facts, NF-87/10-77.
5. "Magellan to Venus," Spaceflight, Apr 1989.
6. "Our Planets at a Glance," NASA Information Summaries, PMS 010 (KSC), July 1986.
7. Lewis, J. S. and Prinn, R. G., *Planets and Their Atmospheres.* Academic Press, Inc., Orlando, FL, 1984.
8. King-Hele, D., *Satellite Orbits in an Atmosphere.* Black and Son Ltd, Bishopbriggs, Glasgow, 1987.
9. *Aviation Week and Space Technology*, 9 Oct 1989, pp. 44–118.

Planetary Environments

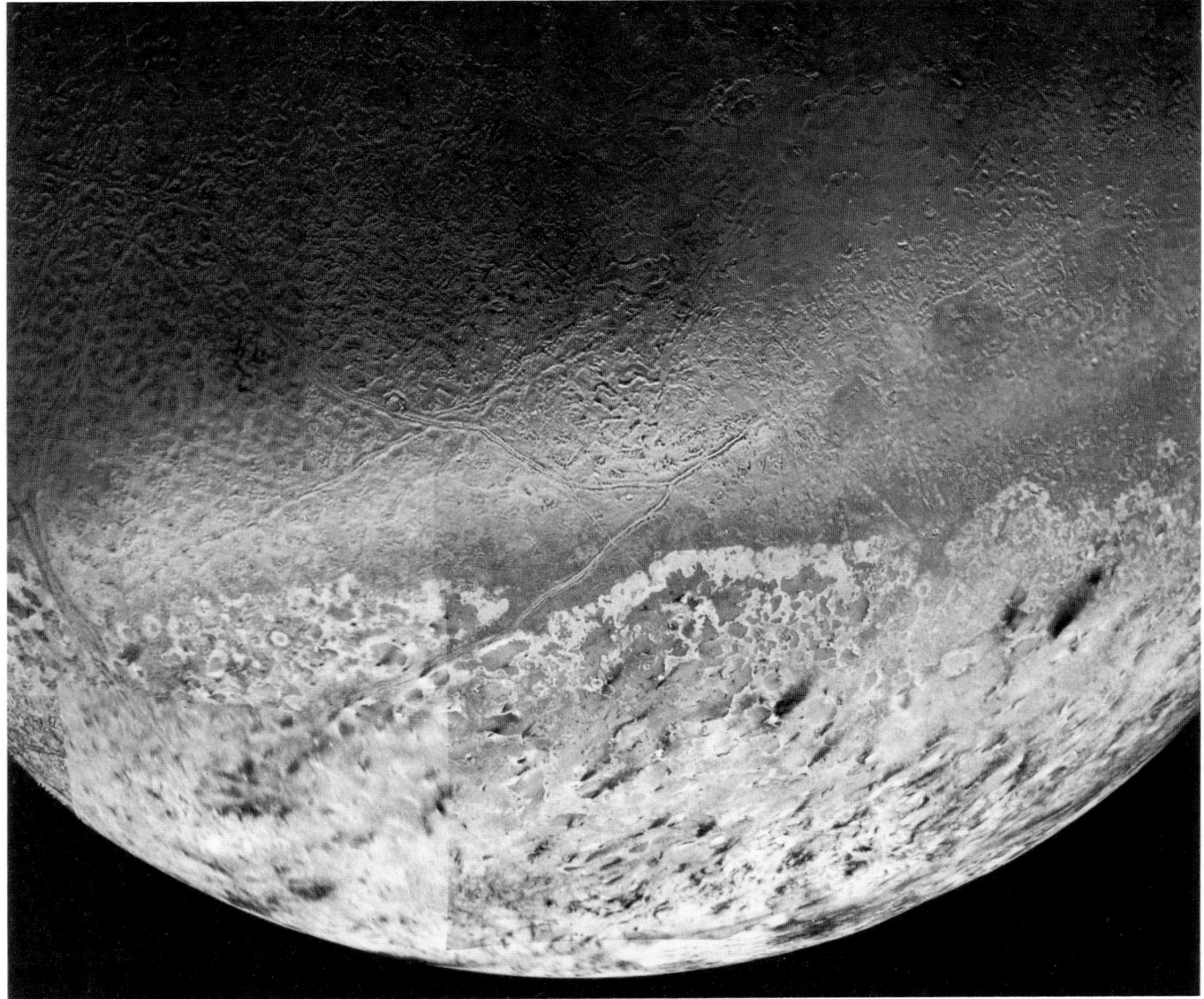

Figure 4.20 Triton, one of Neptune's moons, shows evidence of past fluid flow even though its temperature is estimated as $-240°C$ ($-400°F$), possibly the coldest object observed in the solar system. *Compliments of NASA.*

Table 4.6 Key Parameters of Planets [10]

	Mercury	Venus	Earth	Mars	Jupiter	Saturn	Uranus	Neptune	Pluto
Mean Distance from Sun (millions of km)	57.9	108.2	149.6	227.9	778.3	1,427	2,870	4,497	5,900
Period of Revolution	88d	225d	365d	687d	11.9y	29.5y	84y	165y	248y
Equatorial Diameter (km)	4,880	12,100	12,756	6,794	143,200	120,000	51,800	49,500	3,000
Main Components of Atmosphere	None	CO_2	N_2, O_2	CO_2	H_2, He	H_2, He	He, H_2, CH_4	H_2, He, CH_4	None
Satellites	0	0	1	2	14	12	5	6	1

Figure 4.21 This artist's conception shows Pluto and its moon Charon. *Compliments of NASA.*

10. "A Look at the Planets," NASA PAM107, July 1985.
11. "Mars: The Viking Discoveries," NASA EP-146, October 1977.
12. *Aviation Week and Space Technology*, 1 May 1989, p. 19.
13. "The Voyager Mission: Jupiter, The Giant of the Solar System," NASA Facts, NF-89, 1979.
14. "Encounter with Uranus," NASA Facts, NF-146/12-85.
15. *The New Solar System*. Ed. by Beatty, J. K., O'Leary, B. and Chaikin, A., Sky Publishing Corporation, Cambridge, Mass, 1982.
16. Kliore, A. J., et al, "The Venus International Reference Atmosphere," Adv. in Sp. Res., No. 5 (11), 1986, p. 1–305.
17. *Aviation Week and Space Technology*, 26 June 1989, p. 87.

Discussion Questions

1. What are the major differences between the "inner" and "outer" planets (terrestrial versus Jovian)?

2. Interplanetary exploration is needed to help unlock the mysteries of the creation and evolution of our solar system.
    a. Which planet(s) will provide the most insight into the formation of the solar system?
    b. Which planet or planets will assist researchers most in determining the future of Earth's environment and atmospheric evolution?

3. The composition of planetary atmospheres has been determined by remote observations and in situ measurements.
    a. Which planet, other than Earth, has been visited and analyzed the most?
    b. Which planet has never been visited by an interplanetary probe? Do you feel that a mission to this planet would yield significant information? Explain.

4. The Sun has a great effect on the inner planets, especially their atmospheres.
    a. Why doesn't Mercury have a substantial atmosphere?
    b. What atmospheric constituent of Venus contributes most to its high surface temperature?
    c. What is the major constituent of the Martian atmosphere?

5. Why have some astronomers in the past thought that there was intelligent life on Mars?

6. The Jovian planets are massive and gaseous while the terrestrial planets are small and rocky.
    a. Which planet is so massive that it actually accounts for a majority of the mass in the solar system?
    b. The atmospheres of Jupiter and Saturn consist mainly of which element?
    c. Why does the atmosphere of Jupiter appear to be banded?

7. The United States and the former U.S.S.R. (now called the Commonwealth of Independent States) have been the major space powers in exploring the solar system even though Japan orbited a probe about the Moon in 1990.
    a. What major technological concerns must be addressed by spacecraft designers as the planets closer to the Sun are visited?
    b. The exploration of the Jovian planets will pose different challenges to the satellite builders. What are these?

8. Propose a theory why Saturn has a visible set of rings but Neptune and Uranus do not. (Does the recent verification of a less massive ring system about Neptune support or refute your theory?)

Problems

1. If the orbital lifetime of a satellite in a circular orbit of 180 km altitude about Earth is 6 days:
    a. What is its lifetime in a similar orbit about Venus?
    b. What is its lifetime in a similar orbit about Mars?

Planetary Environments

Figure 4.22 This photographic collage shows the relative size of Earth compared to Jupiter and Saturn. *Compliments of NASA.*

2. When solar activity increases, the atmosphere about Earth expands, creating increased densities at the same geometric altitude and thus reducing satellite lifetimes.

    a. Does this same effect apply to the Venusian and Martian atmospheres? Support your answer with data from available figures.

3. Equation 4.10 describes the photolysis of CO_2 in the Martian atmosphere. Energy with wavelengths less than 204 nm will effect this reaction.

    a. To what frequency does this radiation correspond? (Hint: $c = \nu\lambda$)

    b. Does this frequency of radiation have the same effect in Earth's atmosphere?

4. The average molecular weight of Earth's atmosphere at sea level is approximately 29 atomic mass units (amu). By using weighted averages, determine the average molecular weight for the atmospheres of:

    a. Venus

    b. Mars

    c. Jupiter

Chapter 5

Spacecraft Materials

Selection of the appropriate material for a particular application poses a difficult challenge to spacecraft designers. A number of criteria including strength, weight, volatility, elasticity, reflectivity, thermal stability, and chemical reactivity must be considered. These numerous constraints dictate the usage of a wide variety of materials since no single one possesses all the necessary properties. For example, aluminum is preferred for structural members because of its high strength/weight ratio [1, 2, 3]. Unfortunately, its high coefficient of thermal expansion causes problems during thermal cycling. Composites represent an improvement over aluminum because of their lighter weight and lower thermal expansion. However, their low thermal conductivity makes disposal of waste heat through radiation very difficult. Such trade-offs are routine in the selection process as designers seek those materials possessing the overall greatest advantage for a particular application.

Historically, spacecraft have been constructed mainly of metals and polymers. The use of certain precious metals has been essential to the development of reliable, safe space operations. For example, gold film has been used on spacecraft exteriors for thermal protection and is used on the interior of astronauts' visors to deflect harmful solar radiation [4]. The use of composites in space is, however, growing and their physical and chemical properties improve with the expanding knowledge of the field. So too is the role of ceramics, as evidenced by their utilization on the Space Shuttle for thermal protection. In this chapter, the chemical makeup of these four classes of materials (metals, polymers, composites, and ceramics and glasses) and their interaction with the space environment are examined.

Metals

Metals are a basic building material in spacecraft structures as well as in all subsystems [3]. Examples of their uses include structural members, membranes, fuel tanks, bumper shields, batteries, and thermal control systems. The performance of metals in space is well documented, as evidenced by such references as the Marshall Space Flight Center *Handbook 527* [5]. This materials selection guide lists the metals and alloys acceptable for particular spaceflight systems. Metallurgical principles for spaceflight were mainly borrowed from knowledge gained in the area of aeronautical design. After all, concerns about high strength/weight, flexibility, and corrosion resistance are generic to any type of flight vehicle. Still, space is its own unique environment and is quite different from the realm of air-breathing craft. Critical issues involving the effects of the near Earth environments upon metals and materials in general are now just coming to light. If such man-made objects as Space Station Freedom are to attain their projected 20 to 30 year lifetimes, then a basic understanding of chemically destructive processes must be obtained.

This section will discuss the major primary and secondary structural metals. We will not focus upon the metallurgy, as it is accepted that the properties of a metal are strongly dependent upon its mechanical and thermal history [3]. Rather, emphasis will be placed upon the effect of the space environment upon the basic chemical nature of the material and the attempts to thwart harmful degradation.

Atomic Structure and Radiation Effects

Recall in Chapter 2, the atomic structure within a metal was described basically as a large number of positively charged nuclei surrounded by an amorphous cloud of negatively charged electrons. This description is called the free electron theory and was first put forth by Drude (1902) and Lorentz (1916) [1]. Electrostatic attractions between the nuclei and electrons hold the atoms together in what was previously described as metallic bonding. Metallic bonding is isotropic and therefore the arrangement of metal atoms is only restricted by geometry. Metal atoms assume an ordered arrangement very similar to the close packing of hard spheres. The very simple crystal structures produced by this packing are divided into three classes, shown in Figure 5.1.

These crystal or unit cells are the basic arrangement of the atoms, which is repeated in all three dimensions of a solid crystal. Table 5.1 gives the common crystal structures of metals. Note that a metal can have more than one crystal structure based upon such external conditions as temperature and pressure. Packing density is greatest for close-

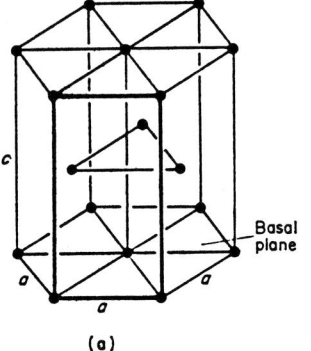

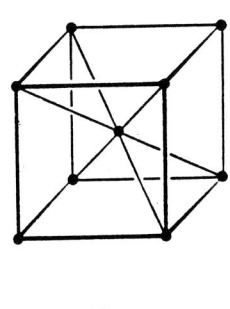

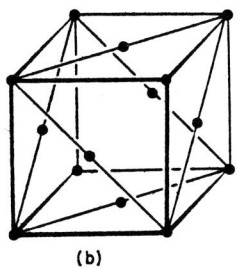

Figure 5.1 Three common crystal cell structures in metals: (a) close-packed hexagonal, (b) face-centered cubic, (c) body-centered cubic [1].

packed hexagonal and face-centered cubic and least for body-centered cubic.

The unsaturated nature of metallic bonding allows for insertion of small atoms into the crystal structure and alloying of dissimilar metals [1]. Atoms with small atomic radii like H, N, and C are able to fit into the interstitial voids of the crystal lattice, thereby altering a metal's physical and mechanical properties. Alloys formed by the unspecific reaction between two metals are possible because of the amorphous electron cloud each nucleus commonly shares. These substitutional solutions are simply the replacement of one metal atom by another within the electron cloud.

As previously discussed, observable mechanical and physical properties like ductility, malleability, and reflectivity are easily explained by the free electron theory. Another important consequence of the free electron structure is the immunity of metals to radiation damage by ionization [1]. Radiation in the forms of high speed charged particles (e.g., protons and electrons) and ionizing rays possesses sufficient energy to remove electrons from the shells of an atom (Figure 5.2). Since electrons are responsible for atomic bonding, the results under ionizing conditions could degrade a material's integrity. Such harmful effects do not occur in metals subjected to moderate amounts of radiation due to a free electron cloud that fills any hole left by a departing electron. Electrons are just redistributed within the cloud while the overall bonding between nuclei and electrons is maintained.

Vacuum Effects

Disadvantages, however, do exist regarding the free electron nature of metals when operating in space. Recall that the absolute pressure in space is nearly that of a vacuum with values ranging from 10^{-7} to 10^{-8} torr for LEO to 10^{-10} torr for GEO [6]. The localized space environment about a satellite does not, however, produce uniform vacuum con-

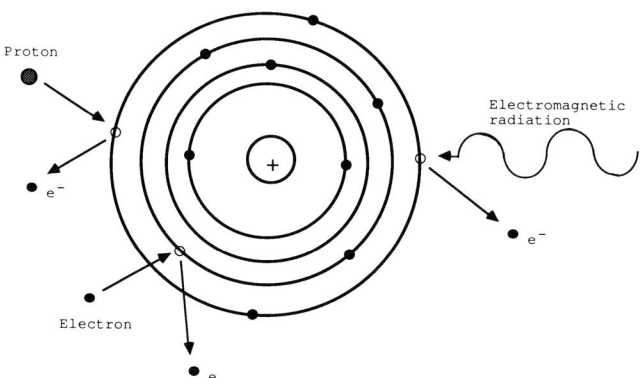

Figure 5.2 Ionization of an atom by radiation (photons) and incident charged particles (such as electrons and protons).

Table 5.1 Common Crystal Structures of Metals [1]

Close-packed hexagonal:

| Be | Cd | Co | Hf | Mg | Os | Re | Tl | Ti | Zn | Zr |

Face-centered cubic:

| Ag | Al | Au | Ca | Ce | Co | Cu | Ir |
| Ni | Pb | Pd | Pt | Rh | Se | Sr | Th |

Body-centered cubic

| Be | Cr | Fe | K | Li | Mo | Na |
| Nb | Ta | V | W | | | |

Spacecraft Materials

ditions due to its relative motion with respect to the atmosphere. On the wake side (opposite side to satellite motion) the partial pressures for hydrogen and helium may drop as low as 10^{-14} torr. The pressure for oxygen and heavier species will be significantly lower than this [7]. Adsorbed gaseous species occupying interstitial sites on the surface of the metal will desorb, thereby interferring with the performance of optical instrumentation.

A spacecraft may be several weeks in orbit before outgassing ceases. Gases such as H_2, H_2O, CO, CO_2, and hydrocarbons can be released through outgassing from aluminum and stainless steel that drastically affect the performance of optical sensors [2, 8].

Reduced pressure also causes an increase in the rate of sublimation of metals. Recall that sublimation is the process by which a solid enters directly into the gas phase without passing through the liquid state. The vacuum of space causes the equilibrium between the solid and gaseous metal to shift, thereby increasing the metal's vapor pressure. Gas phase metal atoms are very mobile which presents problems for spacecraft designers in the forms of redeposition and cold welding. Volatile metals, like zinc and cadmium, sublimate readily at temperatures of 100–150°C [3]. These temperatures, which are easily obtained on surfaces exposed to the Sun, provide the atoms on the surface with sufficient energy to break their metallic bonds and escape. The gaseous atoms can then redeposit on insulators, optical components, and other sensitive surfaces.

Cold welding is the interdiffusion of atoms between two metal surfaces in contact with one another. This occurrence is an example of substitutional solutions formed by the random replacement of one metal atom by another within their crystal structures. Thermodynamically, cold welding is favored when the two metals are of similar size and packing arrangement [1]. For this reaction to occur Gibbs's free energy (ΔG) is negative, and the reaction is spontaneous due mainly to the increase in entropy (ΔS) as randomness among the atoms becomes greater; namely,

$$\Delta G = \Delta H - T\Delta S < 0 . \quad (5.1)$$

The rate of this spontaneous reaction is quite small on Earth, but increases greatly in the vacuum of space. Thus, precautions, such as lubrication, must be taken to prevent cold welding of metals in contact with one another.

Oxidation and Atomic Oxygen Attack

Almost all metals oxidize to one extent or another when exposed to an oxygen rich atmosphere. One need only drive past the local junkyard to see automobiles coated with rust or iron oxide (Fe_2O_3, Fe_3O_4). Corrosion is usually thought of as a harmful, costly process which must be prevented. After all, 1984 estimates put the annual loss in the United States due to corrosion at $80 billion [9]. Although corrosion may be a slow, nonenergetic phenomenon, its presence in space structures may have severe effects. Seven Delta second stage rocket bodies exploded in orbit during the 1970s due to inadvertent mixing of propellants. The corrosion of an aluminum bulkhead separating the fuel and oxidizer is believed to have caused these fragmentations that littered LEO with thousands of pieces of space debris [10].

Despite the obvious disadvantages and problems associated with oxidation, it can have a positive side effect. Namely, oxide layers on metals can act as barriers to further oxidation of the underlying metal. The key is the production of a dense, chemically stable oxide film with good protection properties [1]. Aluminum oxide or alumina (Al_2O_3) is probably the best known example of a protective oxide coating. The formation of alumina takes place as

$$4Al + 3O_2 \rightarrow 2Al_2O_3 . \quad (5.2)$$

With a ΔG^o of -1582 kJ/mole, this spontaneous reaction occurs whenever a piece of aluminum comes in contact with oxygen [11]. The formation of a dense, adherent, amorphous Al_2O_3 layer with a thickness of 2.5–3.0 nm prevents further oxidation from taking place [2, 12]. Alumina is relatively inert chemically and will regenerate on the surface if it is physically penetrated. Increasing film thickness can be accomplished through anodization according to the reaction

$$2Al + 3O^{2-} \xrightarrow[\text{electricity}]{H_2SO_4(aq)} Al_2O_3 + 6e^- . \quad (5.3)$$

This electrochemical process can increase Al_2O_3 thickness to 100,000 nm (100 μm) [12]. By making the film thicker, oxygen will take longer to diffuse to the bare metal surface and therefore corrosion resistance is enhanced.

Other common spacecraft metals also exhibit good corrosion resistance through protective layer formation. For example, chromium, titanium, beryllium, and magnesium respectively undergo the following reactions:

$$4Cr + 3O_2 \rightarrow 2Cr_2O_3 \quad \Delta G^o = -1018 \text{ kJ/mole} \quad (5.4)$$

$$Ti + O_2 \rightarrow TiO_2 \quad \Delta G^o = -866 \text{ kJ/mole} \quad (5.5)$$

$$2Be + O_2 \rightarrow 2BeO \quad \Delta G^o = -585 \text{ kJ/mole} \quad (5.6)$$

$$Mg + 2H_2O \rightarrow Mg(OH)_2 + H_2 \quad \Delta G^o = -597 \text{ kJ/mole} \quad (5.7)$$

Stainless steel gains its corrosion resistance through alloying iron with chromium to form Cr_2O_3 on its surface. Silver and copper are protected by alloying them with 1% aluminum. In all cases, the oxide films are formed here on Earth and remain, to varying degrees, intact in space. Unfortunately, the films cannot regenerate due to the smaller proportions of oxygen, which makes layer thickness and anodization all the more critical.

Concerns about the negative effects of corrosion could be neglected if spacecraft exposure was limited to the near

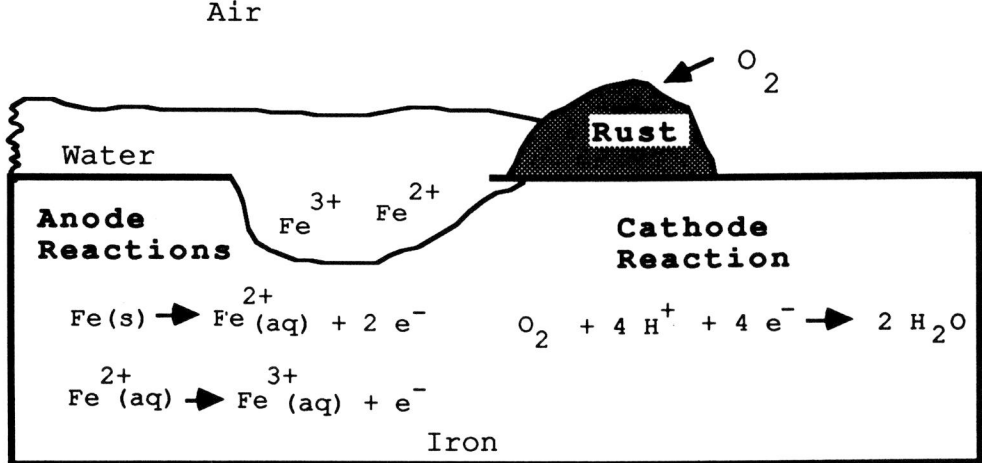

Figure 5.3 Electrochemical corrosion of iron in the presence of air and water. [9]

vacuum of space. After all, corrosion is the same basic process occurring in a galvanic cell or battery. The metal acts as the anode and is oxidized while oxygen is reduced in the cathodic reaction (Figure 5.3.) Corrosion layers of metals such as iron do not adhere very well and are quite porous. The end result is continued oxygen attack of the metal surface and subsequent erosion through flaking.

Missing in the space environment is the aqueous electrolyte necessary for ionic conduction between the two electrodes. Spacecraft, however, are built in the Earth's atmosphere and may be stored for long periods of time before they are launched. Protection of easily oxidized metals is therefore necessary if their natural layers are inadequate. Even aluminum must be protected when alloyed with certain metals like copper [13]. Protective coatings such as paints act as barriers to oxygen and prevent corrosion [12].

Avoidance of contact between dissimilar metals is also important in the prevention of corrosion. Recall Table 1.7 which lists the reduction potentials for the various metals. The more negative the reduction potential versus hydrogen, the more easily that metal is oxidized. Two dissimilar metals in contact with moisture will immediately become a galvanic cell and the metal with the lowest reduction potential will be oxidized; that is, it will lose electrons. This is illustrated in Figure 5.4.

The severity of corrosion is greatest when the metals are widely separated in the activity series. For example, contact between aluminum or magnesium with Fe, Ni, Cu, Ag, or Pb must be avoided when making connections [13]. Al and Mg alloys are, however, fairly stable as is expected by their relative proximity in the activity series. The presence of oxide layers distorts the relative positions of the metals within the series. Therefore, a second series, Table 5.2, is necessary to show the true, relative positions of the metals with their oxide layers. Notice that zinc is now more reactive than aluminum due to the presence of Al_2O_3.

Oxygen attack of metals is not, however, limited to the

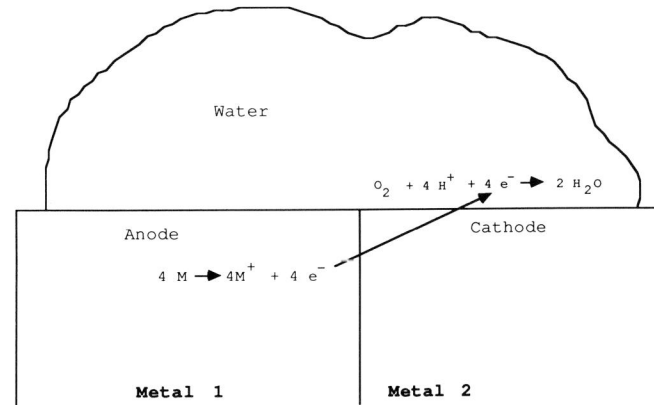

Figure 5.4 Galvanic corrosion between two dissimilar metals in contact with one another in the presence of water.

troposphere. In fact, the dominant constituent from 200 km to 1,000 km altitude is atomic oxygen [14, 15]. Atomic oxygen's great reactivity is due to the presence of two unpaired electrons which previously made up the double bond of molecular oxygen (Figure 5.5).

Recall that atoms and molecules are most stable when their valence or outer electron shell is filled (octet rule). Molecular oxygen achieves an octet around both atoms by sharing two electron pairs. When the covalent, double bond is split between the two atoms, each atom takes with it its six outer shell electrons. In the near vacuum of space, the chances of these atoms colliding and recombining are very slim. Any spacecraft travelling through this region surrounding the Earth will be exposed to an atomic oxygen flux of approximately 3×10^{14} atoms/cm²-sec [14]. Possessing a kinetic energy (8×10^{-19} J or 5 eV) in the region of chemical bonds and activation energies, atomic oxygen reacts easily with spacecraft materials by satisfying its valence shell electron requirement. At the same time, critical bonds within the material are broken, leading to surface degradation and possible structural failure. Such reactivity

Spacecraft Materials

Table 5.2 Galvanic Series (in Seawater) in Order of Decreasing Tendency to Lose Electrons [12]

Corroded end (anode)
- Magnesium
- Magnesium alloys
- Zinc
- Aluminum alloys 7072, Alclad 7075, Alclad 3003
- Aluminum alloys 5083, 5086
- Aluminum alloys 1100, 3003, 5052, 6061, 6063, Alclad 2014, 2017, 2024
- Cadmium
- Aluminum alloy 7075
- Aluminum alloys 2014, 2017, 2026
- Mild steel, cast and wrought iron
- Lead-tin solders
- Lead
- Tin
- Brasses
- Copper
- Bronzes
- Monel, inconel
- Nickel
- Stainless steels (passive)

Protected end (cathode)

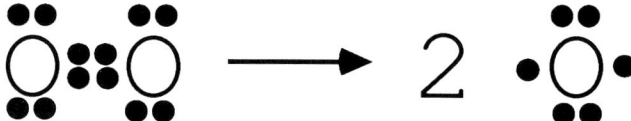

Figure 5.5 Lewis dot structures showing the dissociation of molecular oxygen into atomic oxygen.

Table 5.3 STS-8 Atomic Oxygen Exposure Data [16]

Payload Bay Forward Facing:	$t = 41.2$ hrs
Altitude:	225 km (120 nautical miles)
Velocity:	7.8 km s^{-1}
Mean Oxygen Atom Density: (Calculated)	2.65×10^9 cm^{-3}
Surface Impact Frequency:	2.07×10^{15} cm^{-2} s^{-1}
Integral Fluence:	3.5×10^{20} atoms cm^{-2}

Table 5.4 STS-8 Atomic Oxygen Effects Upon Metals [16]

Material	Nominal Thickness (nm)	Thickness of Metal Converted to Oxide (nm)	Thickness of Metal Lost (nm)
Al film	3.4	0.8	none
Au film	35.5	none	none
Ir film	32.3	none	2.5
Ni film	54.2	0.7	none
Nb film	16.8	1.3	none
Pt film	10	none	none
Os bulk	bulk	none	1100
Ag bulk	bulk	>100, variable	none
Cu bulk	bulk	~3.5	none
Si bulk	bulk	no measurable effect	

is not surprising since the corresponding temperature for an O-atom colliding with the Space Shuttle at 8 km/sec is approximately 41,000 K! The effect of atomic oxygen on most spacecraft materials is simply

Atomic Oxygen + Spacecraft Material
→ Degradation and Failure.

The type and degree of degradation, of course, depends upon the chemical nature of the spacecraft material. As far as metals are concerned, the impact of atomic oxygen is relatively benign. Tables 5.3 and 5.4 show the exposure data and test results for an atomic oxygen experiment flown on STS-8.

Most metals tested showed little or no measurable increases in oxide thickness compared to that obtained under normal atmospheric conditions. In the case of aluminum, atomic oxygen's greater reactivity and diffusivity compared to molecular oxygen allow it to form an even thicker Al_2O_3 protective layer [17]. The behavior of aluminum is characteristic of those metals forming dense, adhesive oxide films on their surfaces. However, preliminary results from the Long Duration Exposure Facility experiment do show atomic oxygen corrosion along the grain boundaries of aluminum.

Two metals which did not perform well in space were osmium and silver. Osmium eroded with a reaction efficiency of 2.6×10^{-26} cm^3/O-atom to form the high vapor pressure compound OsO_4 [14]. Reaction efficiencies for silver were not calculated, but oxide layer thicknesses of 0.4×10^{-6} m have been measured. With a density only 0.68 times that of Ag, Ag_2O adheres poorly to the metal surface and flakes off easily.

Currently, the most popular material of innovative spacecraft designers is beryllium. It is lighter than aluminum and stiffer than steel while also possessing a great capacity to absorb heat. Its melting point is a high 1,285°C. Beryllium also maintains its size, shape, and structure even under severe dynamic loading and extreme temperatures. This material is now being used in inertial guidance systems, missile interstages, reentry vehicles, small rocket nozzles, and a variety of spacecraft structures. Precious metals such as gold, platinum, palladium, and rhodium are also being used in a variety of applications, despite their high cost [4].

Another alternative to either the basic aluminum/steel structure or precious metals is aluminum alloys. The Wright Research and Development Center is pursuing alloys that can withstand temperatures as high as 482°C with good corrosive and rupture characteristics. One unique approach is called mechanical alloying which entails grinding the alloy into "tiny powder particles" which are then attached via surface oxide films. This process is still in the development phases, but classic aluminum-lithium alloys are

Table 5.5 Common Spacecraft Polymers

Chemical Designation	Repeating Unit	Selected Trade Names (RTM)
Polytetrafluoroethylene (PTFE)	$-CF_2-CF_2-$	Teflon, Fluor, Halon
PTFE Copolymers (PFEP)	copolymers with $-CF_2-CF_2-CF_2-$	
Polyvinyl Fluoride (PVF)	$-CH_2-CHF-$	Tedlar, Kynar
Polyacrylonitrile (PAN)	$-CH_2-CH(CN)-$	Orlon, Acrilan
Polymethyl methacrylate (PMMA)	$-CH_2-C(CH_3)(CO_2CH_3)-$	Plexiglas, Lucite
Polycarbonate (PC)	$-[C_6H_4-C(CH_3)_2-C_6H_4-O-C(O)-O]-$	Lexan, Merlon
Polyethylene terephthalate (PET)	$-CH_2CH_2-O-C(O)-C_6H_4-C(O)-O-$	Dacron, Mylar
Nylon-6,6 or Polyamide 66	$-NH(CH_2)_6NH-C(O)(CH_2)_4C(O)-$	Zytel
Polysulfone (PSU)	$-[C_6H_4-SO_2-C_6H_4-O]-$	Ultrason, Victrex
Silicone	$-SiR_2-O-$ R = e.g., CH_3	Silastic, RTV
Poly-p-phenyleneterephthalamide	$-HN-C_6H_4-NHC(O)-C_6H_4-C(O)-$	Kevlar
Polyimide (general name)	(imide–phenyl–O–phenyl repeat)	Kapton

much closer to being ready for use in spacecraft. Martin Marietta Laboratories has developed a special alloy, Weldalite™ 049, for use in aerospace tankage. The new aluminum-lithium alloy has exhibited greater strengths at cryogenic temperatures than any other available aluminum alloy while still maintaining its high temperature (up to 204°C) properties [18]. Research continues to optimize the use of metals in spacecraft construction.

Polymers

Besides metals, polymers are the other class of materials most widely used in spacecraft. Some of their applications include adhesives, lubricants, films, coatings, sealants, structural composites, ablators, and windows. Some of the more common spacecraft polymers are shown in Table 5.5. All spacecraft subsystems utilize the diverse physical and chemical properties available from polymers to perform a variety of tasks. Generally speaking, their relatively light weight and ability to be produced in almost any form make polymers extremely useful to spacecraft designers. The desired physical and chemical properties for a particular application are attained through the proper selection and processing of one of the numerous polymers available. It is the spacecraft designer's responsibility to be aware of the types of polymers qualified for spacecraft use. Such information is contained within MSFC *Handbook 527* which details those acceptable for space operation [5].

Vacuum

Although the reduced pressure of space poses some problems for metals, the effects upon polymers are much more pronounced. These effects are of critical importance to future space platforms and space stations that will be exposed to a vacuum for several years. The most prolific effects are outgassing of low molecular weight constituents, light reaction products, and absorbed gases [3].

Recall from Chapter 1 that polymers are large macromolecules formed through the repetition of repeating units. These large, covalently bonded molecules interact with neighboring molecules through entanglement, crosslinking, and intermolecular forces. Their large size gives these molecules a very low vapor pressure which is typical of high

Spacecraft Materials

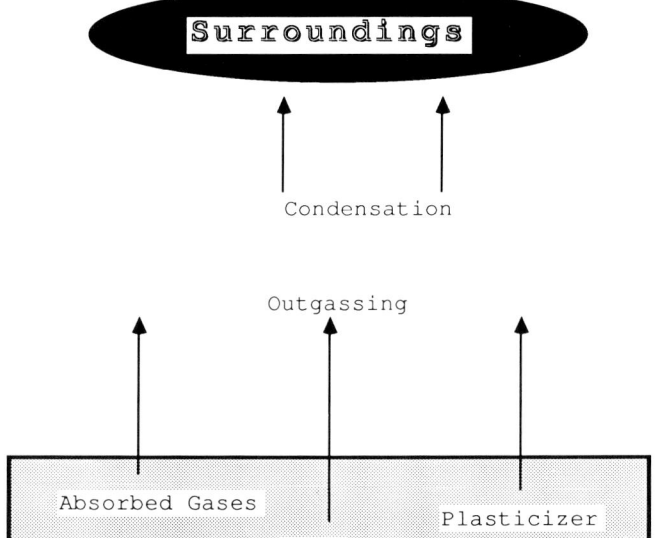

Figure 5.6 Impurities and absorbed gases within polymers can outgas and subsequently condense upon surrounding surfaces.

purity polymers. Unfortunately, many impurities can and do exist within a polymer's structure. First, there is the presence of low molecular weight polymer (LMWP) that did not react fully to attain macromolecular size. The LMWP remains within the polymer until it is exposed to reduced pressure. At that time, the volatile species evaporates and becomes potentially harmful if it condenses upon surrounding surfaces as shown in Figure 5.6.

Another low molecular weight constituent often found in polymers is the plasticizer. Plasticizers are intentionally added to the polymer to impart greater flexibility. They act like lubricants in that the polymer chains now slide more easily past one another, lessening the stress on the covalent bonds. Evaporation of the plasticizer is most apparent to us in the appearance of a hazy film on the interior surface of automobile windows. Vinyl (e.g., polyvinyl chloride, PVC) seats and dashboards must periodically be treated with protectant liquids to replace the lost plasticizer and to retain flexibility. For this reason, PVC (Figure 5.7), which is used in electrical insulation, is not stable enough to be used in space. Other thermoplastics also to be avoided in space applications include certain polyamides, polyvinyl acetates, and butyrates [3].

The final polymer constituents of concern are absorbed gases that are attached to the polymer chains through in-termolecular forces. For example, consider nylon, the repeating unit of which is shown in Figure 5.8. A portion of nylon's strength is gained through hydrogen bonding between nitrogen, oxygen, and hydrogen on adjacent chains. Atmospheric gases, such as O_2, CO_2, N_2, and H_2O, can also hydrogen bond to the nylon molecules. As with other materials, the low pressure of space will cause outgassing of these molecules to occur.

Besides contaminating surrounding materials, outgassing can cause several other detrimental effects. Evolving gases can cause adhesives to bubble and lose contact with the adjoining materials. The loss of plasticizer causes an adhesive to harden and renders it inoperative. Films and coatings may crack during gas evolution. Lubricants, especially silicones, creep from their positions, causing friction and welding between parts. Gaskets become brittle and lose their sealing ability [3]. All these problems can generally be solved by carefully selecting the proper polymer and ensuring that it is of the highest quality. Polymers containing a high degree of additives should not be used if possible. If, however, they are necessary, one should ensure they are sealed (e.g., lubricants), placed far away from critical sensors, and are not adversely affected by loss of volatiles.

Radiation Effects

Ionizing radiation in the form of high energy photons (x rays, gamma rays) and energetic particles (electrons, protons, neutrons) can have detrimental effects upon polymers. While working with nuclear reactors, Charlesby first noticed that polyethylene became infusible, insoluble, and rubber-

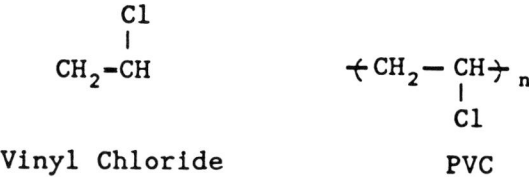

Figure 5.7 The structural formulas of PVC and its monomer.

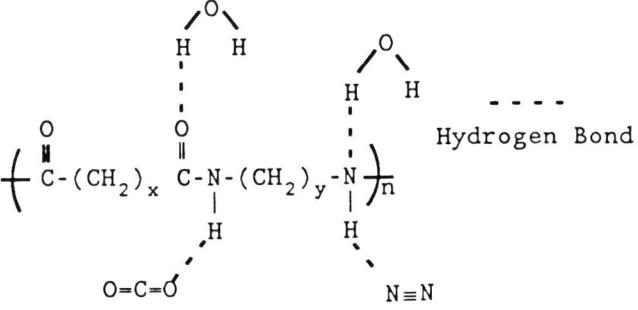

Figure 5.8 Hydrogen bonding of absorbed gases in nylon.

like after exposure to radiation [19]. Subsequent testing showed similar behavior in other polymers including nylon, polystyrene, polyvinyl chloride, and polyvinyl bromide [20, 21]. Polymethyl methacrylate (PMMA) became a crumbly white powder and Teflon was extensively degraded after gamma ray and neutron exposure [22, 23].

Through a series of discrete interactions, radiation affects polymers by depositing enough energy to create a variety of short-lived, reactive species such as ions, free electrons, radicals, and excited electronic states [24]. Depending upon the quantity, type, and kinetics of the intermediates formed, the chemical and physical changes of the polymer fall into two categories: crosslinking and chain scission. The energy of the incident radiation determines the pathway through which these processes occur.

Incident radiation generally imparts energy to atomic electrons, which causes them to transition from their ground state configurations to higher energy states. High energy gamma rays, x rays, electrons, and protons ionize atoms, quickly ($<10^{-12}$s) creating high concentrations of cations, anions, and electrons as follows [25]:

$$P \text{ (polymer)} + \text{radiation} \rightarrow P^+ + e^- . \quad (5.8)$$

These charged particles are quite unstable in their excited states and will quickly ($<10^{-9}$s) undergo dissociation through covalent bond breaking to form free radicals:

$$P^+ \rightarrow A \cdot + B^+ . \quad (5.9)$$

A free radical ($A \cdot$) is a chemical radical or molecular fragment containing an unpaired electron. A second free radical can be formed by the attraction of the initially ejected electron to the cation [26]:

$$B^+ + e^- \rightarrow B \cdot . \quad (5.10)$$

Free radicals are longer lived (hours to days) than ions and migrate throughout the polymer structure causing various reactions to occur [24].

UV radiation causes basically the same chemical changes as its higher energy counterparts, but through a different pathway. Instead of ionizing atoms through electron removal, electrons are excited into higher energy levels. The unstable, excited atoms undergo subsequent homolytic bond cleavage to form two free radicals, namely,

$$R_2 \xrightarrow{UV} R_2^* \rightarrow R \cdot + R \cdot . \quad (5.11)$$

where the asterisk (*) indicates that the molecule R_2 is in an excited state.

The effects of high energy protons on most materials requires additional considerations due to the relatively massive size of the proton. Such protons produce relatively high densities of ionization and electron excitation as well as a greater degree of atomic displacement in the target material from nuclear collisions. [24].

As was mentioned previously, the main responses of polymers to radiation are crosslinking and chain scission. The response is of course dependent upon the type of polymer and the irradiating energy, but the common factor is the formation of free radicals. A free radical's high reactivity stems from the existence of an unpaired electron in its valence shell. Since a full valence shell with all electrons paired within atomic orbitals is the most stable arrangement, the free radical will seek to react and form a covalent bond. Table 5.6 lists the free radicals generated during the photolysis of several polymers.

The least detrimental reaction to the polymer is the recombination of free radicals:

$$R \cdot + R \cdot \rightarrow R_2 . \quad (5.12)$$

Liang et al. [28] showed this reaction in Kapton to be second order; that is,

$$\frac{-d[R \cdot]}{dt} = k[R \cdot]^2 . \quad (5.13)$$

By integrating both sides of the equation, the formula for a second order reaction is obtained as

$$\frac{1}{[R \cdot]_t} = \frac{1}{[R \cdot]_o} + kt \quad (5.14)$$

where

$[R \cdot]_t$ = radical concentration at time t
$[R \cdot]_o$ = initial radical concentration
k = rate constant
t = time

Simple algebraic manipulation to solve for k, using the half-life of radical decay, yields

$$k = \frac{1}{t_{1/2}[R \cdot]_0} . \quad (5.15)$$

The free radicals also react with polymer molecules as covalent bonds are broken. In the cases of polymethyl methacrylate and Teflon, bond cleavage is extensive as the macromolecules "unzip" or depolymerize [23]. PMMA experiences three major scission reactions. one of which occurs at the main-chain carbon-carbon bonds [29] as follows:

$$-CH_2-\underset{\underset{CO_2CH_3}{|}}{\overset{\overset{CH_3}{|}}{C}}-CH_2-\underset{\underset{CO_2CH_3}{|}}{\overset{\overset{CH_3}{|}}{C}}- \xrightarrow{h\nu} -CH_2-\underset{\underset{CO_2CH_3}{|}}{\overset{\overset{CH_3}{|}}{C}}\cdot + \cdot CH_2-\underset{\underset{CO_2CH_3}{|}}{\overset{\overset{CH_3}{|}}{C}}- \quad (5.16)$$

End chain depolymerization to form the parent monomer then proceeds as

$$-CH_2-\underset{\underset{CO_2CH_3}{|}}{\overset{\overset{CH_3}{|}}{C}}-CH_2-\underset{\underset{CO_2CH_3}{|}}{\overset{\overset{CH_3}{|}}{C}}- \rightarrow -CH_2-\underset{\underset{CO_2CH_3}{|}}{\overset{\overset{CH_3}{|}}{C}}\cdot + CH_2=\underset{\underset{CO_2CH_3}{|}}{\overset{\overset{CH_3}{|}}{C}} \quad (5.17)$$

Table 5.6 Free Radicals Generated During Photolysis of Select Polymers [27]

Polymer	Radical	Wavelength of Incident Light (nm)	Temperature during Irradiation (°C)
Polycarbonate	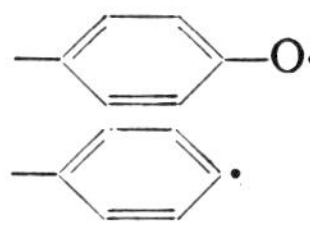 —⟨C₆H₄⟩—O· —⟨C₆H₅⟩·	UV*	−196
Polyethyleneterephthalate	—CH_2—CH_2—O·	313	25
Polyvinylpyrrolidone	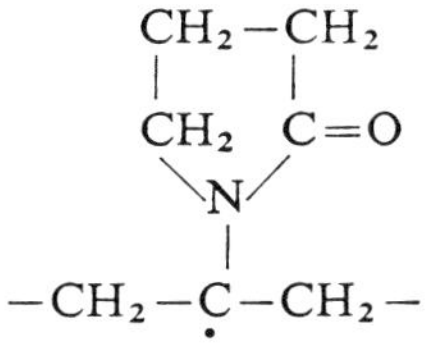 CH_2—CH_2 / CH_2 C=O / N / —CH_2—Ċ—CH_2—	UV	−196
Polyamide-6	—CH_2—ĊH—CH_2— —CH_2—Ċ=O	250–600	−196
Poly(2,6-dimethylphenylene oxide)	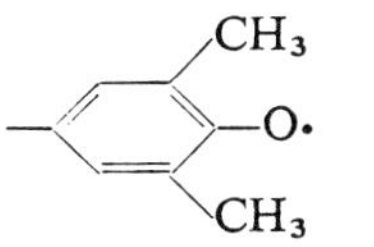 (2,6-dimethylphenyl)—O·	300	room temperature in benzene
Polystyrene	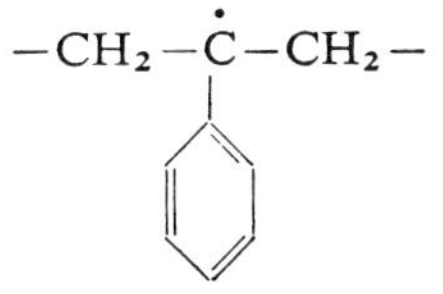 —CH_2—Ċ(C₆H₅)—CH_2—	250	−185
Polymethylmethacrylate	—CH_2—Ċ(CH_3)—O=C—OCH_3	253.7	25
	·CH_3, ·CHO, ·C(=O)—O—CH_3	253.7	−196

*) high pressure mercury lamp

The result is a drastic reduction in the molecular weight of the polymer. PMMA samples with initial molecular weights of 500,000 to 800,000 g/mole contain no material of greater than 20,000 g/mole molecular weight after exposure to 7,500 Mrads (1 rad = 10^{-5} J/g of absorbed energy) of proton radiation [24]. The main anaerobic products of photodegradation are volatile methyl formate, methanol, and methyl methacrylate.

$HCOCH_3$ — methyl formate

CH_3OH — methanol

$CH_2={C}({CH_3})-CO_2CH_3$ — methylmethacrylate

In addition to depolymerization, free radicals can also promote crosslinking between adjacent polymer molecules. Free radicals on adjacent chains bind together to form a covalent bond as can be seen with nylon and silicone [29]:

$$\begin{array}{c}-CH_2-\overset{O}{\overset{\|}{C}}-NH-\overset{}{C}H-CH_2 \\ + \\ -CH_2-\overset{}{C}-NH-\overset{\cdot}{C}H-CH_2 \\ \overset{\|}{O}\end{array} \longrightarrow \begin{array}{c}-CH_2-\overset{O}{\overset{\|}{C}}-NH-CH-CH_2 \\ | \\ -CH_2-\overset{}{C}-NH-CH-CH_2 \\ \overset{\|}{O}\end{array} \quad (5.18)$$

$$\begin{array}{c}\overset{CH_3}{|}\\ -\overset{\cdot}{Si}-O-\\ \\ O-\overset{\cdot}{Si}-\\ |\\ CH_3\end{array} \longrightarrow \begin{array}{c} \overset{}{|} \overset{O}{|} \\ CH_3-Si-Si-CH_3 \\ | \quad | \\ \quad O \end{array} \quad (5.19)$$

The macroscopic effects of radiation exposure can be quite severe depending upon the type and dosage of absorbed radiation. Clouding of thin film surfaces and cracking of elastomeric sealants are just two of the harmful outcomes. Adhesives and sealants are crosslinked to some degree as a means of maintaining both the strength and flexibility essential to their functions. Additional crosslinking imparts greater strength, but unfortunately at the expense of flexibility since the long chains cannot distort as easily as before. Thermal cycling places excessive stress upon the polymer, causing it to eventually crack and fail. For elastomers, such as natural rubber, neoprene, and silicone, increased crosslinking is the predominant response to UV. Chain scission and depolymerization, on the other hand, occur most readily in butyl, polysulfide, and fluororubbers [30].

Perhaps more pervasive than bulk property effects are those occurring at the polymer surface. As will be discussed in a later chapter, surface properties of thermal control materials are quite important to their performance. Surface discoloration in thin films, like the polyimide Ultem^R, is due to the presence of free radicals [31]. Recall these free radicals have lifetimes on the order of hours to days so their discoloring effect can be a continual problem. The general effects of UV on polymers are given in Table 5.7.

Damage to materials by ionizing radiation is a function of the penetration depth with the order being UV < protons < electrons. UV affects the surface properties since its energy is quickly absorbed by surface electrons.

Table 5.7 Effects of Ultraviolet Radiation on Polymer Stability [30]

Polymer	Effect of Exposure to Ultraviolet Energy[a]
Plastics	
Mylar	Decreases tensile strength and elongation
Polyamide (nylon)	No significant effect
Polymethyl methacrylate	Surface discoloration and crazing
Polyethylene	Embrittlement
Polypropylene	Embrittlement
Polyimide	No significant effect
Polystyrene	Yellows
Plasticized polyvinyl chloride	Develops tacky and discolored surface
Teflon	No significant effect
Elastomers	
Butyl	Increases tensile strength and elongation
Hypalon (chlorosulfonated polyethylene)	No significant effect
Neoprene	Increases tensile strength, decreases elongation
Nitrile	Decreases tensile strength and elongation
Styrene-butadiene (SBR)	Decreases tensile strength and elongation
Silicone	Surface crazing
Viton A	No significant effect

(a) Relative effects only, since spectral distribution and total exposure in equivalent Sun hours are not specified.

Protons can penetrate further into the bulk before imparting their energy through collisions. Due to their smaller size than protons, electrons can penetrate even deeper. This ranking, of course, is all dependent upon radiation energy. Protons, in fact, do more damage because they achieve a higher density of deposited energy. For the same incident dose of radiation, 3 MeV protons penetrating 0.01 cm deposit 100 times more energy in the same volume than 3 MeV electrons penetrating 1.3 cm. Thus, heavy particles pose more of a threat to surface coatings while lighter particles can degrade bulk materials [24]. Considering the anticipated 10 to 30 year lifetime requirement for space station materials, both surface and bulk effects must be considered.

Heavy particles, like protons and alpha particles, present another problem and that is localized heating. Not only can they ionize, but these heavy particles can also excite and displace atoms. Molecular rotation, vibration, translation, and bond breakage all generate heat which is absorbed in the surrounding areas. Significant local heating produces much more heterogeneous damage than is produced by gamma or electron radiation [24].

Table 5.8 Degradation of Select Polymer Films in GEO Radiation Environment [32]

Material	Most Damaging Radiation	Environmental Effect
Teflon (fluoropolymer)	Low energy protons, near UV, high energy electrons	Yellowing, becomes brittle
Kapton (polyimide)	Low energy protons, near UV, high energy electrons	Becomes black, some reduction in elongation
Mylar (polyester)	Low energy protons, high energy electrons	Major surface degradation

Increased use of GEO entails operation in a very hostile environment where accumulated radiation surface dosage is approximately 10^9 to 10^{12} rads, well above the threshold of 10^5–10^8 rads for polymer films. Bulk penetration depths of 2.5 μm to 250 μm still accumulate up to 10^9 rads, making most polymer films quite susceptible to complete degradation. Test results of several polymer films in the GEO radiation environment are summarized in Table 5.8 [32].

Obviously the harmful effects of radiation must have been countered to some extent because polymers have successfully operated in space for years. In fact, there are several methods available to scientists and engineers for protecting against radiation degradation. These include screening, absorption, and quenching [29].

In screening, a pigment opaque to UV light is added to the polymer. The pigment acts as a reflector, limiting UV penetration into the polymer. The degree of protection is dependent upon the pigment and the polymer. Reflective pigments work best as surface coatings rather than being interdispersed throughout the polymer matrix. Some common stabilizers of this type are Al, Fe_2O_3, Fe_3O_4, Cr_2O_3, Pb_3O_4, ZnO, TiO_2, and carbon black. Zinc oxide is particularly attractive due to its low cost and wide range of protection (240–380 nm) [29].

Photochemical stabilization by absorption, like screening, involves adding a compound to the polymer. However, absorbers do not reflect UV, but rather, as their name implies, absorb the energy, thus preventing degradation of the polymer itself. The absorber's response to the UV light is different for various classes of materials. Some reemit the light at longer, less energetic wavelengths. Others simply dissipate the energy in the form of heat by vibrating and rotating. Several classes of absorption photostabilizers are aromatic salicylates, benzophenones, and benzotriazoles (Table 5.9).

In the case of quenchers, the mode of operation is passive. The polymer (P) itself absorbs the UV and in turn transfers the energy to the quencher (Q) before bond scission occurs to form radicals ($R\cdot$).

$$P \xrightarrow{UV} P^* \longrightarrow Q \begin{array}{l} \longrightarrow R\cdot + \cdot R \\ \longrightarrow P + Q^* \end{array} \quad (5.20)$$

The excited quencher (Q^*) then harmlessly relaxes back to its normal, ground state. Transition metal chelates, such as the nickel chelate shown in Figure 5.9, are commonly used.

It must be pointed out that all three of these approaches to photostabilization are effective only against UV radiation since high energy gamma rays, x rays, electrons, and protons are nonspecifically absorbed by the entire polymer. The additives also present a problem in terms of outgassing. External shielding is necessary in areas of high radiation flux, such as would be encountered with nuclear reactors. For the lower flux densities of the space environment, scientists have developed polymers with higher radiation stability. Incorporation of aromatic units into the backbone renders these polymers relatively immune [27]. A prime example is Kapton which remains stable to absorbed doses in excess of 10,000 Mrad [24].

An interesting example of changes in satellite design, driven by the interaction of polymers with the space environment, is found in the satellite cabling business. The seemingly simple task of wiring is complicated by outgassing and durability needs. The material GORE-TEX® was developed from the insulator Teflon to improve its outgassing characteristics. Optimum outgassing aspects are low loss at a fast rate. GORE-TEX is made by stretching Teflon quickly, resulting in a strong yet porous material. The porosity of GORE-TEX and similar material has changed satellite operations. At the beginning of the space age, the start of satellite operations was routinely delayed to allow for outgassing of air and other gases. The advent of GORE-TEX has reduced this concern.

Cabling must also be made durable. GORE-TEX and PTFE are already highly resistant to UV degradation, but may be strengthened by the addition of Kapton, as already mentioned. The effect of gases being released from a section of wiring may seem unimportant, but all satellites contain a vast amount of cabling. For example, the European spacecraft EURECA is presently designed to use 75 km of cabling [33].

Atomic Oxygen Effects

Unique to the LEO environment is the presence of atomic oxygen. Examination of Shuttle and LDEF surfaces and particularly those at or near the leading edge revealed corrosion and erosion. Some surface degradation was expected due to contact with debris and ablation, but a chemical change from interaction with a reactive species was apparent. That reactive species turned out to be highly reactive atomic oxygen, the most prevalent constituent in the 200 km to 1,000 km altitude range.

Table 5.9 Selected Examples of UV Screening and Absorbing Photostabilizers [27]

Compound class	Compound		Principal Mode of Action
Pigments	Carbon Black, ZnO, MgO, $CaCO_3$, $BaSO_4$, Fe_2O_3		UV Screener
2-Hydroxy-benzo-phenones	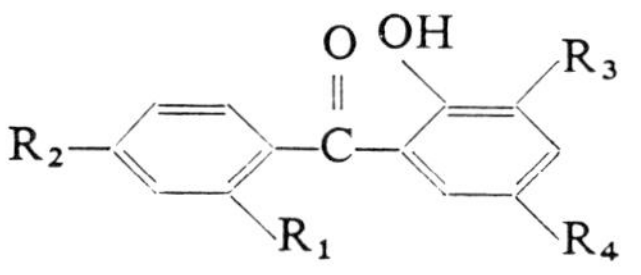 (structure)	R_1 = H, alkyl R_2 = H, alkyl, phenyl R_3 = H, butyl R_4 = H, butyl	UV Absorber
	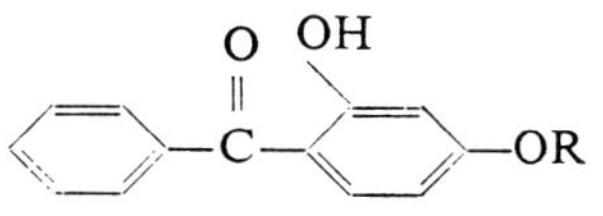 (structure)	R = alkyl	
Phenylsali-cylates	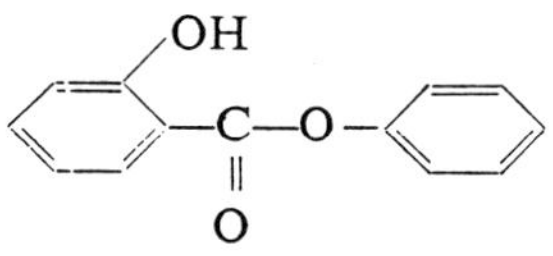 (structure)		UV Absorber
Benzo-triazoles	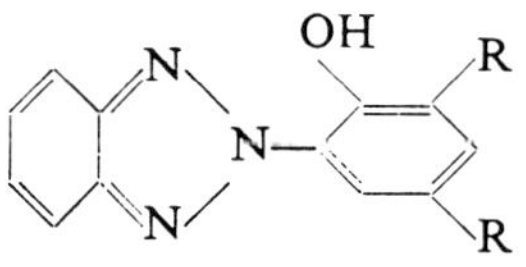 (structure)	R = H, alkyl	UV Absorber

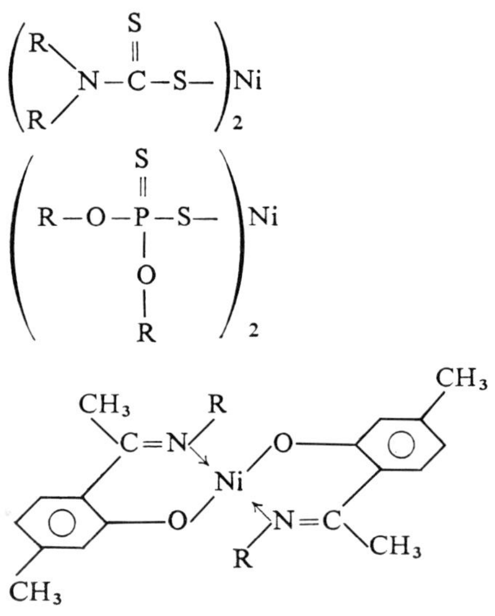

Figure 5.9 Chemical structures of nickel chelates used as quenching photostabilizers [27].

Since the discovery of atomic oxygen, numerous tests and experiments have been conducted to measure its effect upon spacecraft materials [34–41]. Knowing these effects is crucial to proper materials selection for Space Station Freedom since its 30 year lifetime will be spent inhabiting the atomic oxygen rich altitudes. The focus of this work has been mainly in the area of polymers, since, as was discussed in the metals section, metals are generally stable to atomic oxygen. An exception is silver, used for interconnects in solar cell arrays. Polymers, on the other hand, show varying degrees of reactivity based upon their chemical composition.

Early investigations had to first determine whether or not the observed weight loss was due simply to ablation. The actual chemical processes behind ablation will be discussed later, but the important point for now is that the erosion rates for all organics undergoing ablation are approximately the same. This clearly was not observed during Space Shuttle experiments and in fact was evident as early as 1965 in

Spacecraft Materials

Table 5.10 Rates of Weight Loss of Selected Polymers in Atomic Oxygen [42]

Type of polymer	Rate of weight loss (grams/cm^2/min)*
Low-density polyethylene	0.500×10^{-4}
Irradiated low-density polyethylene (1 Mrad)	0.572
Irradiated low-density polyethylene (10 Mrad)	0.705
Irradiated low-density polyethylene (105 Mrad)	0.851
Chemically crosslinked low-density polyethylene	0.684
Low molecular weight highly branched polyethylene	0.675
High-density ethylene butene copolymer	0.638
Polypropylene	0.713
Polybutene-1	0.736
Chlorinated high-density polyethylene	1.03
Chlorinated polyethylene plus 10% polysulfide polymer	0.599
Natural rubber	0.700
Natural rubber sulfur raw stock	0.248
Natural rubber sulfur vulcanizate	0.033
Natural rubber peroxide raw stock	0.618
Natural rubber peroxide cured	0.345
Commercial hard rubber	0.560
Vulcanized ethylene propylene rubber	0.041
Polystyrene	0.260
Poly-3-phenyl-1-propene	0.295
Poly-4-phenyl-1-butene	0.345
Polyvinylcyclohexane	0.471
ABS polymers, several types	0.554
Unplasticized poly(vinyl chloride) copolymer	0.973
Poly (vinyl fluoride)	0.525
Polytetrafluoroethylene	0.13
Perfluorinated ethylene-propylene copolymer	0.091
Poly (methyl methacrylate)	0.442
Polyimide	0.246
Polycarbonate	0.535
Poly (ethylene terephthalate)	0.376
Nylon 6	0.572
Nylon 610	0.669
Formaldehyde polymers	1.19 – 1.62
Polysulfide (chloroethyl formal disulfide)	4.02
Cellulose acetate	1.03

*Low power, Tracerlab LTA-500A

experiments performed at Bell Laboratories [42]. Table 5.10 shows the weight loss rates of numerous polymers in laboratory generated atomic oxygen atmospheres. The reactions are obviously linear with time, which translates to zero order chemical kinetics, as follows

$$\frac{d}{dt}[P] = k[P]^o = k . \quad (5.21)$$

Table 5.11 gives the reaction efficiencies for common spacecraft polymeric materials. Efficiencies are given in volume of reactive material per impinging oxygen atom. A more complete table of reaction efficiencies for both metals and polymers is in the appendix.

Clearly from these results much more complex chemistry than simple ablation is occurring. Interaction of atomic oxygen with a particular material can result in one of the following chemical changes [14, 43]:

1. Abstraction—Oxygen abstracts an atom from the compound.
2. Addition—Oxygen is added to the compound.
3. Replacement—Oxygen replaces an atom on an atomic group.
4. Insertion—Oxygen is added between two bound atoms.

Of these four reactions, abstraction and addition seem to be the most prevalent in polymers.

Table 5.11 Reaction Efficiencies of Selected Materials with Atomic Oxygen in Low Earth Orbit [14]

Material	Reaction efficiency (cm^3/atom)
Kapton	3.0×10^{-24}
	3.4
Tedlar (Clear)	3.2
Polyethylene	3.7
Polysulfone	2.4
Bold Graphite/Epoxy	
1034C	2.1
5208/T300	2.6
Epoxy	1.7
Polystyrene	1.7
Polybenzimidazole	1.5
25% Polysiloxane/ 45% Polyimide	0.3
Polyester 7% 7% Polysilane/93% Polyimide	0.6
Polyester	Heavily attacked
Polyester with Antioxidant	Heavily attacked
Teflon, TFE	<0.05
Teflon, FEP	<0.05
Carbon (various forms)	0.5–1.3
RTV-560	$0.02^* \times 10^{-24}$
DC6-1104	0.02*
T-650	0.02*
DC1-2577	0.02*
Black paint Z306	0.3–0.4*
White paint A276	0.3–0.4*
Black paint Z302	2.03*
Perfluorinated polymers	

*Units are mg/cm^2 for STS-8 mission. Loss is assumed to occur in early part of exposure; therefore, no assessment of efficiency can be made.

Saturated organic molecules (those containing carbon atoms with single bonds) undergo hydrogen abstraction to form polymer radicals:

$$-\underset{\underset{H}{|}}{\overset{\overset{H}{|}}{C}}- + O \longrightarrow -\underset{\underset{H}{|}}{\overset{\overset{H}{|}}{\dot{C}}}- + \dot{O}H \quad . \quad (5.22)$$

If the polymer molecule contains unsaturated carbons (presence of multiple bonds around carbon) O can add across the double bond and create a biradical [43]:

$$O + -\underset{\underset{H}{|}}{\overset{\overset{H}{|}}{C}}=\underset{\underset{H}{|}}{\overset{\overset{}{}}{C}}- \longrightarrow -\underset{\underset{H}{|}}{\overset{\overset{}{}}{C}}-\underset{}{\overset{\overset{\dot{O}}{|}}{\dot{C}}}- \quad . \quad (5.23)$$

A generic model of O/polymer interaction was developed on the basis of data gained from Space Shuttle experiments [35]. The proposed mechanism is as follows:

$$\begin{aligned} P + O &\xrightarrow{k_1} PO \\ PO + O &\xrightarrow{k_2} PO_2 \\ PO + O &\xrightarrow{k_3} V_1 + P \\ PO_2 + O &\xrightarrow{k_4} V_2 + P \end{aligned} \quad (5.24)$$

where

k_i = reaction rate constant
V_1, V_2 = volatiles (e.g., CO, CO$_2$, H$_2$O, HCO)

An intermediate oxide (PO, PO$_2$) is formed on the polymer surface (P) upon exposure to O. The intermediate PO can encounter another O atom and become either PO$_2$ or a volatile product (V_1) and the original polymer. PO$_2$ can also react with another O to form a volatile product (V_2) and the original polymer. The mass loss incurred by polymers is due to the release of volatile products V_1 and V_2 and the extent of degradation is determined by the amount of oxide. The oxide depletion rate (k_3 and k_4), or how quickly the oxide leaves the surface, impacts the amount of weight loss. For example, a rapid reaction to form V_1 and V_2 means little oxide accumulation and a large weight loss. Bulk properties in this case are not affected by the oxide since it does not remain long enough to cause chemical change. In contrast, a slow reaction rate for volatile product formation means longer oxide contact time and low weight loss. Bulk properties, however, may change due to oxide induced chemical reactions.

The extent of degradation, as apparent from Tables 5.10 and 5.11, is dependent upon the chemical composition of the polymer and any additives. Those polymers least affected are perflourinated (e.g., Teflon), sulfur vulcanized rubber, silicones, and highly aromatic polymers [14, 42]. The most reactive appear to be highly branched, like polypropylene, or have the —C—O—C— ether linkage, like Kapton.

Teflon's resistivity to atomic oxygen is due to the presence of fluorine, the most electronegative element. Reactive atomic oxygen is basically unable to compete with fluorine in the electron "tug of war" to achieve a filled valence shell. The —CF$_2$—CF$_2$— structure thus remains intact and no abstraction can occur. Contradicting observations of Teflon's inertness were initially seen in samples returned from the Solar Max satellite. Teflon tape samples suffered noticeable damage in the form of yellowing and cracking. However, post-test analysis revealed the tape was not pure fluorocarbon, but a copolymer containing hydrocarbons as well. Degradation was due to atomic oxygen's attack upon the hydrocarbons [40].

Polymers containing aromatic groups also seem to be highly resistant to atomic oxygen degradation, although Kapton is a most notable exception. Aromatic groups are generally unreactive due to the presence of delocalized, multiple bonds within their cyclic structures. Electrons are held tightly to atoms within these ringed structures and will not react with the electrophilic oxygen atoms. The general trend indicates the greater the amount of aromaticity, the greater the restivity to atomic oxygen degradation. Presence of relatively weak single covalent bonds should thus be minimized.

Besides degrading polymeric materials, atomic oxygen is also responsible for the phenomenon known as "Shuttle glow." The glow is induced by the interaction of the atmosphere with the orbiting spacecraft. This phenomenon is potentially bothersome since it could interfere with sensors operating in the visible and infrared portions of the spectrum.

Several explanations exist for the occurrence of glow and the reader is directed to reference 45 for complete coverage of the area. One model of particular interest is the formation of oxidized intermediates on the material surface which subsequently undergo volatilization. In one possible mechanism, 5 eV oxygen atoms (O) collide with the material (M) surface and produce the following reaction:

$$M + O(5eV) \rightarrow [OM] \rightarrow OM^* \rightarrow OM + h\nu . \quad (5.25)$$

The volatile intermediate $[OM]$ first becomes electronically and vibrationally excited (OM^*) then relaxes back to its ground state through the emission of photons. A second mechanism involves formation of the same excited intermediate OM^*, but this time it interacts with the significant numbers of 10–100 eV hot electrons present in LEO to become even more electronically excited (OM^{**}):

$$M + O \rightarrow OM^* + e^- \rightarrow OM^{**} \rightarrow OM + h\nu . \quad (5.26)$$

The composition of oxidized volatiles is of course dependent upon the material surface. However, likely species include OH, CO, and even CN [46].

The effects of atomic oxygen, UV rays, and thermal cycling all contribute to the peeling of paint from operational satellites. Paints or thin coatings are applied for esthetic reasons and for thermal control. This seemingly unimportant degradation of old rocket body and payload surfaces may actually create a significant space debris hazard. A case in point is the replacement of a Space Shuttle window in 1983 due to an impact crater caused by a 200 micron paint chip. The window replacement cost $50,000 [18]. Since the first Shuttle flight, over twenty other windows have been replaced due to impact chips, but the sources of the impacting fragments for these others have not yet been determined [47].

Composites

Polymers have replaced metals in many applications, but not in the area of structural materials. Inadequate stiffness and dimensional stability are the two major reasons polymers are not used for load-carrying members. These limitations, however, are being eliminated through the fabrication of high strength composite materials which, like polymers, offer considerable weight savings over metals. Composites are now being used for helicopter rotors, torpedo hulls, commercial construction projects, and a wide variety of other applications. This new hybrid material may very well be the "mortar" in building 21st century technology.

In its simplest form, a composite is nothing more than a physical mixture of two dissimilar materials. Hull uses the following three points to define a composite [48]:

1. It consists of two or more physically distinct and mechanically separable materials.
2. It can be made by mixing the separate materials in such a way that the dispersion of one material in the other can be done in a controlled way to achieve optimum properties.
3. The properties are superior, and possibly unique in some specific respects, to the properties of the individual components.

Composites are not the result of space age technology, but in fact occur naturally in the forms of wood, bamboo, bone, and muscle [48]. The first manmade composites based upon polymers occurred around 5000 B.C. (Table 5.12). Strong but flexible papyrus gained the necessary stiffness through mixing it with pitch. The composite hardened into a material suitable for constructing boats. Thus, strength contributed by the papyrus (fiber) and stiffness imparted by the pitch (matrix) resulted in a useful and practical composite building material.

Today, composites have numerous applications with the aerospace industry being their greatest user. The attractiveness of composites is apparent from a relative comparison of their mechanical properties with those of steel and aluminum as is shown in Figure 5.10

Table 5.12 Dates of Development for Some Common Polymer Composites [49]

Date	Material
ca. 5000 B.C.	Papyrus/pitch (boats)
ca. 1500 B.C.	Wood veneer
1909	Phenolic composite
1928	Urea formaldehyde composite
1938	Melamine formaldehyde composite
1942	Glass reinforced polyester
1946	Epoxy resin composites
1946	Glass reinforced nylon
1951	Glass reinforced polystyrene
1956	Phenolic-asbestos ablative composite
1964	Carbon fiber reinforced plastics
1965	Boron fiber reinforced plastics
1969	Carbon glass fiber hybrid composites
1972	Aramid fiber reinforced plastics
1975	Aramid/graphite fiber hybrids

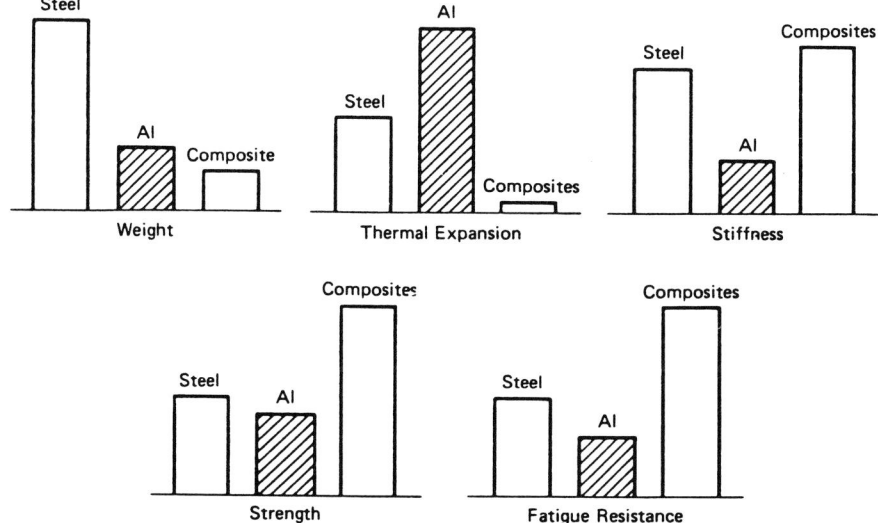

Figure 5.10 Comparison between the physical properties of composites and metals [50].

Since aircraft and spacecraft greatly benefit from high strength/weight ratio materials, composites are quite attractive for heavy structural members. A prime example is the payload bay doors on the Space Shuttle. With a total weight of 1452kg (3,200 pounds), these graphite/epoxy doors weigh approximately 408kg (900 pounds) less than those made from an equivalent alloy structure. Composites will find even greater use in space with the construction of Space Station Freedom. Graphite/epoxy tubular struts are part of the proposed base line design. The unique characteristics of graphite/epoxy elements are also being capitalized on by Hughes Space and Communications Company. They have built a precision antenna reflector for use on defense meterological satellites which far exceeds previous standards for surface contour accuracy. This greater ability to hold a parabolic shape has improved the beam efficiency of the energy used to relay information to and from the satellites and ground facilities [47].

Composites consist basically of a continuous matrix (polymer) containing a fibrous and/or particulate filler. Figure 5.11 shows examples of various composite geometrical arrangements. The desired physical and mechanical properties are determined to a great extent by the geometrical arrangement. However, bonding at the polymer-filler interface is what ultimately dictates these properties. Four mechanisms occurring either together or in isolation at the interface are responsible for adhesion [48]:

1. Adsorption and wetting: The degree to which a polymer wets or adsorbs onto a solid is a function of surface tension or surface free energy.

Surface tension is denoted by γ (N/m) and the contact angle by Θ. The contact angle is a measure of the degree the liquid wets the solid. The work of adhesion (W_A) is given by the Young-Dupré equation:

$$W_A = \gamma(1 + \cos\Theta) . \quad (5.27)$$

Ideal wetting and maximum adhesion between the liquid and solid therefore occur when Θ is zero. W_A represents the physical bonding caused by strong intermolecular forces occurring at the interface. The rougher the solid surface the more area available for bonding with the liquid. Contaminates, however, reduce the effective surface area of the solid and lead to decreased adhesion.

2. Interdiffusion: Physical entanglement is possible when a polymer filler is embedded into a polymer matrix.

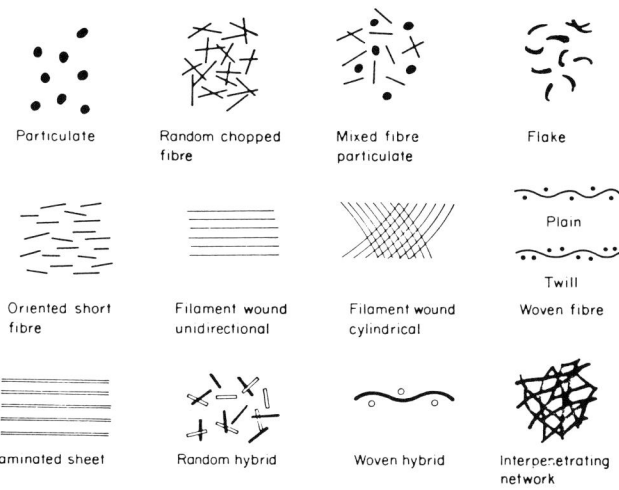

Figure 5.11 The common geometrical arrangements found in composites [49].

Spacecraft Materials

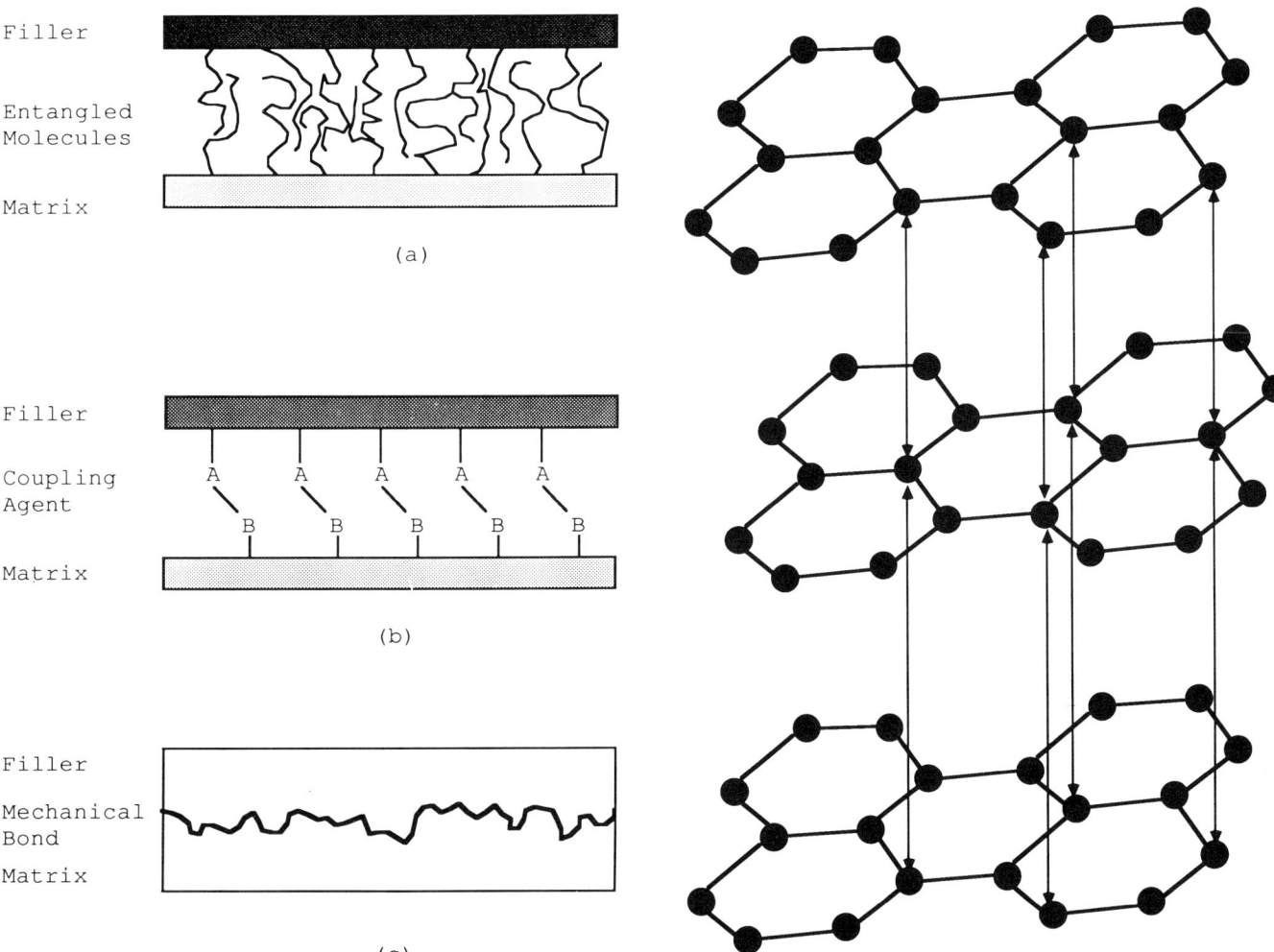

Figure 5.12 The types of bonds formed between the matrix and filler within composites: (a) Bond formation through molecular entanglement preceded by interdiffusion; (b) Bond formation through a coupling agent; (c) Bond formation by physical contact between a liquid polymer and a rough solid surface [48].

Figure 5.13 The layered crystalline structure of graphite.

The two polymers can interdiffuse, causing physical bonding through entanglement (Figure 5.12a).

3. Chemical bonding: Coupling agents are additives which form chemical bonds between the filler and matrix (5.12b). One end of the coupling agent chemically bonds to the matrix while the other end bonds to the filler. Bond strength depends upon the number and types of bonds formed.

4. Mechanical adhesion: This type of adhesion is due to the mechanical interlocking of the two surfaces (5.12c). Again, good wetting is crucial to achieving high strength.

The choices for both the filler and matrix are almost endless, but in fact very few combinations are used for spacecraft design. Graphite fiber, with its low density and high strength-to-weight ratio, is a preferred filler material [50].

Graphite is one of the allotropes of carbon, consisting of layers of two dimensionally bonded atoms. Each carbon atom is covalently bonded to three adjacent atoms, in the same plane, with intermolecular forces between the planes (Figure 5.13). While cotton, cellulose, flax, and hemp were used as graphite starting materials, polyacrylonitrile (PAN) is the most widely accepted precursor. Pyrolyzation of PAN under controlled atmospheric conditions yields graphite over 99.5% pure in carbon [50].

Matrix materials, at least for low temperature applications, consist of both thermosetting and thermoplastic polymers. Thermosetting epoxy resins are the most pervasive for aerospace applications. Epoxies are low molecular weight polymers possessing epoxide rings as end groups. The uncured form of the epoxy is a viscous liquid. Addition of a curing agent, like a diamine, opens the epoxide ring and causes the chains to crosslink into a strong, three dimensional network [51].

$$\begin{array}{c}\text{+CH}_2\text{C—CH}_2\text{ + R} \\ \text{Epoxy Resin}\end{array} \begin{array}{c}\text{H}_2\text{N} \\ \text{R} \\ \text{H}_2\text{N} \\ \text{Curing Agent}\end{array} \longrightarrow \begin{array}{c}\text{Cured Epoxy}\end{array} \quad (5.28)$$

For service temperatures above 300°C, organic polymers are unsuitable as matrix materials. Two alternatives are metals and carbon. Metals, such as aluminum and magnesium, can be used with graphite and boron fibers to make composites comparable to those previously discussed with the added advantage of high temperature operation. Carbon matrices are made from pitch or phenolic resins which have been pyrolyzed. Coupled with graphite, these carbon/carbon composites have found numerous applications in rocket nozzles, nose cones, and cases [50].

Environmental Effects

Composites exhibit basically the same behavior as polymers in the space environment. Outgassing, radiation degradation, and atomic oxygen attack all present problems for composites due to their organic nature. The atomic oxygen environment in LEO is especially harsh on composites. Carbon/epoxy and Kapton/epoxy samples flown on Space Shuttle missions showed extensive erosion of both the matrix and filler. Epoxy showed a reaction rate of 2×10^{-24} cm^3/O-atom. Graphite and Kapton fibers originally covered by epoxy were eroded into a corduroy pattern. The fibers themselves appeared as porous ridges with little apparent residual strength or stiffness [52]. Apparently the same surface oxidation–vaporization process that occurs in polymers is taking place in composites.

Since composites will be used extensively in the space station, the long-term effects of atomic oxygen must be dealt with in order to prevent surface degradation and even structural failure of those load carrying members. The most feasible method of protection available is coating the exposed composite surface with resistant materials. Selected metal foils, like Al for example, are an obvious choice due to their impermeable oxide coatings. Oxides themselves, including those of Al, Ti, and Si, also give excellent protection [53]. Surface coatings are, however, susceptible to pitting from particulate impacts. Once the coating is penetrated, the degradation may eventually result in a very large defect, just as a scratch in a car's paint can lead to a very large corroded area due to exposure of the bare metal to the environment.

Ceramics and Glasses

Ceramics and glasses possess unique chemical, physical, and mechanical properties which enable them to be used for specialized applications aboard spacecraft. The electrical and thermal insulating properties of ceramics make them a preferred material in electrical components and thermal protection systems. Glasses possess the optical transparency necessary for windows and optics. Although their applications are quite different, their chemical compositions are similar.

Ceramics

Ceramics are a class of compounds consisting mainly of inorganic nonmetallic materials which are brittle and undergo fracture with little or no deformation [54]. The raw materials found in nature are formed from complex geologic processes to produce the clays needed to make ceramics. Since oxygen, silicon, and aluminum are the three most abundant elements in the Earth's crust, it is not surprising that silicates (Si_xO_y) and aluminum silicates ($Al_xSi_yO_z$) are the dominant minerals. Traditional ceramics consist of these materials, but new processing techniques have expanded their capabilities. For example, the white thermal tiles developed for the Space Shuttle are silica, SiO_2 [55]. In addition, new ceramics containing nitrides, borides, and carbides are also in use for more demanding applications.

Ceramics can have either ionic or covalent bonding in their crystal structure depending upon the electronegativity relationship of the atoms. Ionic solids consist not of discrete molecules like those of polymers, but rather extended arrays of anions and cations. Covalent solids consist of covalently bonded atoms of a certain geometry extending in three dimensions. In ceramics, the three dimensional array is generally ordered and possesses a crystalline structure. Figure 5.14 shows the covalent structure for one form of silica, SiO_2. Note that each Si is surrounded by four oxygen atoms and each O has two Si neighbors.

Ceramic materials make excellent electrical and thermal insulators due to the nature of their bonding. The strong ionic or covalent bonds prevent the electron movement necessary for thermal and electrical conduction. High melting points are also achieved for the same reason. Silica tiles on the Space Shuttle withstand temperatures in excess of 1093°C (2,000°F) during reentry [55]. The success of the Shuttle demonstrates the effectiveness of this reusable thermal protection system.

The development of a national aerospace plane, fueled by hydrogen, presents another unique opportunity for the

Spacecraft Materials

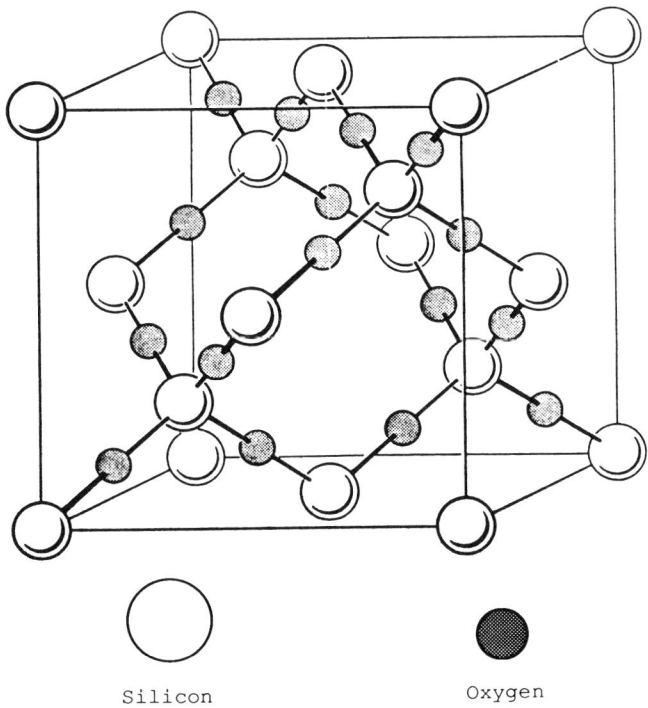

Figure 5.14 Crystal structure of SiO$_2$, silica [54].

use of ceramics. Metals become brittle when exposed to hydrogen while ceramics are very hydrogen-tolerant. A ceramic silicon nitride (SiN) was tested for several hours in the presence of hydrogen at 1,000°C (1832°F) and it retained its strength [47]. As new propulsion and mission scenarios are developed, the use of new materials will undoubtedly increase.

Glasses

Similar to ceramics, but lacking in long-range atomic ordering is the class of materials known as glasses. A schematic representation of the ordering difference is depicted in Figure 5.15.

Most inorganic glasses employ silica as the base material. Depending upon the particular application, any of a number of other compounds may be added. Table 5.13 gives the composition of some commercial glasses.

These same compounds are also used for space applications with silica again being the most prevalent. Sapphire (Al$_2$O$_3$ + FeO, TiO$_2$) is another stable optical material used routinely on spacecraft [30].

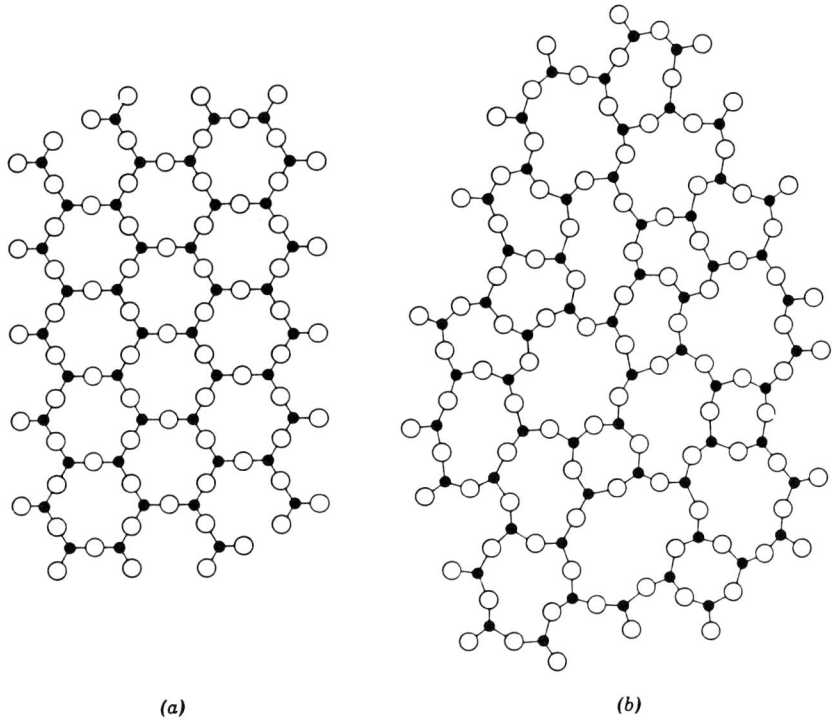

Figure 5.15 Schematic representation of molecular arrangements of the same composition in an ordered crystalline form (left) and a random network glassy form (right).

Table 5.13 Approximate Percentage Composition by Weight of Some Typical Commercial Glasses [54]

Glass	SiO_2	Al_2O_3	Fe_2O_3	CaO	MgO	BaO	Na_2O	K_2O	SO_3	F_2	ZnO	PbO	B_2O_3	Se	CdO	CuO
Container flint	72.7	2.0	0.06	10.4		0.5	13.6	0.4	0.3	0.2						
Container amber	72.5	2.0	0.1	10.2		0.6	14.4	0.2	S-0.02	0.2						
Container flint	71.2	2.1	0.05	6.3	3.9	0.5	15.1	0.4	0.3	0.1						
Container flint	70.4	1.4	0.06	10.8	2.7	0.7	13.1	0.6	0.2	0.1						
Window green	71.7	0.2	0.1	9.6	4.4		13.1		0.4							
Window	72.0	1.3		8.2	3.5		14.3	0.3	0.3							
Plate	71.6	1.0		9.8	4.3		13.3		0.2							
Opal jar	71.2	7.3		4.8			12.2	2.0		4.2						
Opal illumination	59.0	8.9		4.6	2.0		7.5			5.0	12.0	3.0				
Ruby selenium	67.2	1.8	0.03	1.9	0.4		14.6	1.2	S-0.1	0.4	11.2		0.7	0.3	0.4	
Ruby	72.0	2.0	0.04	9.0			16.6	0.2		Tr.*						0.05
Borosilicate	76.2	3.7		0.8			5.4	0.4					13.5			
Borosilicate	74.3	5.6		0.9		2.2	6.6	0.4					10.0			
Borosilicate	81.0	2.5					4.5						12.0			
Fiber glass	54.5	14.5	0.4	15.9	4.4		0.5			0.3			10.0			
Lead tableware	66.0	0.9		0.7		0.5	6.0	9.5				15.5	0.6			
Lead technical	56.3	1.3					4.7	7.2				29.5	0.6			
Lamp bulb	72.9	2.2		4.7	3.6		16.3	0.2	0.2				0.2			
Heat absorbing	70.7	4.3	0.8	9.4	3.7	0.9	9.8	0.7		Tr.*			0.5			

Source. F. V. Tooley, *Handbook of Glass Manufacture*, Ogden Publishing Co., New York, N.Y.
*Trace.

Environmental Effects

For the most part, ceramics are fairly impervious to the space environment. Some outgassing does occur, but nothing to the extent exhibited by polymers. Radiation and atomic oxygen can affect surface properties to some extent, however, these effects are generally not critical to operation.

Glasses, on the other hand, are much more susceptible to damage from the space environment. Outgassing products from metals and adhesives can and do condense on glass surfaces, altering their optical properties. Radiation is probably the most damaging source to glasses since it can cause discoloration. Energetic bombarding particles dislodge electrons from the atoms, creating electron-hole pairs. The trapped electrons act as color centers which possess the appropriate energies to absorb particular frequencies of light. Depending upon the glass, coloration can occur with as little as 600 rad exposure with nearly all glasses showing saturation at 10^{10} rad. Solutions to this problem include additives which occupy free electron sites and substituting inorganic glass with a polymer (e.g., polystyrene) [30]. Atomic oxygen presents no problem for silica and in fact SiO_2 is used as a protective coating for solar cell arrays and composites [53, 56].

Long Duration Exposure Facility

The Long Duration Exposure Facility (LDEF) was launched in 1984 on STS-13 with 57 space experiments on board. Many of the experiments are concerned with spacecraft material/environment interaction. The titles of applicable experiments are [57]:

1. Atomic-Oxygen-Stimulated Outgassing
2. Interaction of Atomic Oxygen With Solid Surfaces at Orbital Altitudes
3. Influence of Extended Exposure in Space on Mechanical Properties of High-Toughness Graphite-Epoxy Composite Material
4. Effect of Space Environment on Space-Based Radar Phased-Array Antenna
5. Space Exposure of Composite Materials for Large Space Structures
6. Effect of Space Exposure of Some Epoxy Matrix Composites on Their Thermal Expansion and Mechanical Properties
7. The Effect of the Space Environment on Composite Materials
8. Microwelding of Various Metallic Materials Under Ultravacuum
9. Evaluation of Long-Duration Exposure to the Natural Space Environment on Graphite-Polyimide and Graphite-Epoxy Mechanical Properties
10. The Effect of Space Environment Exposure on the Properties of Polymer Matrix Composite Materials
11. Space Environment Effects on Space Materials

Spacecraft Materials

12. Balloon Materials Degradation
13. Thermal Control Coatings Experiment
14. Exposure of Spacecraft Coatings
15. Thermal Control Surfaces Experiment
16. Investigation of Critical Surface Degradation Effects on Coatings and Solar Cells Developed in Germany
17. Space Aging of Solid Rocket Materials
18. Exposure to Space Radiation of High-Performance Infrared Multilayer Filters and Materials Technology Experiments
19. Effect of Space Exposure on Pyroelectric Infrared Detectors
20. Effects of Solar Radiation on Glasses
21. Study of Factors Determining the Radiation Sensitivity of Quartz Crystal Oscillators
22. Investigation of the Effects of Long-Duration Exposure on Active Optical System Components
23. Space Environment Effects on Fiber Optics Systems
24. Space Environment Effects

Preliminary Findings on LDEF

LDEF returned to Earth aboard the Space Shuttle in early January 1990 and was disassembled at Kennedy Space Center. The authors were members of an investigation team responsible for one of the experiments (#701 Space Environment Effects on Fiber Optic Systems in tray M0004) [58]. Their impression of LDEF as a whole was that it was in good shape, having suffered no major damage from its almost six years in space. Perhaps the most notable feature was the severe atomic oxygen degradation of thermal blankets made of aluminized Mylar and Kapton. Atomic oxygen entered into the spaces between the multilayer insulation (MLI) through the existing holes and subsequently reacted with the polymer. This resulted in the complete erosion of the polymer film backing which produced thin flakes of aluminum. Film taken in space during recovery showed numerous aluminum flakes coming from LDEF as it was being rotated. During deinteration, investigators found these flakes attached to the interior and exterior surfaces of many experiments on LDEF. Atomic oxygen also degraded many of the adhesive tapes used to attach samples to the experiments. The Kapton backing virtually disappeared, leaving only the adhesive. Material samples secured with Kapton tape were missing from several experiments.

Another noticeable effect was the darkening of thermal control paint samples by ultraviolet (UV) radiation. The darkening could be due to UV radiation interacting directly with the polymer base of the paint. Disks coated with white, A276 polyurethane based paint were situated throughout the exterior surface of LDEF. Those disks on LDEF's trailing edge were brown while those on its leading edge were still white. The authors postulate that atomic oxygen continually eroded the surface of the paint samples on the leading edge, keeping them white in color. The trailing edge, on the other hand, had little atomic oxygen exposure so the effects of UV darkening remained.

A second cause of the darkening could be UV fixation of outgassing species from the Space Shuttle and LDEF itself. Many polymers used in the wiring aboard LDEF are notorious outgassers. The outgas products could have deposited on the exterior surface where they were fixed by UV radiation. Exhaust products from the Shuttle's attitude control rockets may also have been deposited and become fixed on LDEF. This was observed previously in both the Apollo and Skylab programs (Figure 5.16).

Preliminary Findings of Experiment #701

Experiment #701 was conceived at the Air Force Systems Command Weapons Laboratory (WL) in 1977 and selected to be orbited [59] on the first NASA-developed LDEF mission. The experiment consisted of four active optoelectronic links developed by the WL and the Jet Propulsion Laboratory [60, 61] for assessing the fiber optic link performance factors of bit error rate, signal to noise ratio, and optical fiber attenuation induced by space radiation and temperature cycling. The fiber links were space exposed occupying approximately one-third of a 0.91×1.22 m area. A 1 mm thick aluminum alloy (6061-T6) protective sheet devoid of fiber and coated with white Chemglaze (A276) thermal paint comprised 0.34 m^2 of the space exposed surface. This 0.41×0.91 m surface was used to provide protection of additional thermal coverings below its surface which contained the experiment flight recorders, instrumentation, and radiation dosimeters. The experiment was housed on the LDEF structure at the tray location and designation M0004. This location placed the experiment on the space pointing end of the LDEF with the edges of the tray 15° and 45° off the ram or leading edge direction. Figure 5.17 shows LDEF in orbit with the ram direction to the right of the satellite and the space pointing end facing out from the page.

Particulate Impacts

Particulate impacts were observed over the entirety of the tray's space exposed surfaces. Impacts also occurred on the space exposed cabled fibers and are discussed elsewhere [62]. The protective sheet portion of the tray had 293 impacts with crater diameters greater than 100 microns and 29 impacts exceeding 300 microns in diameter. All of the larger craters were photographed using a Polaroid camera mounted to a 48X microscope. A cumulative frequency distribution by crater size was compiled and is plotted in Figure 5.18 along with two similar flux curves derived from Solar Max data [63].

The Solar Max satellite resided in an altitude range of 500 km decreasing to 462 km during its four year initial

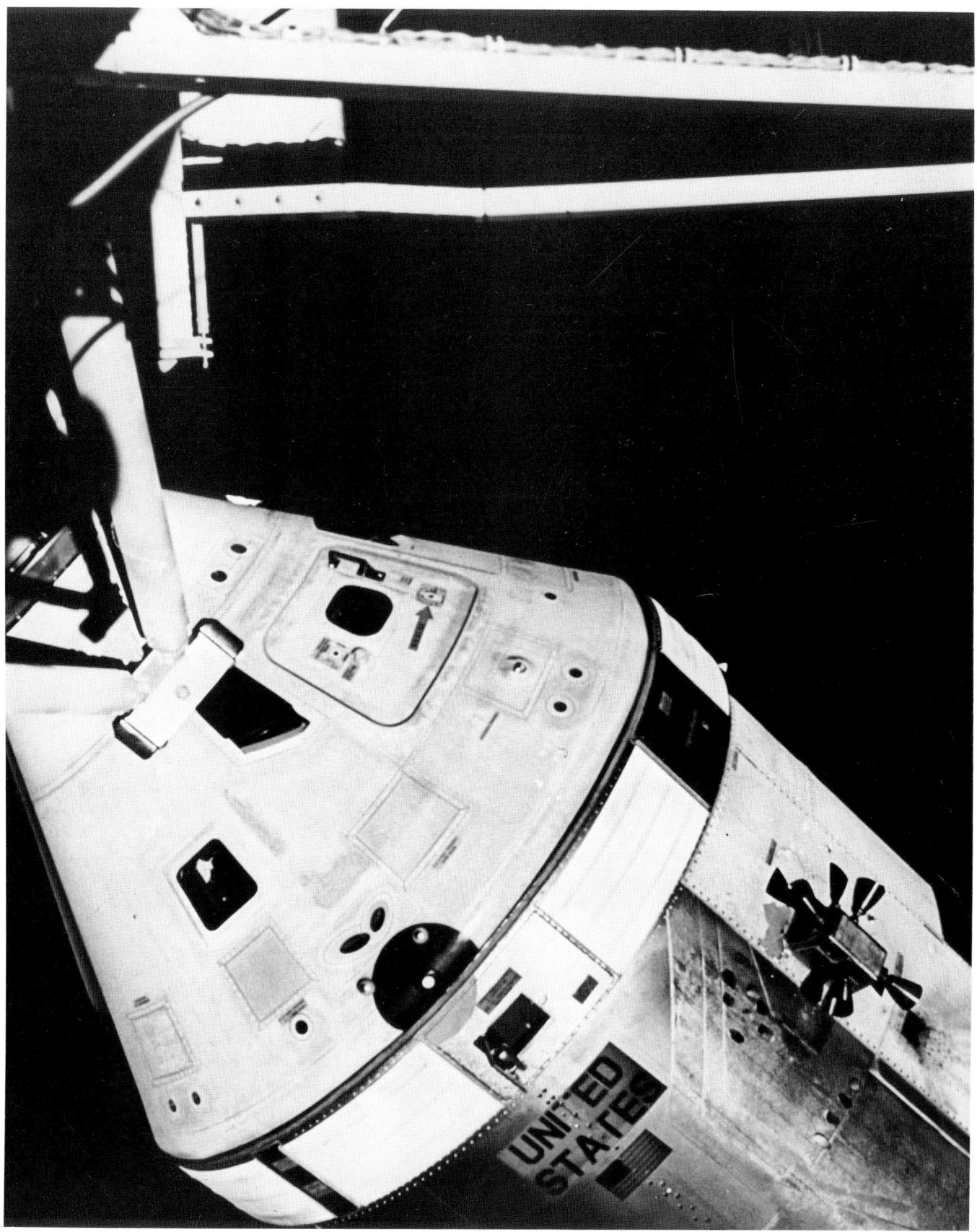

Figure 5.16 A brown residue from thruster rocket exhaust is visible on the surface of the Apollo command and service modules. *Courtesy of NASA*.

Spacecraft Materials

Figure 5.17 Experiment #701 was oriented on the leading edge, just off the velocity vector, of LDEF. *Courtesy of NASA.*

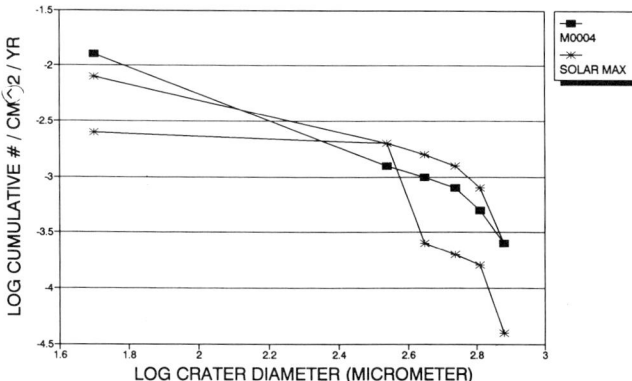

Figure 5.18 The flux data from the Experiment #701 protective sheet is very similar to the data derived from Solar Max material.

stay in orbit while the LDEF drifted from 475 to 335 km in its nearly six year orbital span. Both satellites were in low inclination orbits. The Solar Max was in orbit during a period of decreasing solar activity (1980–1985) while LDEF was exposed to a generally increasing solar activity level (1986–1990). This fact, coupled with the different altitudes and satellite orientations, may be responsible for the differences in flux values.

Impact Distribution

The locations of the impacts on the protective sheet (PS) were logged on a computer-generated grid. This grid divided the sheet into 36 square tiles (4 in × 4 in). The grid is shown in Figure 5.19 which details the sheet's orientation with respect to LDEF and the Earth. The top of the grid was 15° off the ram direction while the grid lower edge was 45° away from the ram. The right side of the PS was space pointing as shown.

In Figure 5.19 the large craters are individually plotted while the smaller craters are tallied on the grid (e.g, tile 26 has 10 small craters, 26 (−10)) on the accompanying table. Each major crater is described by a four digit code: the first two numbers are the x-axis location and the last two are the y-axis location. Thus, Crater 1327 is located approximately 13 inches along the x-axis and 27 inches along the y-axis. Arrows on major impacts indicate the probable direction of the impactor's velocity as evidenced by uneven petaling.

The numbers in parentheses along each row and column represent the number of small impacts in each respective row or column. There is a slight tendency for more small impacts as the ram direction is approached, i.e., moving up the y-axis. This is to be expected if the primary source of particulates is impacting the front of an orbiting object. This is most likely for micrometeoroids impacting the experiment while manmade orbital debris would be expected to hit at 45° off the velocity vector and thus be less numerous as ram is approached.

Crater Analysis

Every impact crater exhibited the classic hemispherical shape except for one observable event. Crater 0604 had a very irregular shape with some petaling. It was one of only two impacts that caused some deformation (near penetration) on the back side of the PS. It is hypothesized that either a relatively large irregular impactor may have struck the plate at a large oblique angle or Crater 0604 resulted from two impact craters overlapping. Figure 5.20 shows Crater 0604.

Each of the impact craters was surrounded by a "paint crater" where the paint had been removed from the aluminum. The diameter of the paint crater on the average was three times the dimension of the crater diameter. Some of the paint craters were almost circular while others were uneven. The characteristics of these paint craters may provide an insight into the age of impacts. It is speculated that at the time of impact a paint crater is uneven but over time the jagged edges break loose until the crater becomes circular.

Around the impact and paint craters a halo is also evident in the paint. The paint in these areas is believed to have been altered by the propagation of the impact shock wave. A number of different types of halos have been found: darkening of paint, a concentric series of rings, or the tearing of paint. Some halos exhibit more than one of these phenomena. Similar features have been noted in previous analyses [60]. Figure 5.21 is a shadowed photograph of the protective sheet taken to highlight the surface characteristics of the many impacts. A simplified grid is overlayed in the photo to help orient these craters.

Chemical Analysis

Surfaces inside the four craters examined to date contain a large number of holes, characteristic of significant melting occurring. Figure 5.22 is taken from Crater 0604 and is representative of the interior of other craters. Chemical analysis using energy dispersive x-ray detection has yielded little insight into the chemical composition of the impactor. The analysis is complicated because 6061 aluminum contains iron, silicon, copper, titanium, manganese, magnesium, chromium, and zinc. The presence of iron, silicon, and magnesium is particularly troublesome because these elements are also the basic materials in many micrometeorites. Examination of Crater 1535 resulted in the observance of crystals of sodium chloride interspersed with the Chemglaze paint (Figures 5.23 and 5.24). The Chemglaze paint itself has the consistency of chalk dust, indicating that the polyurethane base used in the paint was eroded by atomic oxygen leaving the TiO_2 component without a matrix. It was also determined that a brown residue found not only on surfaces of this experiment, but also throughout LDEF, contained a high concentration of silicon (Figure 5.25.)

Spacecraft Materials

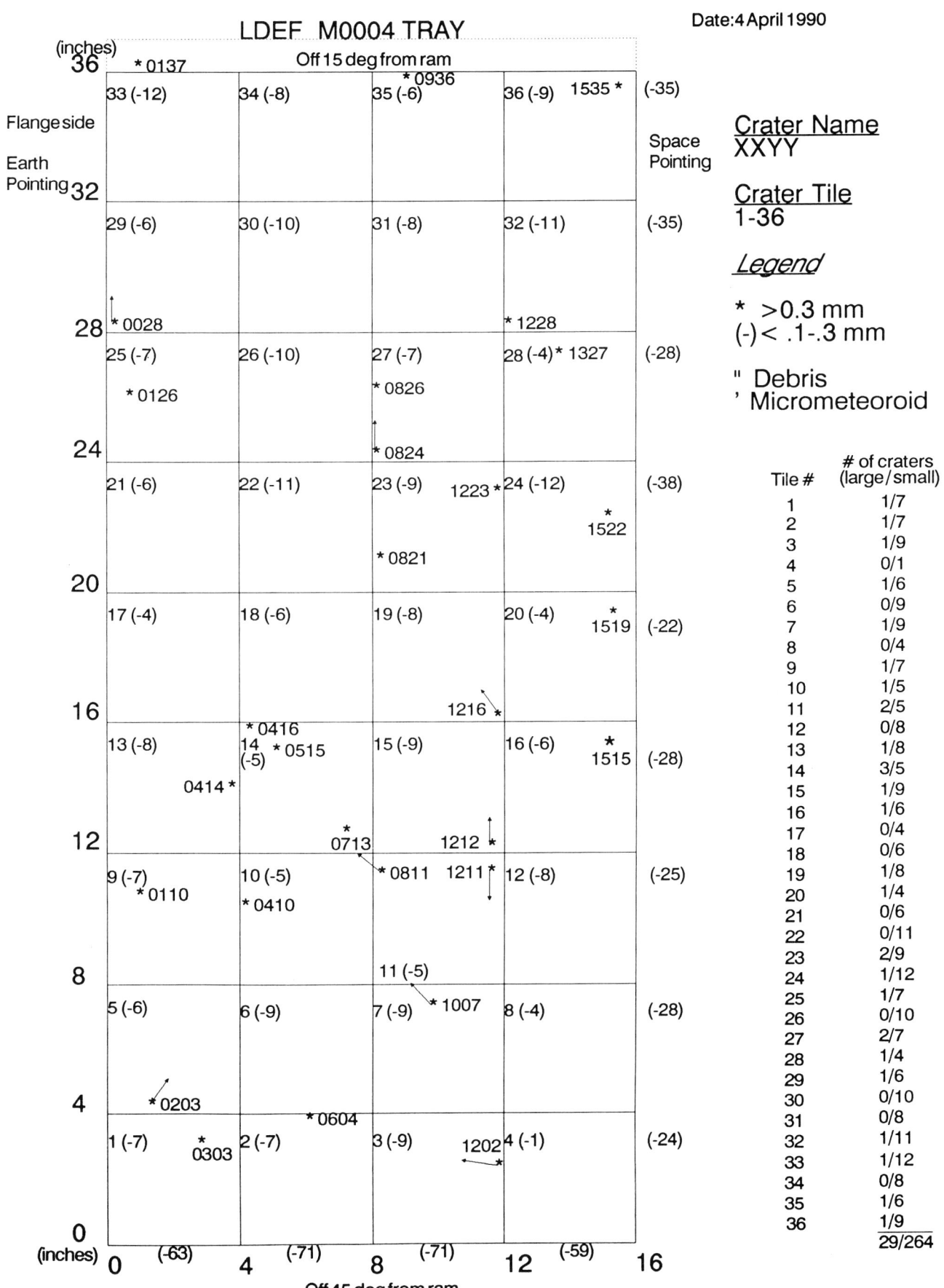

Figure 5.19 The protective sheet from Experiment #701 is depicted on a grid with large craters shown along with directionality vectors.

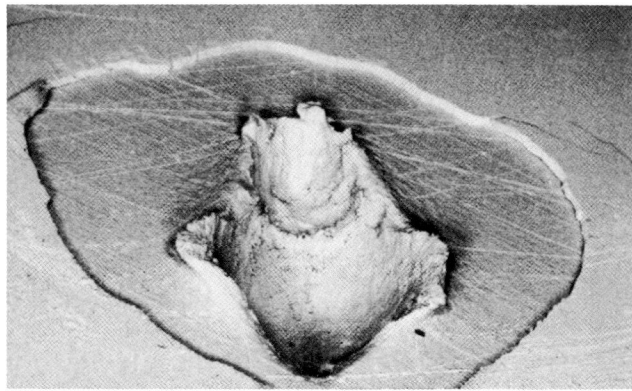

Figure 5.20 LDEF crater 0604 is the only irregular crater on the protective sheet.

Further Study

The interior of each major crater will be analyzed by a scanning electron microscope to determine the nature of the object which created the craters. The brown colored residue (considered to be contaminants) found near the edges of the protective sheet and many other experiments on LDEF will also be analyzed. Additional studies will be conducted to determine if the paint halos may be used to better describe or resolve impact events. Impacts on the fiber optic cabling will be scrutinized and correlated to the operational performance of the active optical links. Analyses will be conducted to determine the particle flux by physical size rather than extrapolating to crater diameters [64]. Adherence to NASA-LDEF analysis procedures [65] will continue to be observed for material and impact analysis. Finally, changes to the fiber optic cabling due to the environment (mostly atomic oxygen, ultraviolet light, and temperature cycling) will be analyzed.

Summary

This chapter has covered the basic chemical nature of four groups of materials and their responses to the space environment. Metals are generally unaffected by the conditions of space except for the occurrences of sublimation by cadmium and zinc and atomic oxygen attack on silver and osmium. Polymers and composites are by far the most chemically reactive groups. Reduced pressure, radiation, and atomic oxygen all have detrimental effects on these classes of materials. Ceramics and glasses respond like metals except for the radiation-induced formation of color centers in glass.

All these factors have been considered by aerospace scientists and engineers. Some problems have been overcome while others still need work. But as one can see from the plethora of research, the areas of chemistry and materials science are of major importance to our space program.

References

1. Cottrell, A. *An Introduction to Metallurgy*, 2 ed. Crane, Russak, & Co. (New York), 1975.
2. *Aluminum: Properties and Physical Metallurgy*, ed. by John Hatch, American Society for Metals (Metals Park), 1984.
3. "Data for Selection of Space Materials," Space Agency, ESA PSS-01-701, Issue 1, Nov 1985.
4. "Special Metals: Beryllium, Gold, Silver Offer Ideal Features for Aerospace Use," Aviation Week and Space Technology, 21 October 1985, pp. 116–118.
5. "Materials Selection List for Space Hardware Systems," MSFC-HDBK-527 Rev F, September 1988.
6. Coulter, D. et al, "O-Atom Degradation Mechanisms of Materials," Proceedings of the NASA Workshop on Atomic Oxygen Effects (JPL Pub 87-14), November 10–11, 1986.
7. "A More Perfect Vacuum," Aerospace America, March 1987, pp. 44–47.
8. McCrary, L. and J. Hueser, "An Apparatus for Weight Loss Determinations in Space Environments," SAMPE Symposium on the Effects of the Space Environment on Materials, April 19–21, 1967.
9. Chang, R., *Chemistry*, 3rd ed., Random House (New York), 1988.
10. Johnson, N. and McKnight, D., *Artificial Space Debris*, Krieger Publishing Company, 1987.
11. Whitten, K. et al, *General Chemistry*, 3 ed., Saunders College Publishing (New York), 1988.
12. Godard, H. et al, *The Corrosion of Light Metals*, The Electrochemical Society, John Wiley & Sons (New York), 1967.
13. Greenwood, N. and A. Earnshaw, *Chemistry of the Elements*, Pergamon Press (New York), 1984.
14. Leger, L. et al, "Review of LEO Flight Experiments," Proceedings of the NASA Workshop on Atomic Oxygen Effects (JPL Pub 87-14), November 10–11, 1986.
15. Tascione, T., *Introduction to the Space Environment*, Orbit Book Co. (Malabar, Fl), 1988.
16. Gregory, J. "Interaction of Hyperthermal Atoms on Surfaces in Orbit: The University of Alabama Experiment," Proceedings of the NASA Workshop on Atomic Oxygen Effects (JPL Pub. 87-14), November 10–11, 1986.
17. Parsons, R. and D. Gulino, "Effect of an Oxygen Plasma on Uncoated Thin Aluminum Reflecting Films," NASA LRC, NAS 1.15: 89882, N87-21999.
18. "Aerospace Materials," Aviation Week and Space Technology, 3 October 1988, pp. 46–76.
19. Charlesby, A., Proc. Roy. Soc. (London), *A215*, 187 (1952).

Spacecraft Materials

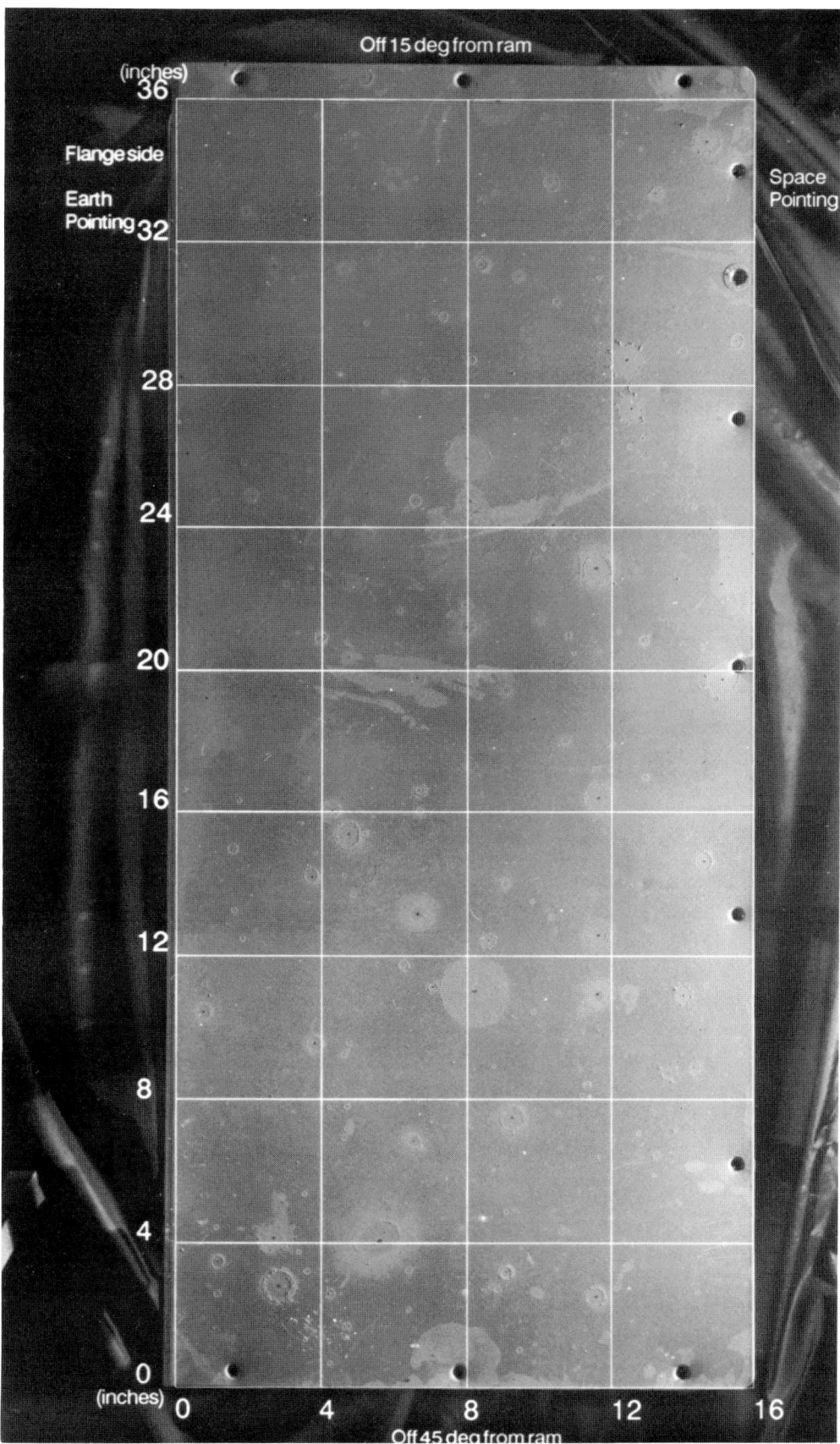

Figure 5.21 A shadowed photograph of the protective sheet shows the distribution of craters and their characteristics much more clearly than a normal photograph. *Photography of Airman Steven Thurow, USAFA/DFSIV.*

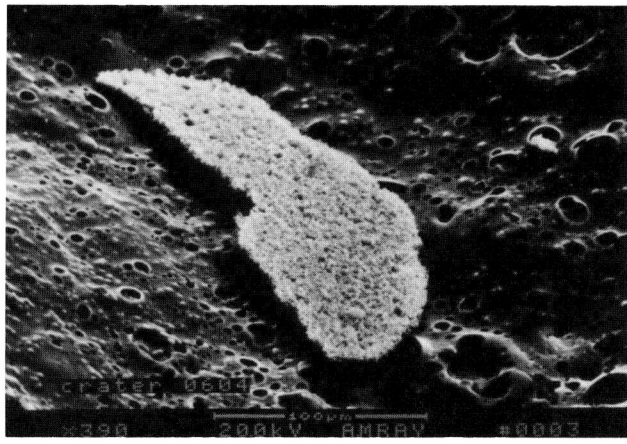

Figure 5.22 The interiors of craters on the LDEF protective sheet show a significant amount of melting. A flake of thermal control paint appears in the center of the crater.

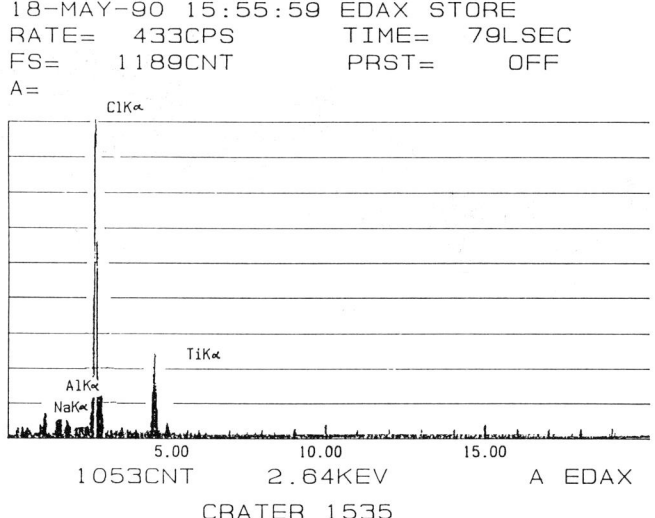

Figure 5.24 The particles of contaminant found in LDEF crater 1535 were identified as sodium chloride by using energy dispersive x-ray analysis.

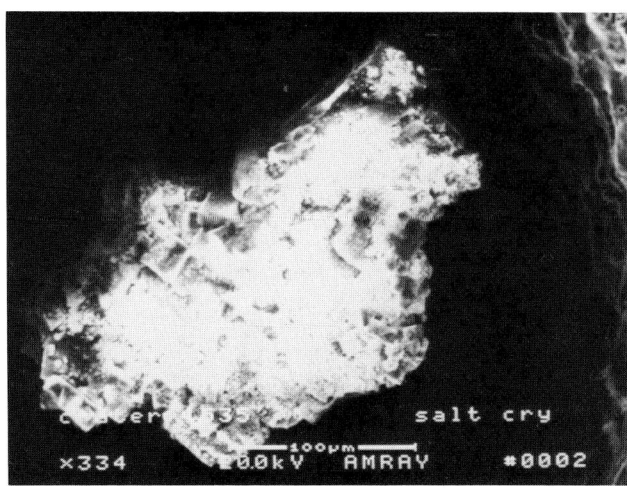

Figure 5.23 The crystal structure of the particulate solid in LDEF crater 1535 is evident upon closer inspection.

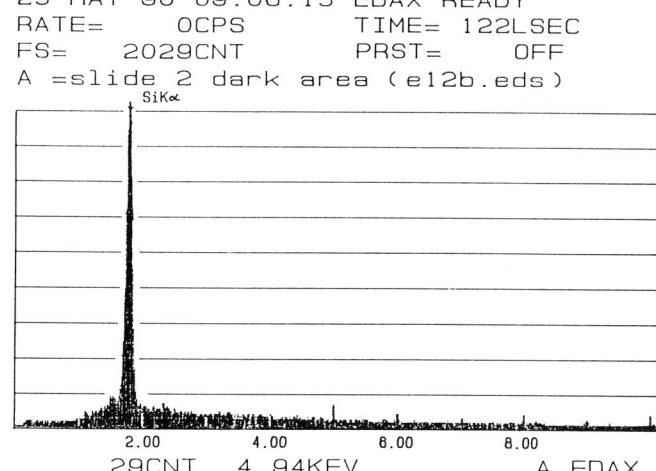

Figure 5.25 Silicon was determined by energy dispersive x-ray analysis to be a very abundant element on the space exposed surfaces of the LDEF.

20. Charlesby, A. and N. Hancock, Proc. Roy. Soc. (London), *A218*, 245 (1953).
21. Wall, L. and M. Magat, J. Chem. Phys., *50*, 308 (1953).
22. Charlesby, A., *Nature*, 171, 167 (1953).
23. Charlesby, A. and M. Ross, *Nature*, 171, 1153 (1953).
24. D. Coulter et al., "The Effects of Energetic Proton Bombardment on Polymeric Materials: Experimental Studies and Degradation Models" JPL Pub 85-101, June 1, 1986.
25. Williams, F., in *Fundamental Processes in Radiation Chemistry*, ed. by P. Ausloos, Interscience (New York), 1968.
26. Odian, G., *Principles of Polymerization*, 2 ed., John Wiley & Sons (New York), 1981.
27. Schnabel, W., *Polymer Degradation*, MacMillan Publishing Co. (New York), 1981.
28. Liang, R. et al, "New Development in Accelerated UV Testing for Life Prediction," NASA Contract NAS 7-918.
29. Ranby, B. and J. Rabek, *Photodegradation, Photo-oxidation, and Photostabilization of Polymers*, John Wiley & Sons (New York), 1975.
30. *Space Materials Handbook*, ed. by J. Rittenhouse and J. Singletary, NASA SP-3051, 1969.
31. Kiefer, R. and R. Orwoll, "Space Environmental Ef-

fects on Polymeric Materials," Final Technical Report NAG-1-678, N88-16879.
32. Schwinghamer, R., "Space Environmental Effects on Materials," NASA TM-78306, N81-10085.
33. Parker, I., "The Vital Connection," Space, January 1989, pp. 41–42.
34. Colony, J. and E. Sanford, "Mechanisms of Polymer Degradation using an Oxygen Plasma Generator," NASA TM-100681.
35. R. Liang et al. "Mechanistic Studies of Polymeric Samples Exposed Aboard STS VIII," JPL Pub 87-25, December 15, 1987.
36. S. Koontz et al, "Materials Selection for Long Life in LEO: A Critical Evaluation of Atomic Oxygen Testing with Thermal Atom Systems," N89-12590.
37. E. Lan et al., "Atomic Oxygen Effects on Candidate Coatings for Long-Term Spacecraft in Low Earth Orbit, "N89-12592.
38. Banks, B., Rutledge, S. and J. Brady, "The NASA Atomic Oxygen Effects Test Program," N89-12589.
39. "Atomic Oxygen Effects Measurements for Shuttle Missions STS-8 and 41-G," NASA TM 100459, N89-14331, September 1988.
40. R. Liang et al, "Degradation Studies of SMRM Teflon," 18th International SAMPE Technical Conference, Vol 18, October 7–9, 1986, pp. 1050–1055.
41. "Proceedings of the NASA Workshop on Atomic Oxygen Effects," ed. by D. Brinza, JPL Pub 87-14, November 10–11, 1986.
42. R. Hansen et al, "Effect of Atomic Oxygen on Polymers," Journal of Polymer Science, Part A, Vol 3, 1965, p 2205–2214.
43. R. Cvetanovic, "Kinetics and Mechanisms of Some Atomic Oxygen Reactions," Proceedings of the NASA Workshop on Atomic Oxygen Effects (JPL Pub 87-14), November 10–11, 1986.
44. "Space Station Atomic Oxygen Effects," Presentation by Lockheed MSC to NASA-LeRC, 11 July 1986.
45. "Second Workshop on Spacecraft Glow," ed by J. Waite and T. Moorehead, NASA Conf. Pub. 2391, May 6–7, 1985.
46. R. Liang et al, "Mechanistic Studies of Interactions of Materials with Energetic Oxygen Atoms in Low Earth Orbit," NASA-JPL.
47. Orbital Debris Monitor, October 1990, P.O. Box 136, USAFA, CO.
48. Hull, D., An Introduction to Composite Materials, Cambridge University Press, 1981.
49. Sheldon, R., Composite Polymeric Materials, Applied Science Publishers (New York), 1982.
50. Delmonte, J., Technology of Carbon and Graphite Fiber Composites, Van Nostrand Reinhold Company, 1981.
51. Billmeyer, F., Textbook of Polymer Science, Inter-Science Publishers (New York), 1966.
52. Zimcik, D. et al, "The Effect of Low Earth Orbit Space Environment on Polymeric Spacecraft Materials," Proceedings of the Third European Symposium on Spacecraft Materials in Space Environment, Oct 1–4, 1985, pp 81–89.
53. Dursch, H. and C. Hendricks, "Protective Coatings for Composite Tubes in Space Applications," NASA-CR-178116, N87-18669.
54. Kingery, W., Bowen, H. and D. Uhlmann, Introduction to Ceramics, 2 ed., John Wiley & Sons (New York), 1976.
55. Banas, R. et al, "Lessons Learned from the Development and Manufacture of Ceramic Reusable Surface Insulation Materials for the Space Shuttle Orbiters," N84-10158.
56. Bilger, K., Gjerde, H. and B. Sater, "Photovoltaic Array Environmental Protection Program," 24th IECEC, August 6–11, 1989.
57. "The Long Duration Exposure Facility," ed. by Clark, L. G. et al, NASA SP-473, 1984.
58. McKnight, D., Taylor, E. and R. Dueber, "Space Debris and Micrometeorite Events Experienced by WL Experiment 701 in Prolonged Low Earth Orbit," to be published in the Journal of Geophysical Research-Space Physics.
59. Taylor, E. W., "Fiber Optic Space Effects," Proc. First Fiber Optic Conference, Chicago, IL, Information Gatekeepers, Sept 6–8, 1978, pp. 245–247, 1987.
60. Taylor, E. W., "Space Environment Effects on Fiber Optic Systems," in The Long Duration Exposure Facility (LDEF) Mission 1 Experiments, Edited by L. G. Clark, et al, NASA SP-473, pp. 182–4, 1984.
61. Johnson, A. R., Bergman, L. A. and E. W. Taylor, "Fiber Optic Experiment for the Shuttle Long Durataion Exposure Facility," SPIE Vol 296 Fiber Optics in Adverse Environments, 1982.
62. Taylor, E. W. et al, "Response of Temperature Cycled Active Fiber Optic Links in Low Earth Orbit," submitted to Opt. Soc. Amer., App. Optics, June 90.
63. Warren, J. L. et al, "The Detection and Observation of Meteoroid and Space Debris Impact Features on the Solar Max Satellite," proceedings of the 19th Lunar and Planetary Science Conference, pp. 641–657, 1989.
64. Westine, P. S. and S. A. Mullin, "Methods of Using Scale Models for Studying Hypervelocity Impact Damage," AFATL-TR-86-81, 1986.
65. See for example: NASA-LDEF Environment Effects on Materials, Special Investigation Group (MSIG) Test Plan, Jan 1990 and; LDEF M/D SIG Test Plan, 1990.

Discussion Questions

1. The ambient pressure in space is generally less than 10^{-6} torr.
    a. What impact does this have on the selection of a material for a particular spacecraft application?
    b. In particular, what metals and polymers should be avoided for space applications?
    c. How can spacecraft outgassing be minimized in space?

2. Corrosion can have both harmful and beneficial effects on spacecraft life and performance.
    a. In terms of activity, which metals in Table 1.7 are most and least likely to oxidize? Tell why?
    b. Why do fairly active metals like aluminum and magnesium resist extensive corrosion?
    c. Explain what anodization does to a metal.

3. Bonding in metals can be explained using the free electron theory.
    a. Using this theory, explain why metals are able to retain their integrity when exposed to radiation.
    b. Are the levels of natural radiation found in space harmful to metals?

4. Ideally, polymers used in space should be pure and of the proper molecular weight. Unfortunately, few polymers meet these criteria.
    a. Name the three impurities which normally outgas from a polymer in space.
    b. Discuss the detrimental effects of outgassing on the polymer itself and other spacecraft subsystems.

5. What is the long lived specie created in polymers exposed to ionizing radiation which is responsible for causing degradation? Discuss the two main responses a polymer will exhibit given the presence of this species.

6. Discuss the methods available for protecting polymers from ionizing radiation.

7. Atomic oxygen is extremely reactive and can cause polymers to quickly erode through chemical processes.
    a. Why is atomic oxygen so much more reactive than molecular oxygen?
    b. Why are some polymers, like polyethylene, so reactive with atomic oxygen while others, like teflon, are fairly unreactive?
    c. Explain the phenomenon known as "Shuttle glow."

8. Metals are being replaced in spacecraft more and more by composite materials. What are the advantages and disadvantages of this substitution?

9. The strength of a composite is due to the binding forces between the filler and matrix. Discuss the mechanisms responsible for adhesion between these two components.

10. Ceramics and glasses are often considered as one general class of materials.
    a. What type or types of chemical bonding occur in ceramics and glasses?
    b. How do the two differ in terms of their atomic structures?

Chapter 6

Spacecraft Power

The electric power subsystem supplies power to the spacecraft during its entire lifetime. Depending upon the type of mission, the power system could be any one of several types. For those spacecraft orbiting Earth or travelling in the vicinity of the Sun, the solar array/battery is the preferred choice. If the mission is to travel away from the Sun toward the outer reaches of the solar system, then a radioisotope thermoelectric generator (RTG) is used. Manned missions beginning with Gemini have relied upon fuel cells for both power and life support. The Soviets have used nuclear reactors on several of their more important satellites and the United States has considered their use on SDI platforms. The space station will have a solar dynamic power system in addition to a solar array. In this chapter the chemical principles behind the design and operation of these power sources are discussed.

Photovoltaics

The physical phenomenon responsible for converting light into electricity is known as photovoltaics. First observed by the French physicist Edmond Becquerel in 1839, the photovoltaic effect was initially utilized by photographers for photometric devices. The first working silicon solar cell was demonstrated in 1954 by Bell Laboratories and its first use in space occurred in 1958 aboard Vanguard I. Vanguard's array consisted of 108 silicon solar cells, generating about 1 W of power with an average cell efficiency of 10%. Since then, solar cells have been the primary energy conversion source of the U.S. space program [2].

Principles of Solar Cell Operation

An understanding of solar cell operation is best attained through the use of an example. Silicon solar cells are chosen since they are the best understood and most widely used. Others, such as gallium-arsenide and indium-phosphide, operate in the same basic manner.

Silicon is in Group 4 of the Periodic Table immediately below carbon. It contains four valence shell electrons and is classified as a semiconductor. Each Si atom covalently bonds to four adjacent atoms (remember the octet rule) to form the tetrahedral crystal lattice shown in Figure 6.1. This particular arrangement is repeated throughout the entire crystal.

Covalently bonded compounds are normally poor conductors of electricity due to electron immobility and a large energy gap between the valence and conduction bands. Silicon, being a semiconductor, has a rather small band gap energy (1.1 eV) which can be "jumped" by exciting the sample. Light incident upon Si possesses enough energy to excite electrons into the conduction band where they are free to travel. At the same time, a positively charged "hole" is created at the vacated site. Adjacent electrons can fill this hole, thereby giving mobility to the hole as well (Figure 6.2). This process occurs quickly and frequently as electrons and holes move throughout the crystal.

The simple random motion is not, however, all that is required to generate current. Electrons would eventually lose energy and fall back into holes unless there was a potential barrier present in the semiconductor preventing recombination from occurring. Solar cells in fact have this barrier in the form of the p-n junction.

Semiconductor doping is performed by adding an impurity to an otherwise pure sample. In the case of Si, any dopant containing more than four valence shell electrons (e.g., phosphorus) would introduce unbound electrons into the crystal lattice. The presence of free negative charges gives rise to the name n-type (Figure 6.3.).

Similarly a dopant containing less than four valence shell electrons (e.g., boron) can be placed within Si to form a p-type semiconductor. The result in this case is the presence of bonds with missing electrons or what are referred to as holes. These positively charged holes are, as before, able to freely move throughout the crystal (Figure 6.4).

The formation of the junction occurs when samples of p-type and n-type are brought together. Excess electrons on the n-side cross the junction to fill holes on the p-side (Figure 6.5a). A potential is now set up across the junction

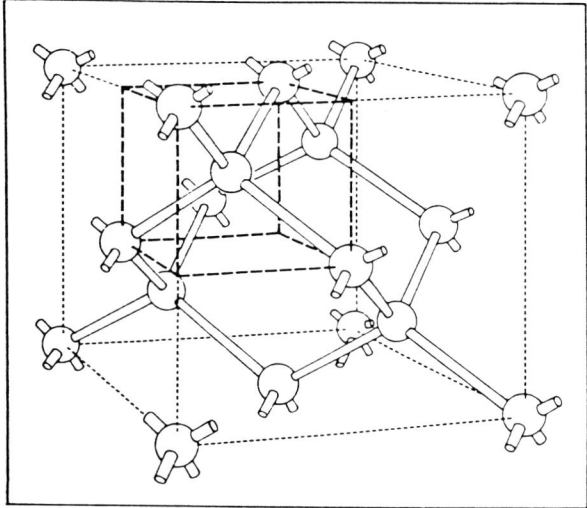

Figure 6.1 The three dimensional network crystal structure of covalently bonded atoms in silicon [2].

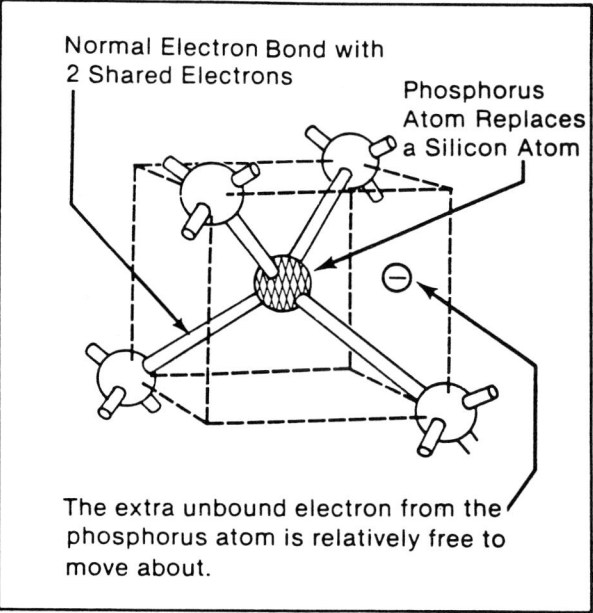

Figure 6.3 Crystal structure of covalently bonded silicon doped with phosphorus to make an n-type semiconductor [2].

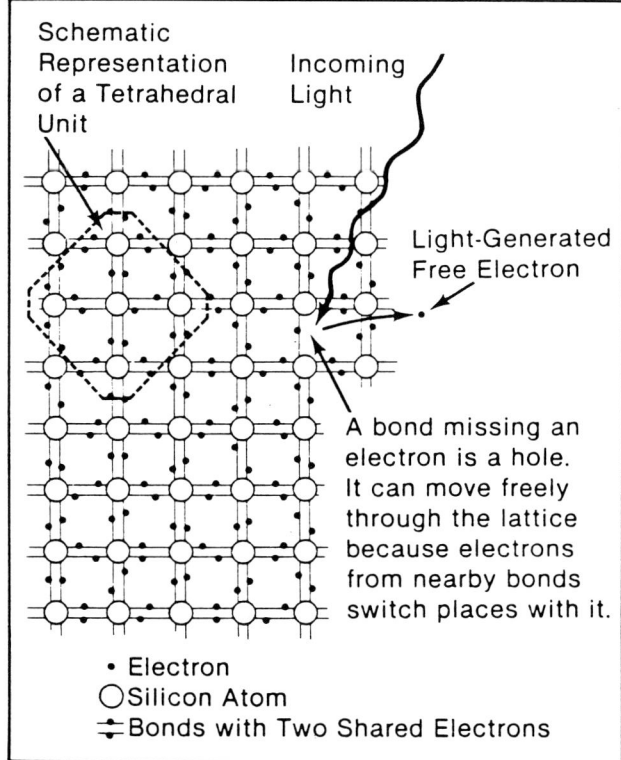

Figure 6.2 The process by which electron-hole pairs are created in silicon by incoming light [2].

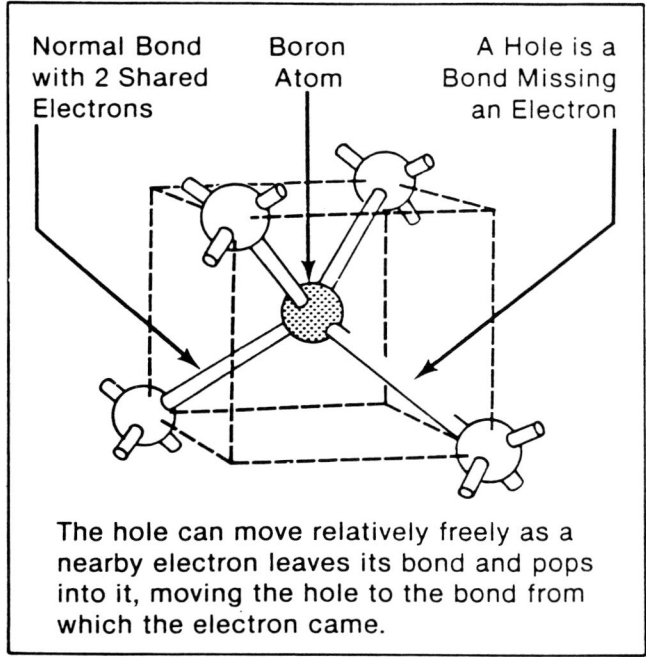

Figure 6.4 Crystal structure of covalently bonded silicon doped with boron to make a p-type semiconductor [2].

because the n-side has surrendered electrons to become positively charged and the p-side has accepted electrons to become negatively charged (Figure 6.5b). Equilibrium is eventually established as electrons from the n-side are now repelled at the junction by the negative charge of the p-side (Figures 6.5c and d).

Formation of this potential barrier is the key to solar cell operation. Light striking the p-side creates an electron-hole pair with the electron attracted to the junction and the hole repelled. The electron in fact crosses the junction to the n-side where it is free to travel, repelled only by the junction itself. Electron-hole pairs formed on the n-side by light

Spacecraft Power

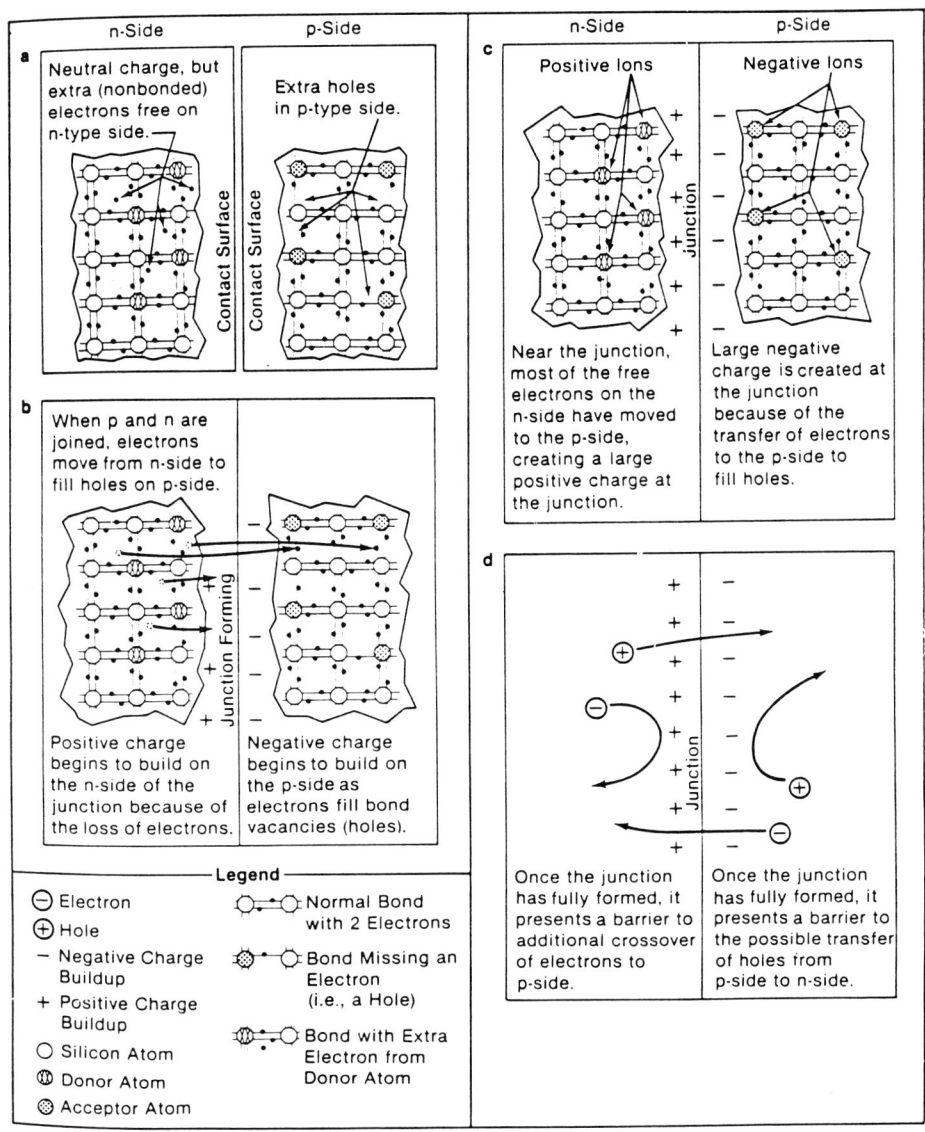

Figure 6.5 Mechanism for the formation of a potential barrier at a p-n junction [2].

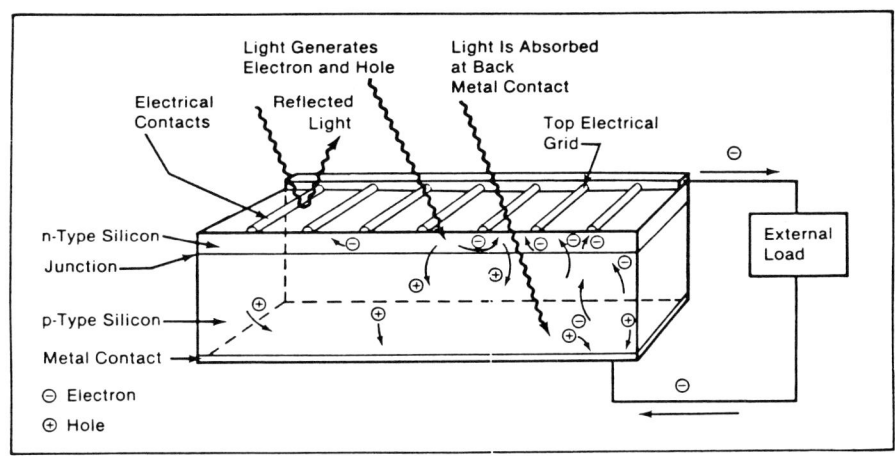

Figure 6.6 Light incident on the solar cell generates electron-hole pairs which are separated by the potential barrier, creating a potential that forces electrons through an external circuit [2].

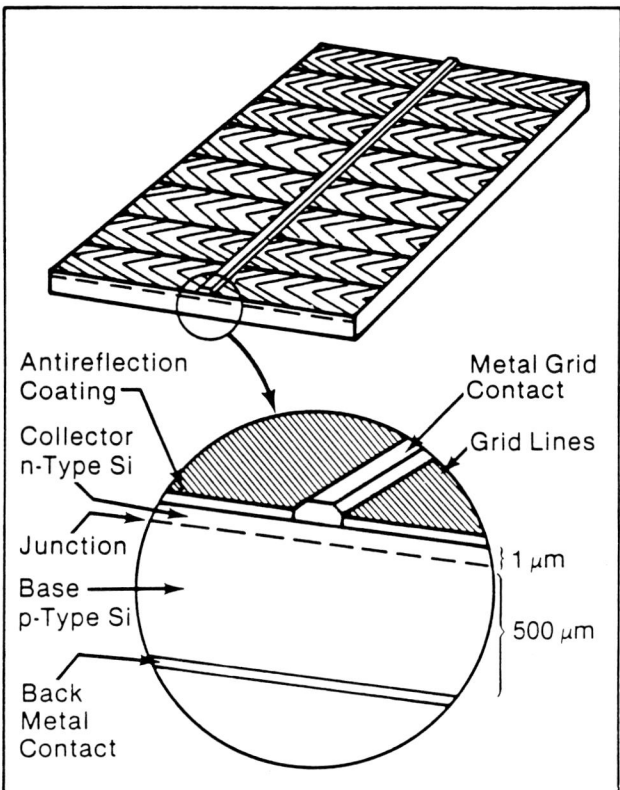

Figure 6.7 The components of a silicon solar cell [2].

undergo the same process, only this time the holes cross the junction to the p-side. Connection of an external circuit enables the electrons to continuously flow from the n-side to the p-side where they recombine with the holes (Figure 6.6).

Solar Cell Design

Solar cells are basically designed to maximize conversion of sunlight into electricity. Currently, silicon solar cells are about 15% efficient and gallium-arsenide cells have 17% efficiency [3]. Figure 6.7 shows a typical Si solar cell.

The optical interface between sunlight and the cell is crucial to cell performance since absorption losses in the wavelengths the cell is responsive to must be minimized. For example, untreated silicon reflects more than 30% of incident light. In order to avoid this unacceptable loss, antireflective coatings are applied to the cell's surface. A double layer of SiO reduces reflection to less than 3% [2] while even lower values can be obtained with tantalum pentoxide (Ta_2O_5) [3].

Current collection is accomplished by metal contacts on the front and back of the cell. The back contact is generally a metal sheet, like aluminum. The front contact is more crucial since it must not shade the cell from sunlight. A grid consisting of many fine fingers of titanium/silver is generally used as a means of minimizing both electrical resistivity and shading.

Cover slides made of fused SiO_2 glass are placed on the surface of the solar cell to protect against physical degradation caused by collisions with particles and debris. Reflection by the cover itself is also a concern so it too is coated with an antireflective layer, generally magnesium fluoride, MgF_2. Silicone adhesive bonds the cover to the cell [3].

As for the silicon cell itself, the amounts of dopants and the position of the p-n junction are quite important. The p-type base contains 10^{16} boron atoms per cm^3. Phosphorus is placed on the surface and heated until it diffuses to a depth of 1.2 μm. The phosphorus atom density varies from 10^{18} at this depth to 10^{21} at the cell's surface. This high concentration at the surface results in the formation of a Si-P alloy and the loss of electron donor behavior in phosphorus. Termed the "dead layer," the surface is highly resistive and is the area of much unwanted electron-hole recombination [2].

Initially, space solar cells were designed with a thin p-layer on top of the thick n-layer. However, this design suffered large losses in output power upon exposure to the charged particle environment of space. Changes led to the current thin-emitter n on p design, which provided higher radiation resistance. Thickness of the top n-type layer is important due to the characteristic absorption of sunlight by silicon. Over 50% of the incident light is absorbed in the first 3 μm of depth, yet almost 300 μm is necessary to totally absorb the light. Ideally, 50% of the light should be absorbed in each of the two layers. This necessitates designing the cell with a thin n-type layer and a thick p-type layer [2].

Solar Array Design

Solar cells for space application range in size from about $2 \times 2 \times 0.036$ cm up to $7.1 \times 7.1 \times 0.10$ cm [4]. These relatively small area cells do not provide nearly enough power by themselves, so they must be combined to form an array. Array sizes vary according to the spacecraft's power needs. Vanguard I used 108 Si cells which provided less than 1 W of total power. Contrast this with Skylab's approximately 225,000 cells 2×2 cm, 2×4 cm, and 2×6 cm supplying over 16 kW total power [5]!

Array shape is also dependent upon the mission and type of spacecraft. Curved arrays are used on spin stabilized satellites while three axis stabilized spacecraft have flat arrays. Cells are fixed to the supporting substrate through some type of organic adhesive or adhesive tape. The substrate itself can be either rigid or flexible. Figures 6.8 and 6.9 show the construction and designs of typical solar arrays. In Figure 6.8, the support is made of aluminum honeycomb covered with graphite/epoxy face skins. The polymer Kapton can be used in place of graphite/epoxy and

Spacecraft Power

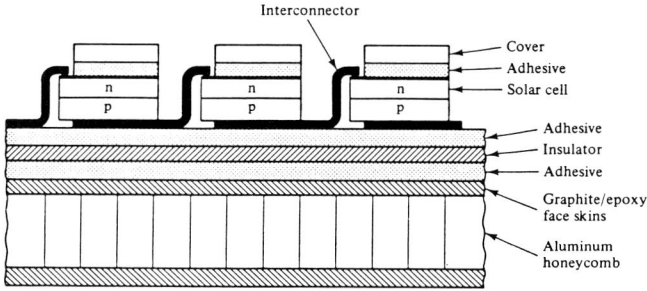

Figure 6.8 Cross section of a typical rigid solar array [3].

in some flexible arrays, a Kapton blanket forms the entire substrate. Electrical connections between cells are made of high conductivity silver.

Environmental Effects

The space environment is particularly harsh on the solar array due to its large size and constant exposure. Outgassing is, of course, a usual problem, but radiation and LEO atomic oxygen are far more damaging.

Like all semiconductor devices, solar cells can suffer permanent electrical degradation when exposed to high energy radiation. Electrons and protons trapped within the Earth's magnetic field do the most damage. Particulate radiation collides with atoms within the cell, dislodging electrons and forming ions. In the case of protons, Si nuclei are displaced from their lattice sites. These displacements restrict electron-hole movement and actually reduce their total number within the crystal. Localized heating from atomic vibration increases cell resistivity as well. In addition, color centers created in cover slides and the darkening of adhesives decrease transmitted incident light by as much as 10% in the first year of a GEO satellite. The combined effects in GEO for a 7-year mission cause solar cell performance to degrade by as much as 25%. For this reason, solar arrays are designed not based upon beginning-of-life requirements, but those at end-of-life [3,5]. Orbital debris is a growing concern which may necessitate changes in overall system design involving larger safety factors and different orientations for the solar arrays.

Atomic oxygen's deleterious effects do not act on the solar cell itself but upon the polymers used in array construction. Of particular interest to designers of Space Station Freedom is the degradation of Kapton, the polyimide film used for cell structural support and electrical insulation in flexible solar array blanket designs [6]. Recall from the previous chapter that Kapton is one of the polymers highly susceptible to chain scission by atomic oxygen. Recession of the material would cause decreases in electrical power output, structural load carrying capability, and array lifetime. Space station arrays have a 15 year design lifetime, equating to a total fluence of 5.5×10^{22} O-atoms/cm^2 [7]. In this environment, unprotected Kapton would virtually disintegrate, leaving little freestanding Kapton on the solar array blanket.

Research in this area has focused upon atomic oxygen

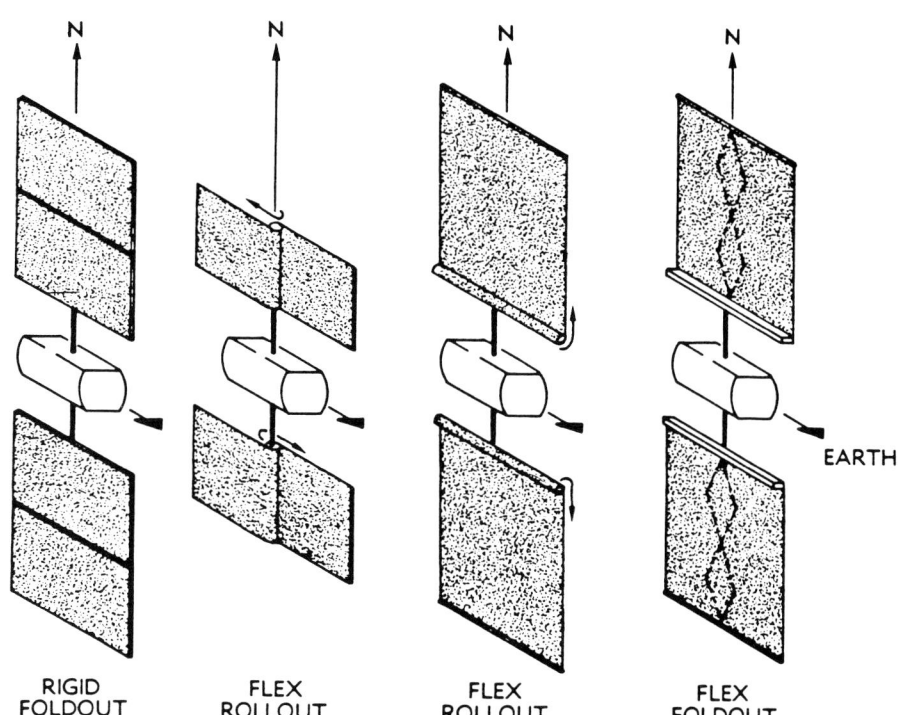

Figure 6.9 Four concepts for the deployment of a solar array on a three axis stabilized satellite [3].

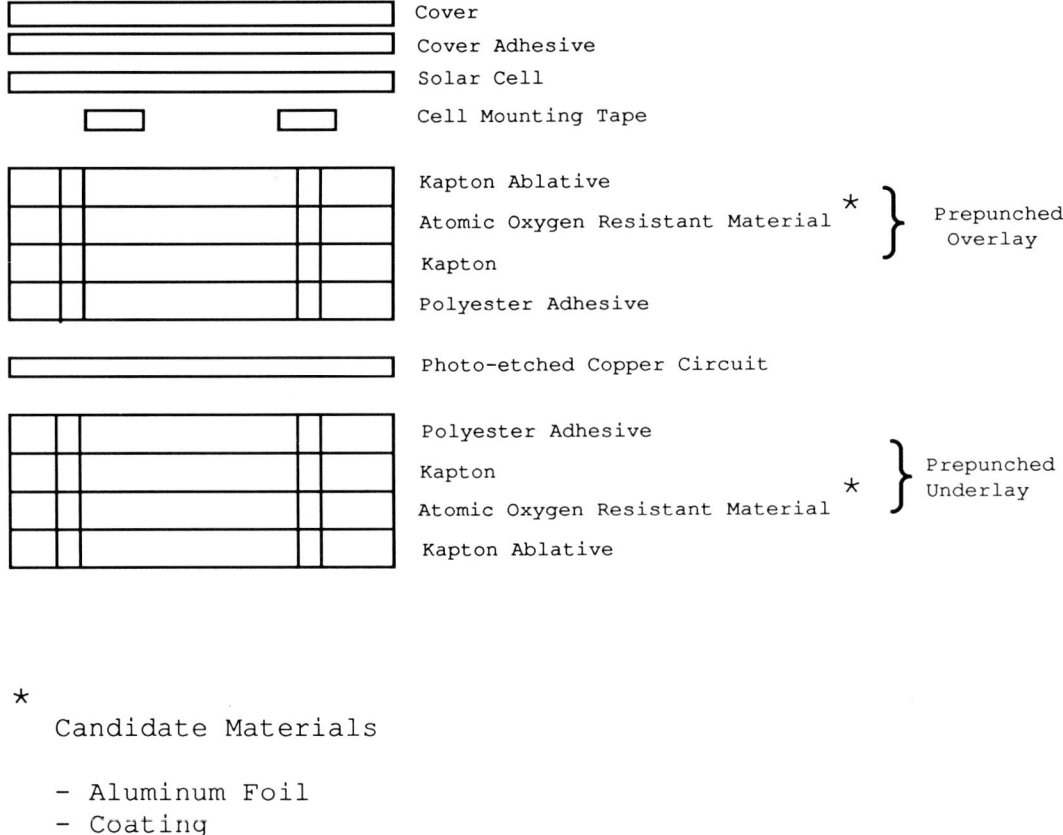

Figure 6.10 Schematic diagram of an atomic oxygen resistant solar array blanket for Space Station [4].

resistant protective layers and coatings. Figure 6.10 shows one particular design being studied by NASA and Lockheed. Aluminum is atomic oxygen resistant because of the presence of Al_2O_3 on its surface. A lighter weight alternative would be a coating. Theoretically, coatings in their highest oxidation state should provide the best protection against oxidation. For this reason, SiO_2 was selected for evaluation. Silicon is in its highest oxidation state of $+4$ and will resist any further loss of electrons. Results to date show it to be quite effective as the relative reactivity of SiO_2 coated Kapton varies from 0.03 to 0.06 compared to uncoated Kapton [8].

Batteries

Secondary batteries are used aboard spacecraft as energy storage devices and provide power during lift-off and eclipse periods. During sunlight periods, a portion of the solar array's current is shunted to recharge the batteries. The type of battery used is dependent upon a number of factors, including mission duration, profile, and altitude. Four secondary batteries used in space programs are silver-zinc, silver-cadmium, nickel-cadmium, and nickel-hydrogen. A fifth battery, high temperature sodium-sulfur, is currently under development by the Air Force with projected availability in the late 1990s. Before looking at these specific systems, a general discussion of battery operations is necessary.

Basic Concepts

Many of the basic concepts of battery operation were covered in the electrochemistry section of Chapter 1. Figure 6.11 is a schematic of the electrochemical operation of a discharging and charging cell. Remember that oxidation (electron loss) occurs at the anode and reduction (electron gain) takes place at the cathode. The driving force pushing the electrons through the load on discharge is the electromotive force (cell voltage). Cell voltage is a function of the particular electrode couple with theoretical values obtained from the half-cell reduction potentials given in Table 1.7. During cell charging, the chemical reactions are reversed by supplying power to the cell. Thus, the cell is returned to its original chemical composition and again able to supply power.

In addition to voltage, several other critical factors must be taken into account when considering batteries. These factors include capacity, energy, power, weight, useful life, and shelf life.

Capacity is the total amount of electricity available from

Spacecraft Power

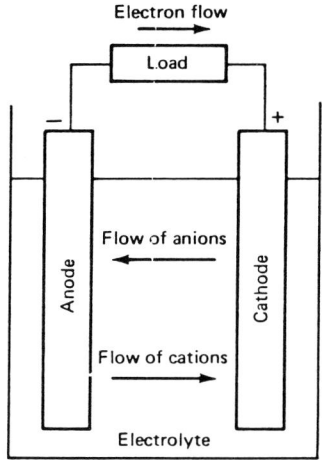

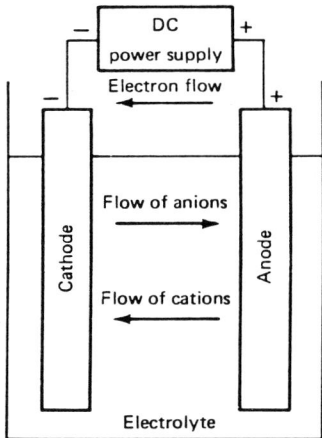

Figure 6.11 Electrochemical operation of a cell during charge and discharge [9].

a cell and can be expressed in units of coulombs (C) or ampere-hours (A·hr). A theoretical capacity can be calculated for any electrode couple on the basis of atomic/molecular weight and number of electrons transferred. The following shows the calculations for the theoretical capacity of a Ni-H$_2$ cell based strictly on cell reactants:

$$H_2 + 2NiOOH \rightarrow 2Ni(OH)_2 \quad (6.1)$$

Recall: 1 faraday(F) = 96,487 C/mole
= 26.8 A·hr/mole

For H$_2$: $\dfrac{1 \text{ mole H}_2}{2.01 \text{ g}} \cdot \dfrac{2 \text{ moles } e^-\text{'s}}{\text{mole H}_2} \cdot \dfrac{26.8 \text{ A·hr}}{\text{mole } e^-\text{'s}}$

= 26.6 A·hr/g H$_2$ or 0.037 g H$_2$/A·hr

For NiOOH: $\dfrac{1 \text{ mole NiOOH}}{91.7 \text{ g}} \cdot \dfrac{1 \text{ mole } e^-}{\text{mole NiOOH}} \cdot \dfrac{26.8 \text{ A·hr}}{\text{mole } e^-\text{'s}}$

= 0.292 A·hr/g NiOOH
or 3.42 g NiOOH/A·hr

Theoretical capacity = 0.037 + 3.42 = 3.46 g/A·hr or, reciprocally, 0.289 A·hr/g

Cell energy relies not only on capacity, but also upon the electromotive force. Multiplying theoretical capacity by theoretical voltage yields the cell's specific energy in W·hr/g.

For Ni-H$_2$: 0.289 A·hr/g × 1.5 V
$$= 0.43 \text{ W·hr/g}. \quad (6.2)$$

To find the cell's power output, simply multiply voltage by current:

$$P(\text{watts}) = E_{\text{cell}}(V)I(A) \quad (6.3)$$

The theoretical values calculated above are not realistic because they don't take into account the mass of electrolyte and cell hardware. Still, the theoretical values given in Table 6.1 are helpful to scientists and engineers in designing and comparing new systems.

Maximization of energy and power is important when related to the fourth critical factor, weight. Lightweight power sources are crucial, given the limitations on payload weight and the astronomical launch costs associated with placement into orbit. Therefore, engineers are concerned with a cell's or battery's specific energy and specific power which are defined as

$$\text{Specific Energy} = \text{energy/mass (W·hr/kg)} \quad (6.4)$$
$$\text{Specific Power} = \text{power/mass (W/kg)}. \quad (6.5)$$

Lighter weight batteries enable spacecraft designers to either increase the energy/power for a given weight amount or use the weight savings elsewhere. Since the batteries can comprise as much as 10% of a satellite's weight, a minor increase in energy density is significant.

The final critical factor in a cell and battery design is longevity in the forms of useful life and shelf life. Orbit altitude dictates the amount of use a battery will receive. For LEO, eclipse periods can occur as many as 16 times within a 24-hour period on a continual basis. GEO is less demanding because eclipse seasons are 45 days long and centered around the vernal and autumnal equinoxes. The spacecraft experiences one eclipse per day with a duration of 12–72 minutes [3]. In a five year period, a LEO battery will have been cycled 29,000 times compared to only 450 times for a GEO battery. This inequity in cycling accounts for the difference in calendar lifetimes between LEO and GEO batteries.

Batteries are often manufactured years in advance of a spacecraft's launch. Ideally, they should be able to supply the same amount of energy at launch as the day they were made. Unfortunately, this is not the case due to a number of spontaneous chemical reactions slowly occurring inside the cells which cause them to discharge. The classic example of the results of self-discharge is the inoperable flashlight kept in the trunk of the car for emergencies. Even

Table 6.1 Theoretical Voltages and Capacities of Major Battery Systems. *Reprinted with permission of McGraw-Hill* [9].

Battery system	Anode	Cathode	Reaction mechanism	V	Theoretical Capacity† g/Ah	Theoretical Capacity† Ah/kg
Primary						
Leclanche	Zn	MnO_2	$Zn + 2MnO_2 \rightarrow ZnO \cdot Mn_2O_3$	1.6	4.46	224
Magnesium	Mg	MnO_2	$Mg + 2MnO_2 + H_2O \rightarrow Mn_2O_3 + Mg(OH)_2$	2.8	3.69	271
Alkaline MnO_2	Zn	MnO_2	$Zn + 2MnO_2 \rightarrow ZnO + Mn_2O_3$	1.5	4.46	224
Mercury	Zn	HgO	$Zn + HgO \rightarrow ZnO + Hg$	1.34	5.27	190
Mercad	Cd	HgO	$Cd + HgO + H_2O \rightarrow Cd(OH)_2 + Hg$	0.91	6.15	163
Silver oxide	Zn	Ag_2O	$Zn + Ag_2O + H_2O \rightarrow Zn(OH)_2 + 2Ag$	1.6	5.55	180
Zinc/air	Zn	O_2 (air)	$Zn + \frac{1}{2}O_2 \rightarrow ZnO$	1.65	1.55	800
Li/SO_2	Li	SO_2	$2Li + 2SO_2 \rightarrow Li_2S_2O_4$	3.1	2.64	379
Li/MnO_2	Li	MnO_2	$Li + Mn^{IV}O_2 \rightarrow Mn^{III}O_2(Li^+)$	3.5	3.50	286
Reserve						
Cuprous chloride	Mg	CuCl	$Mg + 2CuCl \rightarrow MgCl_2 + 2Cu$	1.6	4.14	241
Zinc/silver oxide	Zn	AgO	$Zn + AgO + H_2O \rightarrow Zn(OH)_2 + Ag$	1.81	3.53	283
Secondary						
Lead-acid	Pb	PbO_2	$Pb + PbO_2 + 2H_2SO_4 \rightarrow 2PbSO_4 + 2H_2O$	2.1	8.32	120
Edison	Fe	Ni oxide	$Fe + 2NiOOH + 2H_2O \rightarrow 2Ni(OH)_2 + Fe(OH)_2$	1.4	4.46	224
Nickel-cadmium	Cd	Ni oxide	$Cd + 2NiOOH + 2H_2O \rightarrow 2Ni(OH)_2 + Cd(OH)_2$	1.35	5.52	181
Silver-zinc	Zn	AgO	$Zn + AgO + H_2O \rightarrow Zn(OH)_2 + Ag$	1.85	3.53	283
Nickel-zinc	Zn	Ni oxide	$Zn + 2NiOOH + 2H_2O \rightarrow 2Ni(OH)_2 + Zn(OH)_2$	1.73	4.64	215
Nickel-hydrogen	H_2	Ni oxide	$H_2 + 2NiOOH \rightarrow 2Ni(OH)_2$	1.5	3.46	289
Silver-cadmium	Cd	AgO	$Cd + AgO + H_2O \rightarrow Cd(OH)_2 + Ag$	1.4	4.41	227
Zinc/chlorine	Zn	Cl_2	$Zn + Cl_2 \rightarrow ZnCl_2$	2.12	2.54	394
High temperature	Li(Al)	FeS	$2Li(Al) + FeS \rightarrow Li_2S + Fe + 2Al$	1.33	2.99	345
High temperature	Na	S	$2Na + 3S \rightarrow Na_2S_3$	2.1	2.65	377
Fuel cell						
H_2/O_2	H_2	O_2 (or air)	$H_2 + \frac{1}{2}O_2 \rightarrow H_2O$	1.23	0.336	2975

†Based on active anode and cathode materials only.

though the batteries in the flashlight were never used, they still lost their energy through these unstoppable chemical reactions. Batteries able to sustain their capacity for long periods of time are said to have an excellent shelf life. For spacecraft cells and batteries, shelf life is indefinite as long as the liquid electrolyte has not been added. Once electrolyte is added, however, shelf life can be decreased to less then one year for some systems. Precautions such as refrigerating the batteries, thereby slowing down self-discharge, can help to extend shelf life.

Battery Systems

Silver-Zinc and Silver-Cadmium Batteries

The silver-zinc and silver-cadmium batteries have a long record of space service dating back to the Mercury program. They have since flown on such missions as Mercury, Gemini, Apollo, Pioneer, Mariner, Ranger, Surveyor, and Viking. The cell consists of a Zn or Cd negative electrode, a silver (II) oxide (AgO) positive electrode, and 7.5 M po-

Spacecraft Power

tassium hydroxide (KOH) aqueous electrolyte. Reactions occurring during cycling are as follows:

Positive Electrode:
$$2AgO + H_2O + 2e^- \underset{\text{Charge}}{\overset{\text{Discharge}}{\rightleftarrows}} Ag_2O + 2OH^-$$
$$Ag_2O + H_2O + 2e^- \underset{\text{Charge}}{\overset{\text{Discharge}}{\rightleftarrows}} 2Ag + 2OH^-$$

Negative Electrode:
$$Cd/Zn + 2OH^- \underset{\text{Charge}}{\overset{\text{Discharge}}{\rightarrow}} Cd(OH)_2/Zn(OH)_2 + 2e^-$$

Overall:
$$AgO + H_2O + Cd/Zn \underset{\text{Charge}}{\overset{\text{Discharge}}{\rightleftarrows}} Ag + Cd(OH)_2/Zn(OH)_2$$

(6.6)

Silver batteries are unique in that they have a dual plateau discharge curve due to the multiple oxidation states of silver ($+2 \rightarrow +1 \rightarrow 0$). Figure 6.12 shows that the voltage of Ag-Zn remains at about 1.8 V for the first 30% of discharge, which corresponds to the reduction of AgO to Ag_2O. The remaining 70% of discharge at 1.5 V is the reduction of Ag_2O to Ag along with the preceding reaction.

Figure 6.12 also shows two general characteristics about battery operation. First, there is the decrease in voltage from the initial, equilibrium value as a load is applied. This decrease is known as an overpotential and is caused by three factors:

1. Ohmic resistance in the electrolyte, electrodes, leads, and connections
2. Kinetics of ion exchange occurring at the electrode-electrolyte interfaces
3. Reacting ion depletion at the electrode surface

The second is the decrease in percent-of-nominal capacity as the discharge rate increases. This too is caused by these same three factors and is representative of greater cell inefficiency at higher current rates of discharge.

In addition to the electrodes and electrolyte, silver cells, as do all spacecraft cells, contain a separator that prevents physical contact between the electrodes and keeps the electrolyte in place. Actually, a silver-zinc cell has three different separators as shown in Figure 6.13. Directly adjacent to the silver electrode is a nylon or polypropylene separator used to protect the main separator from oxidation by the highly chemically reactive electrode material. A similar separator is also placed next to the zinc electrode to keep its active material from migrating toward the silver electrode. The main separator is made of cellophane and absorbs most of the electrolyte. It also provides the main barrier to migration of particles and ions from both electrodes. Active material migration is the life-limiting factor in Ag-Zn systems, generally causing failure in less than 150 cycles. For this reason, these batteries are restricted to short-life missions [9].

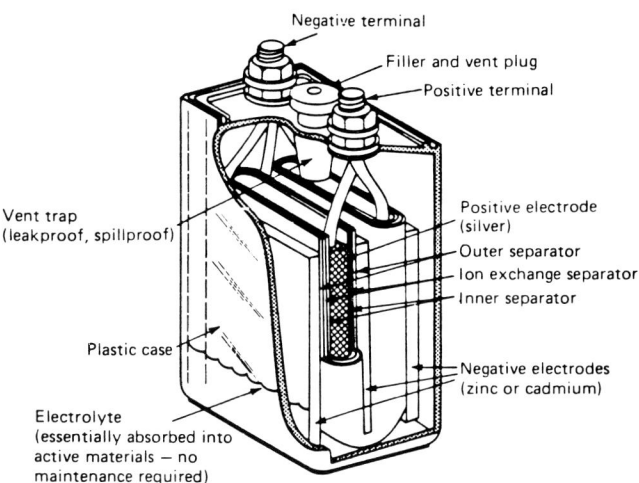

Figure 6.13 Cutaway of a silver-zinc (or silver-cadmium) secondary cell [9].

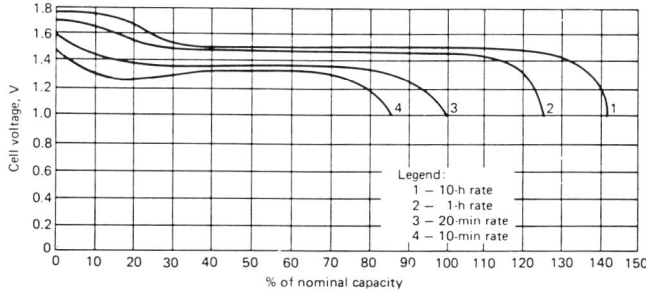

Figure 6.12 Typical discharge curves of silver-zinc cells for various rates at 20°C [9].

Even with a limited cycle life, Ag-Zn continues to be used because of its high specific energy, 99 W·hr/kg. The ability to handle high discharge rates is another one of its advantages. Spacecraft cells are hermetically sealed to prevent electrolyte evaporation, therefore the generation of hydrogen and oxygen gas during cycling is always a problem. So too is the dissolution of AgO during stand time. Lengthy periods of idleness are therefore not recommended. The energy density of silver-cadmium is about 60% that of silver-zinc, but is attractive due to its one to two orders of magnitude longer life. It is also the only system made from completely nonmagnetic materials [9].

Nickel-Cadmium Batteries

The workhorse of spacecraft energy storage systems is the nickel-cadmium battery. Its early inception into the space program and continued dominance are attributed to its ruggedness, dependability, and long cycle life. The system consists of a cadmium impregnated, sintered nickel negative electrode and a NiOOH impregnated, sintered nickel positive electrode. The electrolyte, 6.9 M KOH (*aq*), is completely absorbed into the electrodes and the nylon separator to prevent its unwanted redistribution in zero gravity. Cell reactions during cycling are as follows:

Negative Electrode:
$$Cd(s) + 3OH^-(aq) \underset{\text{Charge}}{\overset{\text{Discharge}}{\rightleftharpoons}} Cd(OH)_3^-(aq) + 2e^-$$
$$\rightleftharpoons Cd(OH)_2(s) + OH^-(aq)$$

Positive Electrode:
$$NiOOH + H_2O + e^- \underset{\text{Charge}}{\overset{\text{Discharge}}{\rightleftharpoons}} Ni(OH)_2 + OH^-$$

Overall:
$$Cd + 2NiOOH + 2H_2O \underset{\text{Charge}}{\overset{\text{Discharge}}{\rightleftharpoons}} Cd(OH)_2 + 2Ni(OH)_2$$
$$E^o = 1.30 \text{ V} \qquad (6.7)$$

Notice that during cycling, the cadmium electrode passes through a soluble intermediate, $Cd(OH)_3^-$, while nickel's conversion is a direct, solid-state mechanism [10].

A characteristic of Ni-Cd batteries is their deliverance of a fairly constant voltage over their entire discharge. This phenomenon can be explained using equation 1.55; namely,

$$E_{cell} = E^o{}_{cell} - \frac{RT}{nF} \ln Q$$

where
- E = cell potential
- E^o = standard cell potential
- F = Faraday's constant (96,487 coulombs/mole)
- n = number of moles of electrons transferred
- Q = reaction quotient.
- R = universal gas constant (8.314 J/mole · K)
- T = temperature (K)

At 25°C this equation simplifies to

$$E_{cell} = E^o{}_{cell} - \frac{0.0257}{2} \ln Q . \qquad (6.8)$$

Recall that Q is the reaction quotient and is expressed as the ratio of the concentrations of products/reactants. For the Ni-Cd cell, Q is

$$Q = \frac{[Cd(OH)_2][Ni(OH)]^2}{[Cd][NiOOH][H_2O]^2} . \qquad (6.9)$$

However, the expression is simplified by treating the concentrations of all pure solids and liquids as constant and replacing them with unity in the expression to yield a value of Q equal to one. Placing this back into equation 6.8 shows why the equilibrium voltage remains constant:

$$E_{cell} = E^o{}_{cell} - \frac{0.0257}{2} \ln 1.00 \qquad (6.10)$$
$$E_{cell} = E^o{}_{cell} .$$

In addition to normal charging and discharging, Ni-Cd cells must be able to tolerate overcharge and overdischarge. Overcharge and overdischarge occur in series connected cells when their individual capacities are not exactly the same. Cells reaching complete charge or discharge before others in the string produce gas from the breakdown of the electrolyte. The operation of a sealed cell is based upon the design being nickel limited, meaning the nickel electrode becomes fully charged before the cadmium electrode does. On overcharge, O_2 formed at the Ni electrode passes through the permeable separator to the Cd electrode where it recombines to form OH^- as shown in Figure 6.14. Separator permeability is key to allowing oxygen to migrate quickly enough to prevent internal pressure increases which can cause the battery to explode, as was the case for a U.S. weather satellite in LEO [11].

Nickel-cadmium cells can tolerate overcharge as long as the rate of recombination is as fast as the rate of generation. Such is not the case on overdischarge, where forced current through a discharged cell sends it into reversal. Both H_2 and O_2 are generated at the Ni and Cd electrodes respectively according to the reactions

$$2H_2O + 2e^- \rightarrow H_2 + 2OH^- \qquad (6.11)$$

and

$$4OH^- \rightarrow O_2 + 2H_2O + 4e^- .$$

Oxygen will recombine in the following charging process, but the H_2 will remain. The increased pressure can significantly degrade cell performance and the mere presence of H_2 presents the possibility of explosion and fire [11]. Proper matching of cell capacities and avoidance of complete discharge are really the only precautions available to prevent cell reversal.

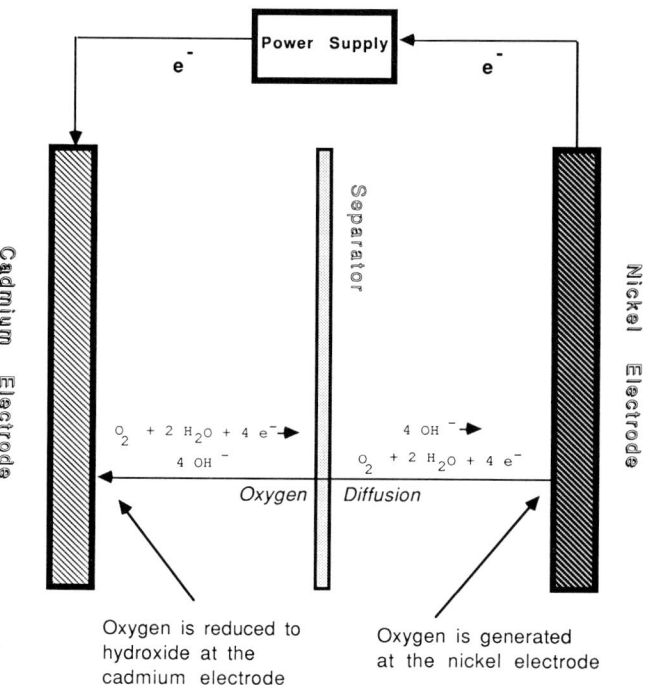

Figure 6.14 Oxygen generation and recombination in a nickel-cadmium cell during overcharge.

Spacecraft Power

Even with the above drawbacks, Ni-Cd cells are still quite durable. Their long cycle life capability is the main reason they are still used. Batteries operated in LEO experience anywhere from 20,000 to 35,000 charge/discharge cycles while those in GEO are subjected to about 1,000 cycles. The depth-of-discharge (DOD) or the amount of capacity removed on a single discharge has a great deal to do with a battery's life. A DOD of 25–35% is permissible in LEO while 50–60% is generally used in GEO (specific energy averages about 28 W·hr/kg) [12].

Still, the battery's performance does degrade with time and cycles for a number of reasons. Evaporation of water through the safety vents of overpressurized cells causes higher internal resistances. Figure 6.15 shows how the soluble intermediate species of the cadmium electrode $(Cd(OH)_3^-)$ is transported by the hydrodynamic force created during the molar volume change of Cd to $Cd(OH)_2$ conversion. Active material precipitates in the separator and eventually bridges the two electrodes to create an internal short circuit [13]. The Ni electrode itself breaks apart from internal stress caused by volume change. Repeated cycling to the same DOD leads to voltage drop and capacity loss in what is termed "memory effect." In some cases, like the memory effect, cell performance can be restored through reconditioning. Essentially, the cell is completely discharged and then slowly recharged. This reforms the Cd and NiOOH crystals in a smaller size with better electrical contact to the current collector. Repeated reconditioning proves to be less effective with increased battery age until a point of failure is reached.

Nickel-Hydrogen Batteries

Future spacecraft under development will require power levels far beyond those presently supplied. Batteries currently comprise about 10% of a spacecraft's weight, but the percentage would grow to 40–50% with Ni-Cd cells. Research and development of the lighter Ni-H_2 battery was therefore begun in the early 1970s. With an energy density of 45 W·hr/kg, Ni-H_2 uses the same positive electrode as Ni-Cd, but has a nickel/platinum catalyst for the reaction of gaseous H_2 as a negative electrode. The circular electrodes are stacked along with an asbestos or zirconia separator and sealed inside a cylindrical nickel alloy pressure vessel. Potassium hydroxide electrolyte is completely absorbed by the cell stack while H_2, at a pressure of 30 to 40 atm, fills the entire vessel. Figures 6.16 and 6.17 show typical nickel-hydrogen cells.

The net reaction for Ni-H_2,

$$1/2 H_2 + NiOOH \underset{\text{Charge}}{\overset{\text{Discharge}}{\rightleftharpoons}} Ni(OH)_2 \quad (6.12)$$

shows no change in KOH or H_2O amounts, but does indicate the consumption of H_2. State-of-charge is difficult to determine in Ni-Cd, but can be easily determined in this system simply from the internal pressure. During overcharge or overdischarge, the following reactions occur:

Overcharge:
$$\text{Ni: } 2OH^- \rightarrow 2e^- + 1/2 O_2 + H_2O \quad (6.13)$$
$$\text{H: } 2H_2O + 2e^- \rightarrow 2OH^- + H_2 \quad (6.14)$$
$$\text{Recombination of } O_2: O_2 + 1/2 H_2 \rightarrow H_2O \quad (6.15)$$

Overdischarge (Reversal):
$$\text{Ni: } H_2O + e^- \rightarrow OH^- + 1/2 H_2 \quad (6.16)$$
$$\text{H: } 1/2 H_2 + OH^- \rightarrow H_2O + e^- \quad (6.17)$$

In both cases, gases formed are recombined to maintain the constant concentration of the electrolyte. The overdischarge protection is a unique characteristic of this system.

Thus far Ni-H_2 has been limited to use in GEO satellites (Figure 6.18). Besides having a higher energy density than Ni-Cd, it also has a longer cycle life because the troublesome cadmium electrode is absent. Cell sizes vary from the 30 A·hr ones used in Intelsat V up to the 81 A·hr cells proposed for Space Station Freedom.

Sodium-Sulfur Batteries

In keeping with its philosophy of always looking to the future, the aerospace community is developing its next generation spacecraft battery, high temperature sodium-sulfur. The driving impetus is again weight reduction, and at 110–220 W·hr/kg, Na-S certainly is attractive (Figure 6.19).

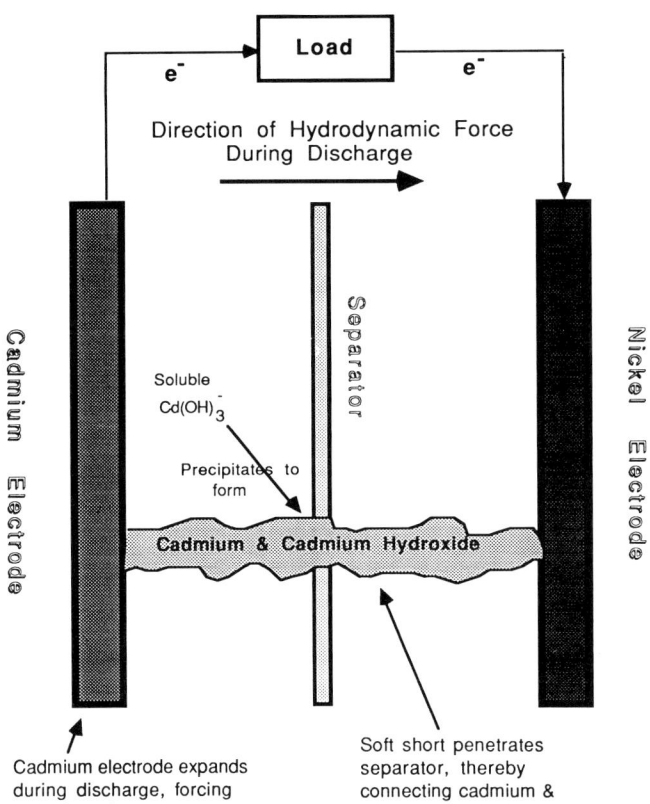

Figure 6.15 Mechanism for the formation of a soft short between electrodes in a nickel-cadmium cell.

NICKEL-HYDROGEN CELL

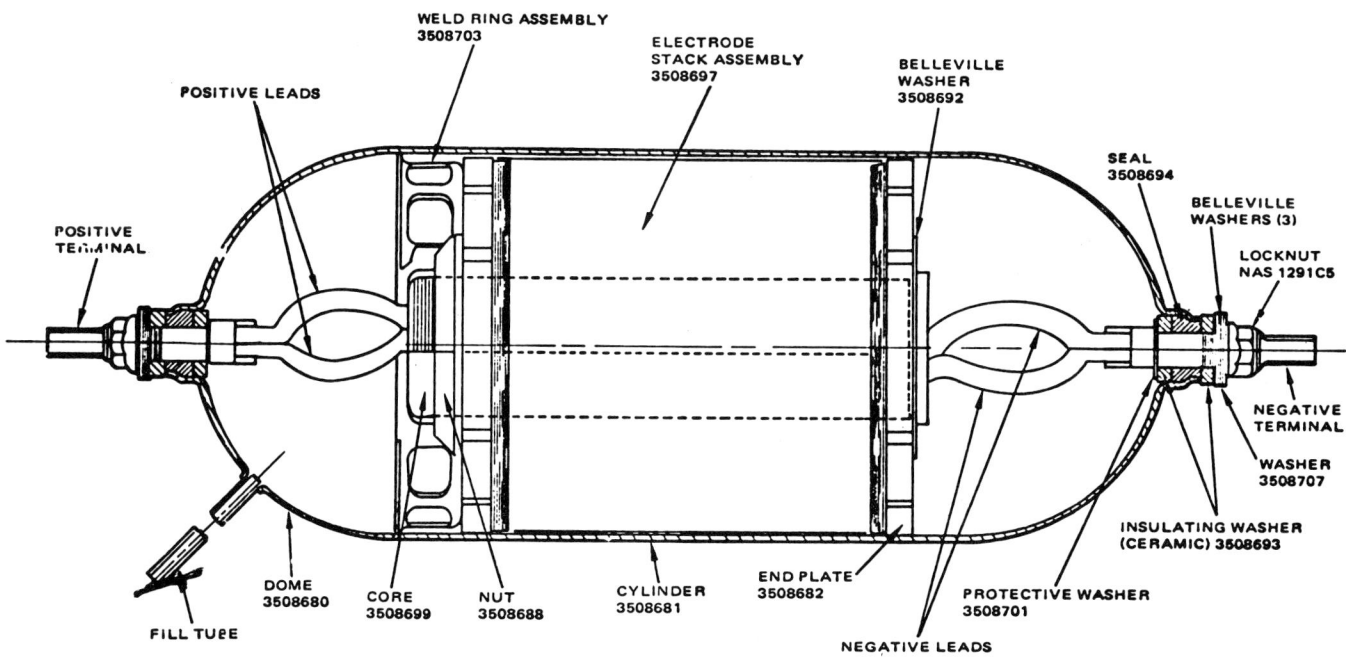

Figure 6.16 Schematic of nickel-hydrogen cell. *Courtesy of Hughes Aircraft Company.*

Figure 6.17 A disassembled nickel-hydrogen cell showing the internal components.

Sodium-sulfur has shown great promise since its invention in 1966 by Ford Motor Company. The Department of Energy has been developing it for electric vehicles and utility load leveling applications. Since 1983, the Air Force has been interested in Na-S for spacecraft applications due to its high energy density and potential for long life. Current projections show Na-S being operational by the late 1990s.

Sodium-sulfur represents a significant departure from traditional batteries with its liquid phase electrodes (at 350°C) and solid phase electrolyte. Figure 6.20 depicts the cylindrical design being developed. The anode is liquid sodium metal contained within a stainless steel tube equipped with a porous metering plug. Surrounding the sodium is the sodium ion conducting ceramic electrolyte, β''-alumina (Na_2O, Al_2O_3). This ceramic is impermeable to solid and liquid phase transport, but allows sodium ions to migrate through its ionic crystal lattice. Outside the electrolyte is the cathode, consisting of graphite felt impregnated with sulfur. The graphite serves as a current collector to transport electrons to and from the chromium coated stainless steel case. Individual half-cell and total cell reactions are

$$\text{Negative electrode: } 2Na \underset{\text{Charge}}{\overset{\text{Discharge}}{\rightleftharpoons}} 2Na^+ + 2e^-$$

$$\text{Positive electrode: } 2Na^+ + xS + 2e^- \underset{\text{Charge}}{\overset{\text{Discharge}}{\rightleftharpoons}} Na_2S_x$$

$$\text{Overall: } 2Na + xS \underset{\text{Charge}}{\overset{\text{Discharge}}{\rightleftharpoons}} Na_2S_x (2 \leq x \leq 5)$$

(6.18)

During discharge, Na is oxidized at the anode-electrolyte interface and the resulting sodium ions migrate through the β''-alumina. At the cathode they initially combine with sulfur to form Na_2S_5. For the first 58% of discharge the cathode is a two phase mixture of immiscible sulfur and Na_2S_5. Past this point, all of the pure sulfur is depleted and sodium

Spacecraft Power

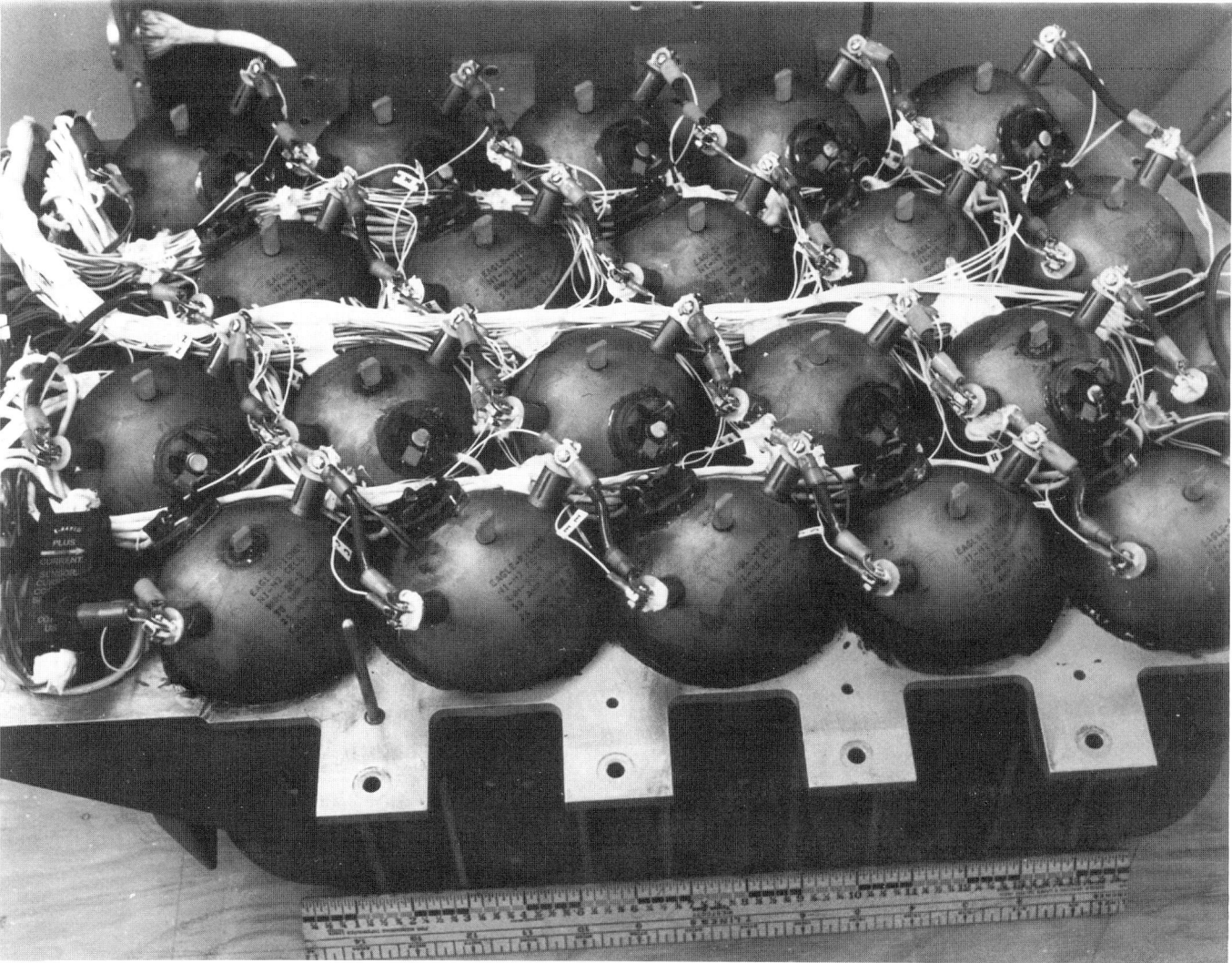

Figure 6.18 Ni-H$_2$ battery [9].

reacts further with the Na$_2$S$_5$ to subsequently form Na$_2$S$_4$, Na$_2$S$_3$, and Na$_2$S$_2$. End of discharge is defined as the cathode being pure Na$_2$S$_3$. Equilibrium voltage at various sodium to sulfur ratios is shown in Figure 6.21.

At first the cell maintains a constant voltage over the two phase region ($L_1 + L_2$). Once all of the sulfur is consumed, the voltage declines throughout the single phase region (L_2) where the polysulfides are completely miscible. A second two phase region begins once solid Na$_2$S$_2$ forms and the voltage again levels off. Formation of a solid can cause structural damage to the cell so overdischarge beyond Na$_2$S$_3$ should be avoided [9].

A high temperature of 300–350°C is necessary for this system to properly operate. Both sodium and sulfur must be in their liquid state and electrolyte able to conduct sodium ions at a sufficient rate. Keeping the battery hot is not a problem however, when one looks at the enthalpy change for the reaction on discharge [14].

$$2Na + 2S \rightarrow Na_2S_2$$
$$\Delta H = -412.640 \frac{kJ}{mole} \text{ at } 600 \text{ K} \quad (6.19)$$

In fact, once the battery is placed within a thermal container, heat must be ejected to prevent degradation of the aluminum components of the cell. Direct mixing of Na and

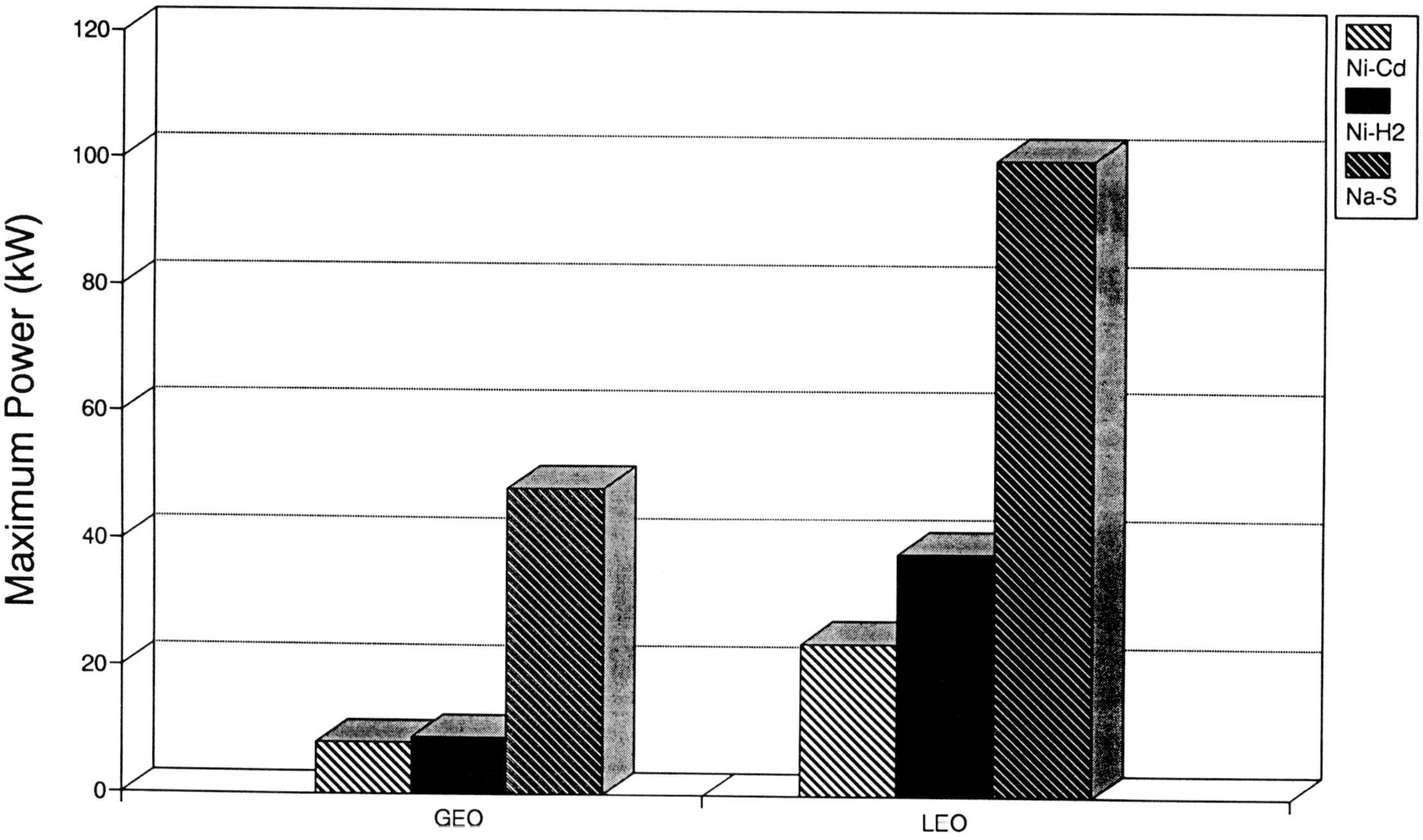

Figure 6.19 Comparison of satellite power system outputs for three different orbits using nickel-cadmium, nickel-hydrogen, and sodium-sulfur batteries.

S caused by complete failure of the ceramic or seals is not likely, but could be catastrophic if it occurred.

Fuel Cells

Similar to batteries in their mode of operation is the class of power systems known as fuel cells. A fuel cell can be thought of as a primary battery with its reactants stored externally. The traditional spacecraft fuel/oxidizer mixture of H_2-O_2 is not combusted, but rather electrochemically reacted to produce low voltage, direct-current electrical energy. The major advantage of doing the reaction electrochemically and not chemically is nonapplicability of the Carnot efficiency limitation. Fuel cells have been supplying power to all manned space missions beginning with Gemini. It was an explosion of the fuel cell oxygen tank that caused near catastrophe aboard Apollo 13.

The basic design of a fuel cell, as shown in Figure 6.22, is quite simple. Gaseous fuel undergoes oxidation at the anode to produce ions and electrons. The ions are transported through the electrolyte and rejoin the electrons at the cathode to reduce the oxidizer. Oxidation products continuously flow out of the cell in order to allow further reactions.

Fuel/oxidizer combinations vary depending upon the application, but for space missions only hydrogen-oxygen is used. The overall reaction itself is the combination of H_2 and O_2 to produce H_2O. However, the actual reaction mechanism is different for acidic and basic electrolytes. In acidic electrolyte, the hydronium ion (H_3O^+) transports current, while in alkaline electrolyte the hydroxide ion (OH^-) performs this function. Table 6.2 shows the reac-

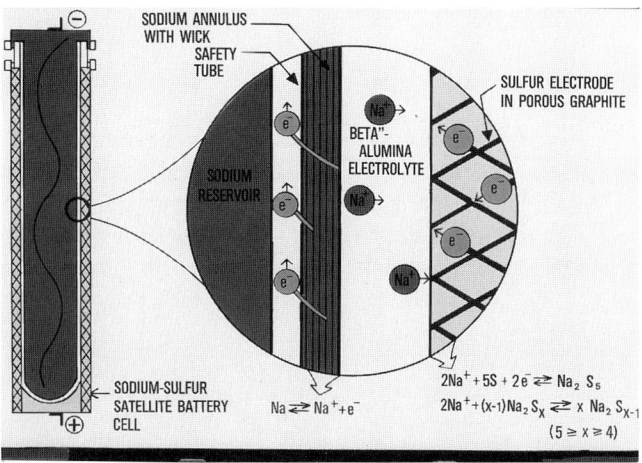

Figure 6.20 Cutaway of a sodium-sulfur cell showing the concentric cylindrical design [9].

Spacecraft Power

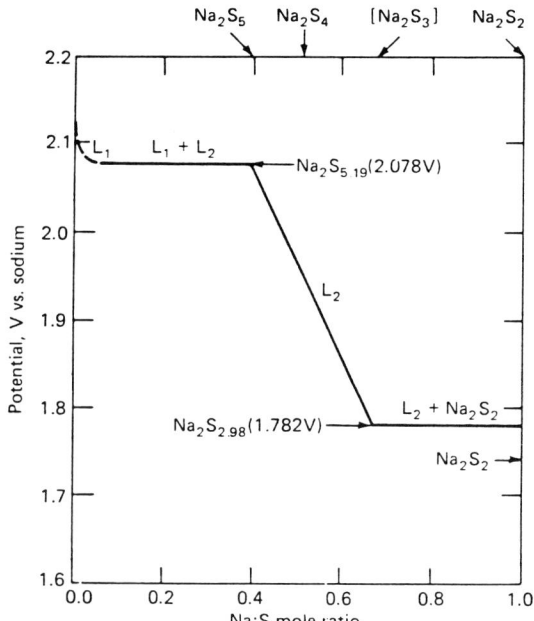

Figure 6.21 Equilibrium potential versus mole ratio of sodium to sulfur for a sodium-sulfur cell [9].

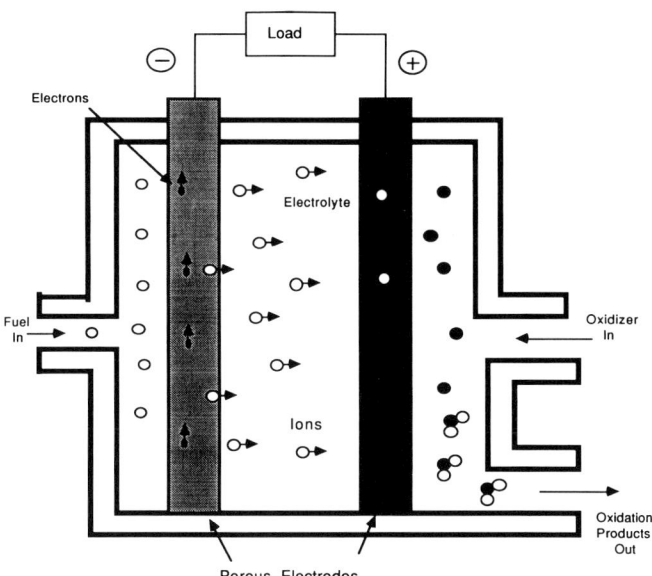

Figure 6.22 Schematic showing the basic operation of a simple fuel cell [15].

tions occurring in each type of fuel cell. Note the difference in location of water production in each design. The theoretical output voltage for a single cell of either design is 1.23 V [9].

Operation of a fuel cell is not, however, as simple as the chemistry may suggest. A number of different parameters including catalyst, temperature, and pressure ultimately determine the cell's output. A catalyst, as defined in Chapter 1, is a nonconsumable substance which participates in a reaction. Its function is to provide a low activation energy pathway for the primary reaction. Platinum is the preferred catalyst used for both the anode and cathode because of its ability to adsorb these gases. Once this chemisorption process takes place, the bonds of H_2 and O_2 are broken and the individual atoms reside on the catalyst surface available for further reaction. Cleanliness of the platinum itself is critical to the operation. Any poisons (e.g., lead) adsorbed onto the surface will retard the chemisorption process and degrade the overall performance of the cell.

Temperature and pressure directly affect cell efficiency through their input to cell voltage. The Nernst equation (equation 1.55) can be expressed in terms of pressure:

$$E_{cell} = E^o_{cell} - \frac{RT}{nF} \ln \left(\frac{\Pi[P_{products}]^x}{\Pi[P_{reactants}]^y} \right) \quad (6.20)$$

$$E_{cell} = 1.229 \text{ V} - \frac{(8.314)(T)}{(2)(96500)} \ln \left(\frac{P_{H_2O}}{P_{H_2} \cdot P_{O_2}^{1/2}} \right) \quad (6.21)$$

For example, if the partial pressures of H_2, O_2, and $H_2O(g)$ are 10 atm, 10 atm, and 1 atm respectively and the temperature is 25°C, E_{cell} will be:

$$E_{cell} = 1.229 - \frac{0.0257}{2} \ln \frac{1}{10\sqrt{10}}$$

$$E_{cell} = 1.273 \text{ V}$$

By varying the fuel cell's temperature and pressure, cell voltage can theoretically be adjusted to the desired value. In practice, however, other considerations must be taken into account. There are limitations on the stability of the materials themselves in high pressure, corrosive environments. A pressure differential between H_2 and O_2 must also be maintained to force product water out of the cell and prevent flooding [15].

Departures from calculated, theoretical voltages do occur as the result of inefficiencies within the cell. The ideal

Table 6.2 Reaction Mechanisms for the Two Types of Hydrogen-Oxygen Fuel Cells

Acid Electrolyte	Alkaline Electrolyte
Anode: $H_2 + 2H_2O \rightarrow 2H_3O^+ + 2e^-$	$H_2 + 2OH^- \rightarrow 2H_2O + 2e^-$
Cathode: $1/2O_2 + 2H_3O^+ + 2e^- \rightarrow 3H_2O$	$1/2O_2 + 2e^- + H_2O \rightarrow 2OH^-$
Overall: $H_2(g) + 1/2O_2(g) \rightarrow H_2O(l)$	$H_2(g) + 1/2O_2(g) \rightarrow H_2O(l)$

Table 6.3 Thermodynamic Values for the H_2-O_2 Fuel Cell at Various Temperatures [15]

T (K)	$\Delta H°$ (kcal/mole)	$\Delta G°$ (kcal/mole)	η_i ($\Delta G°/\Delta H°$)
298	−57.80	−54.64	0.94
400	−58.04	−53.52	0.92
500	−58.27	−52.36	0.90
1000	−59.21	−46.03	0.78
2000	−60.26	−32.31	0.54

efficiency for a cell is the change in Gibbs's free energy divided by the enthalpy of reaction:

$$\eta_{ideal} = \frac{\Delta G}{\Delta H} \quad (6.22)$$

The departure from 100% ideal efficiency is due to the $T\Delta S$ term in the Gibbs-Helmholtz equation. Thermodynamic properties and cell efficiency at various temperatures are given in Table 6.3.

Actual efficiency (η_{AC}) is obtained by substituting $\Delta G°$ with $-nFE_{AC}$:

$$\eta_{AC} = -nFE_{AC}/\Delta H \quad (6.23)$$

Similar to η_{AC} is the cell voltage efficiency (η_v) given by

$$\eta_v = V_{AC}/V \quad (6.24)$$

where V is the theoretical cell voltage.

Still another efficiency used to determine the fraction of electrochemical energy being utilized is the faradaic or current efficiency

$$\eta_F = \frac{I}{nFN_{fu}} \quad (6.25)$$

where

N_{fu} = total number of moles reacted electrochemically per second

I = current.

The causes of inefficiency within fuel cells are exactly the same ones observed in batteries. Lumped together under polarization effects, the following three phenomena account for cell voltage depression (Figure 6.23).

1. Activation or chemical polarization
2. Concentration polarization
3. Resistance or ohmic polarization

1. **Activation or chemical polarization.** This surface related phenomenon, is a function of the kinetic rate of exchange between ions in the liquid phase and atoms adsorbed on the catalyst. Measured in terms of the exchange current density (j_o), the greater the degree of exchange, the smaller the polarization. Figure 6.24 shows chemical polarization at various current densities. Maximization of the catalyst surface area is a key method to enhancing the ex-

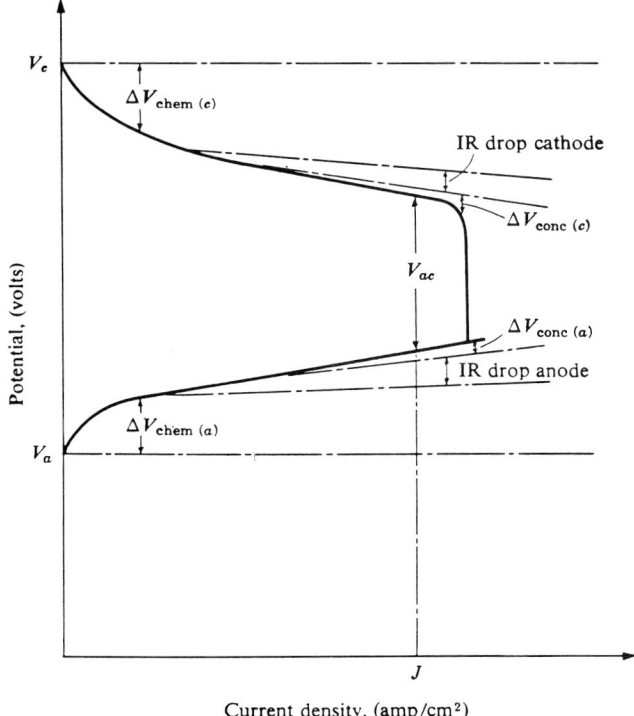

Figure 6.23 Cell potential depression as a function of current density. The graph shows the contribution of the three types of polarization (chemical, concentration, and IR) to voltage depression as the current density increases [15].

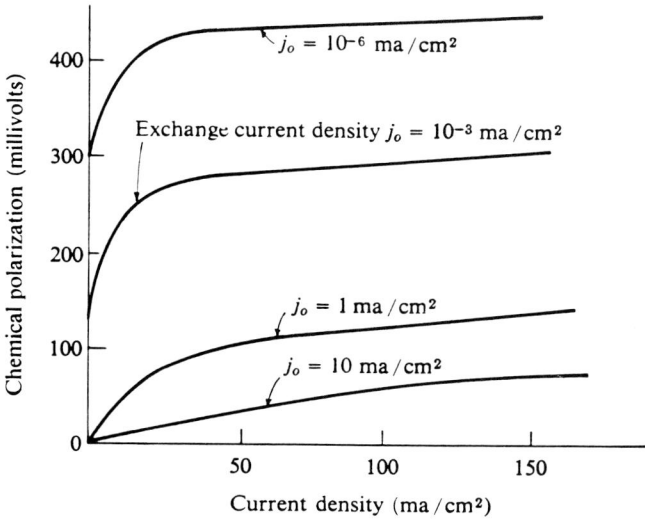

Figure 6.24 Chemical polarization as a function of current density at four different exchange current densities [15].

change process. Electrodes are therefore built with pore sizes ranging from 1 to 100 microns in diameter [15].

2. **Concentration polarization.** Once a load is placed upon a cell, reactant depletion occurs at the electrode-electrolyte interface creating a concentration gradient (Figure 6.25). Using an equivalent concentration profile to model

Spacecraft Power

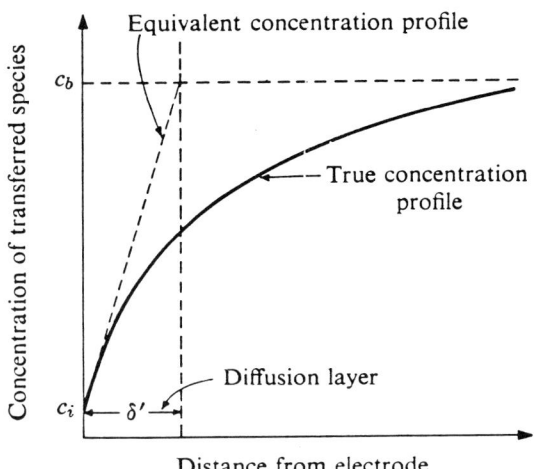

Figure 6.25 Concentration of transferred species as a function of distance from the electrode surface. Profile shows concentration decreases as the species approaches the electrode. The dashed line is a useful approximation of the true profile [15].

the true concentration profile, the maximum or limiting current is given by

$$J_L = \frac{nFD_L}{\delta'} C_b \quad (6.26)$$

where

- J_L = limiting current density
- D_L = diffusion coefficient
- δ' = thickness of diffusion layer of stagnant fluid
- C_b = average concentration in bulk electrolyte

Concentration polarization at the anode and cathode are a function of J_L:

$$\Delta V_{\text{Anode}} = \frac{RT}{nF} \ln\left(\frac{J_L + J}{J_L}\right) \quad (6.27)$$

$$\Delta V_{\text{cathode}} = \frac{RT}{nF} \ln\left(\frac{J_L}{J_L + J}\right) \quad (6.28)$$

$$\Delta V_{\text{conc}} = \Delta V_{\text{anode}} + \Delta V_{\text{cathode}}. \quad (6.29)$$

Prevention of concentration polarization is best accomplished through electrolyte circulation which maintains concentration homogeneity throughout the entire electrolyte [15].

3. **Resistance or ohmic polarization.** Aggregate losses due to resistance of electrolyte, electrodes, and cell hardware.

Fuel cells were developed for the manned space program because the maximum life of batteries was four days [16]. Recharging with solar arrays was not feasible and batteries of equivalent energy to the fuel cell would weigh 10 times more [17]. An added bonus was the production of water to be used for drinking and supplying humidity to the crew cabin. The overall system, as shown in Figure 6.26, is quite complex. With all of the moving parts, fuel cells suffer from a reliability problem and are not especially suited for long duration continuous operation. Space Station designers chose Ni-H_2 batteries over fuel cells for this reason.

Still, fuel cells are ideal for short missions such as Gemini, Apollo, and Space Shuttle. Fuel cell development for space began with the Gemini program which required lifetimes longer than batteries could reasonably attain without the tremendous weight penalty. These first systems were known as solid polymer electrolyte (SPE) fuel cells. Figure 6.27 is a schematic diagram of Gemini's H_2/O_2 cell. Fuel and oxidizer are stored as liquids under high pressure and low temperature in what is termed supercritical storage. The gases are preheated before being transported throughout the fuel cell by manifolds to the individual cells. Gas flows through the trapezoidal channels formed by the current collector and the electrode through which it passes to react at the SPE interface.

The SPE, with its chemical structure shown in Figure 6.28, is an ion exchange membrane (IEM) as opposed to the aqueous electrolyte used in alkaline fuel cells. It consists of a three dimensional network of polymer molecules to which are attached acidic ions. Early IEM networks were copolymers of vinyl compounds or polystyrene-divinyl benzene. Current membranes being developed are made of a fluoropolymer (e.g., Nafion). The network does not contain ions nor does it readily absorb water. Treatment with chlorosulfonic acid ($ClSO_3H$) produces a water absorbent network containing highly acidic sulfonic acid groups ($-SO_3H$). The water provides the medium by which the hydronium ion can move from anode to cathode in a working fuel cell [18]. Wetting of the IEM is crucial to cell operation, especially when one considers the fire hazard it represents if it dries out.

In order to prevent drowning of the cathode by product water, wicks are used to channel H_2O away to a felt pad and then to two tanks. Proper cell temperature 60°C (140°F) is maintained by circulating coolant through ducts. Thirty-two of these cells comprise one stack of which three make up a single fuel cell. Two fuel cells each provided 1 kW of power for the Gemini spacecraft (Figure 6.29). Internal cell pressure was maintained between 2.4×10^5 and 3.1×10^5 Pa (20 and 30 psia) with cell potential averaging 0.78 V [18, 19].

Although the SPE fuel cell performed admirably, NASA chose to develop the alkaline fuel cell for Apollo and Space Shuttle because of its greater efficiency (Figures 6.30 and 6.31). Alkaline fuel cells have an aqueous electrolyte (45% by weight KOH) held in place by a matrix, for example, (asbestos or potassium titanate (K_2TiO_4)). In this case, OH^- and not H_3O^+ is the ionic species transporting

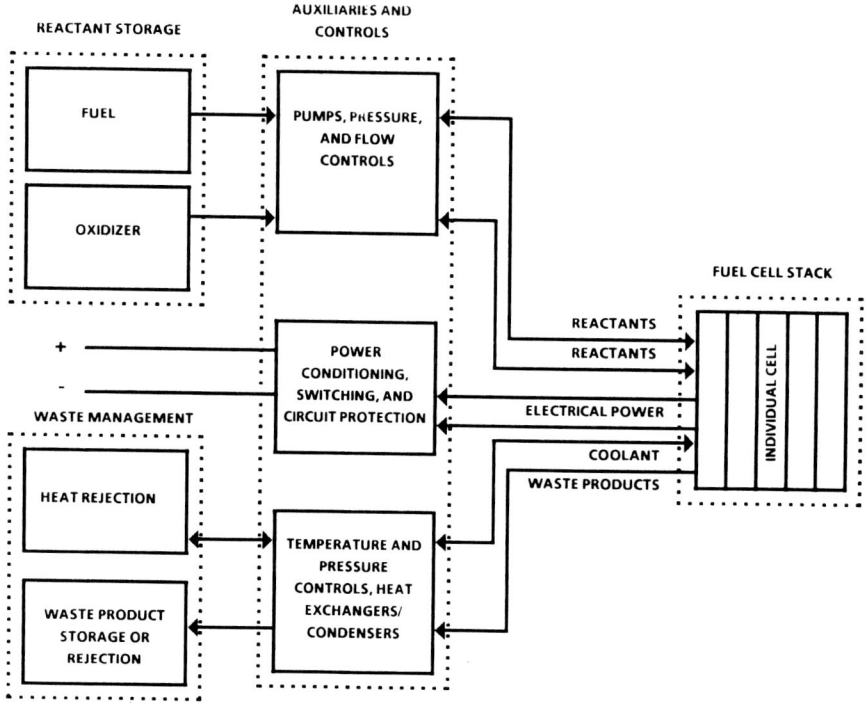

Figure 6.26 Schematic of a simple fuel cell system [17].

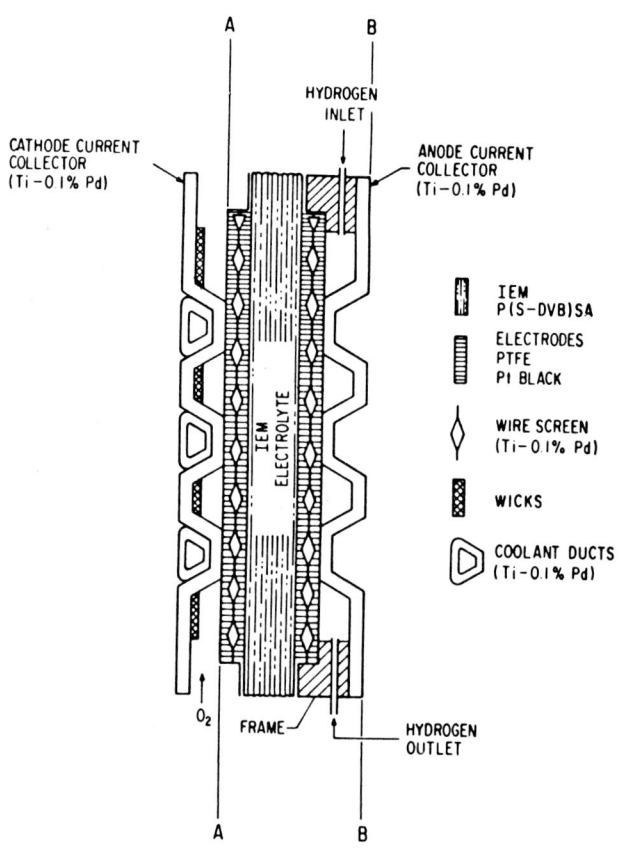

Figure 6.27 The design of an individual cell in General Electric's solid polymer electrolyte acid fuel cell [18].

charge with water produced at the hydrogen electrode. The hydrogen electrode is a platinum/carbon mixture, but the Pt is replaced by gold in the cathode. Reactants are stored in the critical state at pressures between 1.1×10^6 to 2.2×10^6 Pa (150 to 302 psia) for H_2 and 1.5×10^6 to 7.0×10^6 Pa (200 to 1,005 psia) for O_2 (Figure 6.32). Other pertinent parameters about the Space Shuttle's fuel cells are given in Table 6.4. Technological advancements allow the Shuttle's fuel cells to produce six to eight times as much power as those on Apollo, yet weigh 20 kilograms less [9].

Radioisotope Thermoelectric Generators

Spacecraft scientists and engineers realized at the beginning of the space program the limitations of solar cells and batteries for certain missions. Solar cells, with their low conversion efficiency, could not realistically provide enough power at distances far from the Sun. Batteries suffer from limited lifetime and require recharging from an external source. Interplanetary probes, such as Pioneer, Voyager, and Galileo, are examples of missions ill-suited for the use of solar cells and batteries. A self-contained, long-life energy source known as the radioisotope thermoelectric generator (RTG) was therefore developed. With over three decades of flight experience, RTGs have demonstrated characteristics of long life, high reliability, adaptability to extreme mission environments, and safe handling and use [21]. The United States has launched 23 satellites using RTGs since 1961. On a weight basis, they are competitive with solar cells, but are more costly due to their expensive

Spacecraft Power

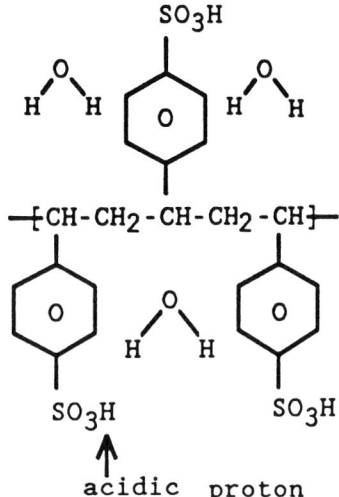

Figure 6.28 Chemical structure of the ion exchange membrane in the acid solid polymer electrolyte fuel cell.

Figure 6.29 Gemini spacecraft SPE fuel cell [20].

Figure 6.30 Apollo spacecraft fuel cell assembly [9].

Figure 6.31 Space Shuttle fuel cell assembly. *Courtesy of International Fuel Cell.*

fuel [20]. Still, NASA has found the RTG to be the best alternative for missions where solar cells are not feasible.

Radioisotope Fuels

The RTG is a self-contained power system which converts heat into electricity. The heat source is an unstable radioisotope fuel which decays through atomic particle emission. Figure 6.33 explains the various types of atomic radiation.

Heat generated by the decay process is converted into electricity by thermoelectric devices (Figure 6.34). On a weight basis, radioisotopes release about 1,000 times more energy than the best chemical fuels [20]. Unfortunately, the conversion process is only 6.5% efficient for current thermoelectric converters. Couple the inefficiency with shielding requirements and vast heat disposal and the weight advantage rapidly disappears.

Several hundred radioisotopes exist, but very few are suitable heat sources. The main criteria for selection are a high specific power, sufficient half-life, and no major shielding requirement. Specific power is the thermal output measured in watts divided by the fuel weight. Half-life is a characteristic of radioisotopes and indicates how quickly the nucleus decays. Equation 6.30 gives the relationship between half-life and the fraction of remaining nuclei left:

$$\log_{10} \frac{N_0}{N_t} = \frac{0.301 t}{t_{1/2}} \quad (6.30)$$

where

N_0 = initial number of nuclei
N_t = number of nuclei present at time t
$t_{1/2}$ = half-life

The decay is exponential and the rate is of great concern

Figure 6.32 Hydrogen and oxygen for the Shuttle's fuel cells are stored in these insulated tanks at 22 K and 89 K respectively *Courtesy of Ball Aerospace.*

Table 6.4 Orbiter Fuel Cell Operating Parameters

Power Output*	H$_2$ Consumption	O$_2$ Consumption	H$_2$O Production
2 kW	0.0784 kg/hr	0.622 kg/hr	0.73 kg/hr
7	0.2873	2.283	2.59
12	0.5193	4.114	4.44

* Constraints: 2–7 kW continuous, 12 kW for 15 min (every 3 hours), 16 kW for 15 min max (contingency).

to mission planners since sufficient fuel must be present for the life of the mission. After one half-life, one-half of the original amount will remain, and after two, only one-fourth remains. This relationship can be expressed by

$$\frac{N}{N_0} = \left(\frac{1}{2}\right)^n \qquad (6.31)$$

where n is the number of half-lives.

Shielding of the fuel entails protecting both man and equipment from the harmful effects of radiation. Alpha particles are the largest and slowest of the decay particles, so emitters of these are potential candidates for fuels. Table 6.5 lists the properties of several radioisotope fuels.

From Table 6.5, only plutonium-238 (in the form of PuO$_2$) has been employed as a fuel in spacecraft. Plutonium is an alpha particle (helium nucleus) emitter which decays in the following way:

$$^{238}_{94}\text{Pu} \rightarrow {}^{234}_{92}\text{U} + {}^{4}_{2}\text{He} \qquad (6.32)$$

Heat is generated by the movement of the alpha particles and the vibration of the recoiling nuclei. The energy range of alphas from emitter isotopes is from 6 to 7 MeV, as compared with only 3 MeV for beta and gamma emissions. Alpha sources therefore possess higher thermal power densities then do other radioisotopes. Also, almost all energy is absorbed within the fuel source itself, thereby minimizing the requirements for shielding [15]. The radioisotope fuel being used in the general purpose heat source (GPHS), shown in Figure 6.35, for Galileo and Ulysses is an isotopic mixture of plutonium dioxide (PuO$_2$) containing 83.5%, ±1%, ^{238}PuO$_2$. Formed into ceramic pellets with a volume of approximately 13 cm^3, the PuO$_2$ is encased by iridium and graphite to prevent release of radiation should it reenter the atmosphere. Each RTG contains 72 fuel pellets providing a total thermal output of 4,410 watts at a weight of 11.03 kg [23].

Thermoelectric Conversion

Conversion of the heat to usable electricity is the job of the thermoelectric converters, which operate on the principle

Spacecraft Power

- **Alpha** particles are the slowest of the three types of ionizing radiation, and despite a speed of about 10,000 miles per second, they can travel only a few inches in the air. Alpha particles lose their energy almost as soon as they collide with anything. They can easily be stopped by a sheet of paper or the skin's surface.

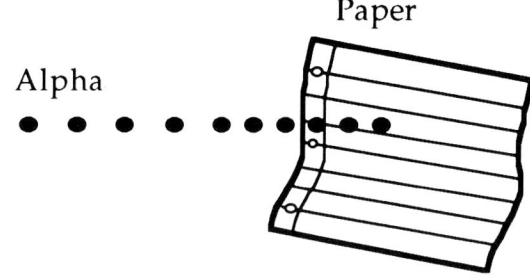

- **Beta** particles are much less massive than alpha particles. They can travel as much as 100,000 miles per second and can travel in the air for a distance of about 10 feet. Beta particles can pass through a sheet of paper, but may be stopped by a thin sheet of aluminum foil or glass.

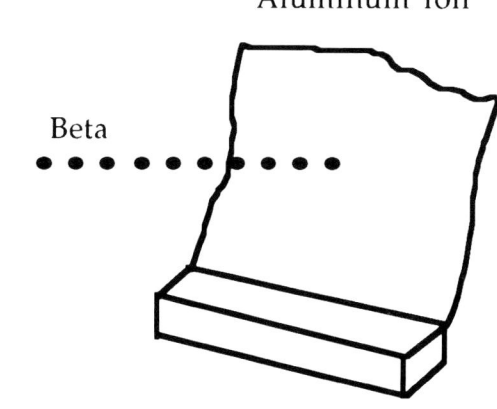

- **Gamma** rays, unlike alpha or beta particles, are waves of pure energy. Gamma rays travel at the speed of light (186,000 miles per second). Gamma radiation is very powerful and penetrating and requires a thick wall of concrete, lead, or steel to stop it.

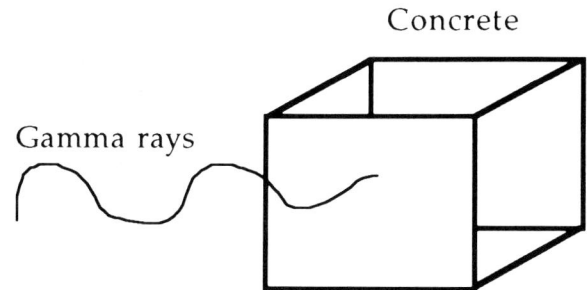

Figure 6.33 Types of ionizing radiation [22].

of the thermoelectric effect. When joined together and then heated at their junction, two dissimilar metals or semiconductors will produce current, giving rise to the thermoelectric effect. The flow of current in semiconductors is similar to that in solar cells. Figure 6.36 depicts what happens when n-type and p-type semiconductors are connected through at heated junction. The free electrons and holes are driven away from the heat source toward the cold junction. Connection of an external circuit at the cool junction allows the electrons to flow and fill holes on the p-side. Simultaneously more holes are generated and the electrons, which are repelled by the negative charge at the cool end, travel through the hot junction to the n-side.

RTGs of today operate at 1,000°C on the hot side (maximum operating temperature for Pu) and 300°C on the cool side. With conversions of only 6%, 94% of the heat is waste which must be radiated through attached fins to prevent the system from melting. The semiconducting material is made of silicon-germanium alloy in atomic percentages ranging from 50% Si and 50% Ge to 67% Si and 33% Ge. Boron and phosphorus are the dopants. Small amounts of gallium phosphide may also be added to decrease thermal conductivity, thereby increasing conversion efficiency [24]. Figure 6.37 shows the individual components of the RTG used in the Viking lander spacecraft.

One major limitation of RTGs is their limited electrical output, which is in the range of tens to hundreds of watts. Four SNAP-19 RTGs, each producing 40 W at beginning of mission (BOM), are aboard Pioneers 10 and 11. Three multihundred watt units with a combined, nominal electrical output of 450 W (BOM) are still operating on the Voyager spacecraft. Galileo, shown in Figure 6.38, has two

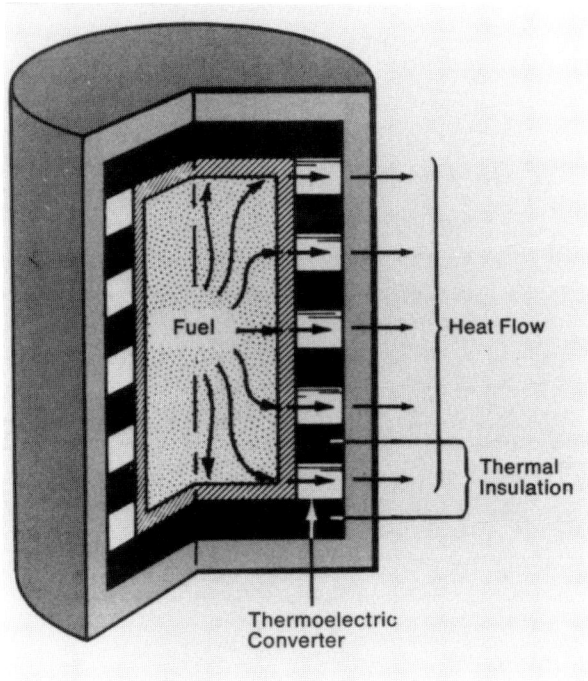

Figure 6.34 Idealized radioisotope thermoelectric generator showing the fuel delivering heat to the thermoelectric converters which in turn generate electricity [20].

GPHS-RTGs which will produce 285 W_e apiece. These power outputs pale next to the multikilowatt solar arrays being developed for Space Station and the 100 kW SP-100 nuclear fission reaction. Still, RTGs have proven their long life and high reliability in space and will continue to be used for special missions [25, 26].

Alternative Power Sources

All four of the preceding energy sources are used on operational spacecraft. They represent the best technology available at this time, but they are by no means the only systems available. Other energy sources and converter technologies have been looked at over the years and several are being considered for future missions. In this section, alternate heat sources and conversion systems are discussed.

Heat Sources

Solar Concentrators

The most readily available source of heat in space is of course the Sun. With photovoltaics, its usable light energy is converted directly into electricity by a solar cell. Concentrators can gather more light and in turn focus them on the solar cell to increase power output. Concentrators can also be designed to collect large amounts of sunlight to produce a tremendous amount of heat at the focal point. This heat is then converted into electricity by a converter (e.g., thermoelectric).

Space Station could be the first to use a solar concentrator like the one in Figure 6.39. In its initial operating configuration, Si solar cells and Ni-H$_2$ batteries will provide approximately 38 kW of electricity. Eventually solar concentrator powered dynamic systems (SDPS) will be added. The function of the concentrator, in this system, is to collect sunlight and reflect it through an aperture and onto a heat engine's receiver. High reflectivity is essential and the materials chosen for this purpose are silver and aluminum. One complete concentrator unit consists of 456 triangular facets (1 m on a side) joined to form 19 hexagonal elements. A single facet is composed of 6.4 mm thick Al honeycomb core to which is bonded 0.19 mm graphite-epoxy composite. A thin layer of Cu is placed over the composite to promote adhesion of the 200 nm thick Ag reflector. Silver, however, is quite susceptible to atomic oxygen attack, as was discussed in the previous chapter. Therefore, an Al$_2$O$_3$ coating is placed over the silver to prevent atomic oxygen penetration. A final coating of SiO$_2$ covers the Al$_2$O$_3$ to ensure moisture resistance and retard radiation degradation as well as provide additional protection against atomic oxygen [12].

Nuclear Fission

The use of nuclear reactors for spacecraft power is anything but a new idea. The SNAP-10A reactor shown in Figure 6.40 was launched on April 3, 1965, and was the first nuclear reactor placed in space. It supplied 500 W (electrical) and weighed a total of 435 kg (960 pounds) with its attached thermoelectric converter [21]. The Soviet space

Table 6.5 Properties of Radioisotope Fuels [15]

Isotope	Fuel form	Decay mode	Compound power density W/cm^3	Compound power density W/g	Half-life years	Shielding required	Maximum operating temp, °C	Availability kW/yr (thermal)
Polonium 210	GdPo	α	820	82	0.38	minor	1600	70
Plutonium 238	PuO$_2$	α	3.7	0.41	86.4	minor	1000	11
Curium 242	Cm$_2$O$_3$	α	1050	98	0.45	neutron	1600	1.5
Curium 244	Cm$_2$O$_3$	α	28.6	2.6	18.0	neutron	1600	8
Promethium 147	Pm$_2$O$_3$	β	2.1	0.28	2.6	minor	1000	2
Strontium 90	SrO/SrF$_2$	β	1.2	0.24	28	heavy	1000	31

Spacecraft Power

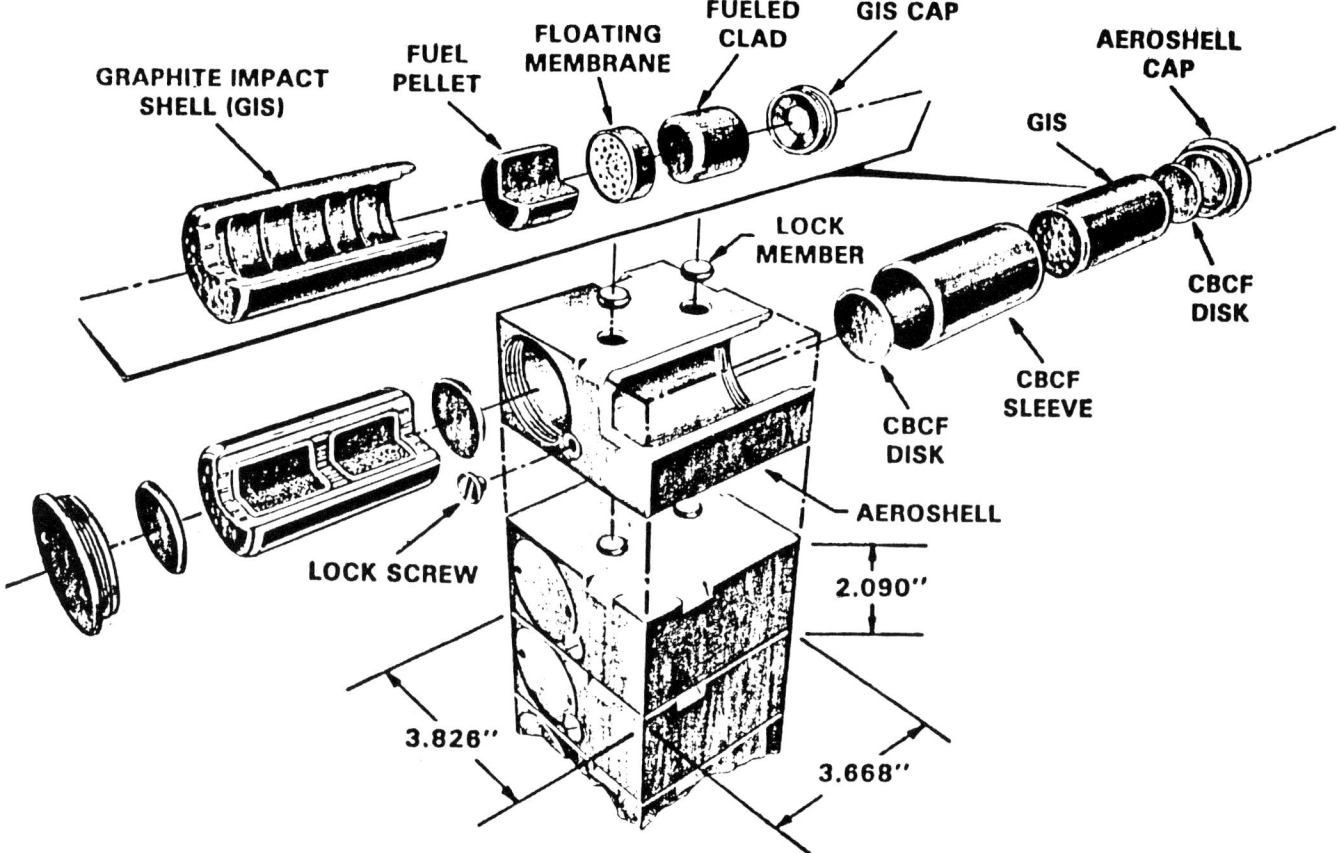

Figure 6.35 Components of the general purpose heat source used in the RTGs for Galileo and Ulysses spacecraft [23].

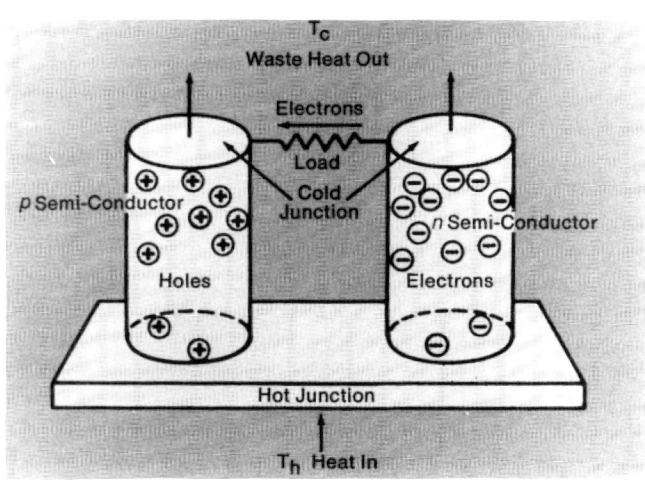

Figure 6.36 Depiction of the thermoelectric effect between n- and p-type semiconductors. Heat applied to the junction forces electrons to flow from the n-type material through an external load to the p-type material [20].

program has employed many more nuclear reactors than the United States, having placed at least 30 nuclear power systems into LEO.

Since these initial successful tests, U.S. interest in nuclear fission reactors dwindled. Only recently, with the huge power requirements called for by SDI has the technology been reborn. The current nuclear power system, SP-100, is a uranium fueled reactor coupled to thermoelectric converters (Figure 6.41).

During fission, the nucleus absorbs a neutron, becoming unstable to the point of splitting into two parts of approximately equal mass. These fission fragments are themselves radioactive and undergo further decay with the emission of alpha and beta particles and gamma radiation. Uranium-235, for example, produces over 80 primary products with mass numbers ranging from 72 to 160. On the average, each of these products progresses through three stages of radioactive decay to produce over 200 isotopes in the final fission products [28]. Three possible reactions are

$$^{235}_{92}U + ^{1}_{0}n \longrightarrow \begin{cases} ^{90}_{37}Rb + ^{144}_{55}Cs + 2^{1}_{0}n \\ ^{87}_{35}Br + ^{146}_{57}La + 3^{1}_{0}n \\ ^{72}_{30}Zn + ^{160}_{62}Sm + 4^{1}_{0}n \end{cases} \quad (6.33)$$

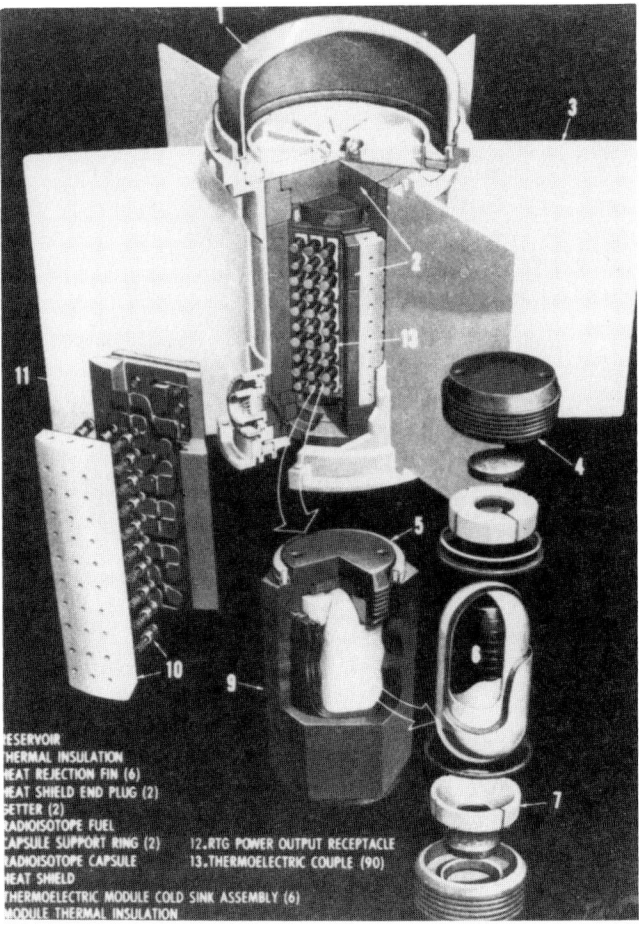

Figure 6.37 Components of the RTG in the Viking lander spacecraft [15].

The tremendous amount of energy released during fission is the result of converting mass into energy according to Einstein's well known equation

$$\Delta E = mc^2 \tag{6.34}$$

where
- ΔE = energy, J
- m = mass, kg
- c = speed of light = 3.00×10^8 m/s.

Take, for example, the mass difference in the fission of one mole of ^{235}U to produce nuclei with mass numbers of 95 and 139.

$$^{235}_{92}U + ^{1}_{0}n \rightarrow ^{95}X + ^{139}Y + 2^{1}_{0}n \tag{6.35}$$

$$\Delta m = 1 \text{ mole}\left(94.945 \frac{g}{\text{mole}}\right) + 1 \text{ mole}\left(138.955 \frac{g}{\text{mole}}\right)$$
$$+ 2 \text{ moles}\left(1.009 \frac{g}{\text{mole}}\right) - 1 \text{ mole}\left(235.124 \frac{g}{\text{mole}}\right)$$
$$- 1 \text{ mole}\left(1.009 \frac{g}{\text{mole}}\right)$$
$$\Delta m = -0.215 \text{ g}$$

Using unit conversions and the basic definition of the joule $\left(\frac{kg \cdot m^2}{s^2}\right)$, the mass-energy conversion shown in equation 6.34 becomes

$$\Delta E = 9.00 \times 10^{10} \frac{kJ}{g} \cdot (-0.215g) \tag{6.36}$$
$$= -1.94 \times 10^{10} \text{ kJ}.$$

Compared to nuclear fission, the energy released by conventional chemical reactions is quite small—about 10 million times less.

A key to sustaining a fission reaction is the continual production of neutrons. For each individual fission reaction in ^{235}U, an average of 2.51 neutrons is released [28]. These neutrons are essential since they initiate fission in other nuclei and therefore sustain the chain reaction. For a controlled reaction, the number of nuclei produced per event must be reduced to around one. Neutron absorbing control rods, consisting of cadmium or boron, are inside the reactor core to prevent reactor runaway. A moderator, which can be H_2O, D_2O, or graphite, is also in the core to slow down the neutrons to a speed where they can be captured by the $^{235}UO_2$ ceramic fuel pellets [29].

Converter Systems

Thermoelectric is the only method used to date to convert heat into electricity in space systems. However, as previously mentioned in the discussion of solar concentrators, NASA has plans to use a dynamic power system aboard Space Station Freedom (Figure 6.42). The inherent advantage of solar dynamic over photovoltaic is its greater efficiency (4% to 7% versus 14% to 22%) [30]. This translates into a much smaller collector area that has less drag at orbit altitude and requires less propellant to reboost [31]. Additionally, a solar dynamic power system is less expensive to expand in response to increased power needs. It has been estimated that it would cost $5 million for each additional kilowatt for a solar array system while a solar dynamic system could be expanded for $2 million/kW. The solar array power system would also require significantly more maintenance, necessitating more Shuttle flights than a solar dynamic power design [30].

The power system itself is the closed Brayton cycle generator shown in Figure 6.43. During operation, helium-xenon gas is heated by the heat source and then expands through a turbine, driving an alternator to produce AC. The heat source could be nuclear, radioisotope, or in this case, solar. During insolation, sunlight is focused from the concentrator through the aperture into a solar receiver. The solar receiver in Figure 6.44 shows a series of heat pipes attached to the inner wall of a 2.6 m (8.5 ft) long cylinder with a diameter of 2.6 m (8.5 ft). Each heat pipe has within it tubes containing the helium-xenon working fluid and a phase change material (PCM). Surrounding these tubes is

Spacecraft Power

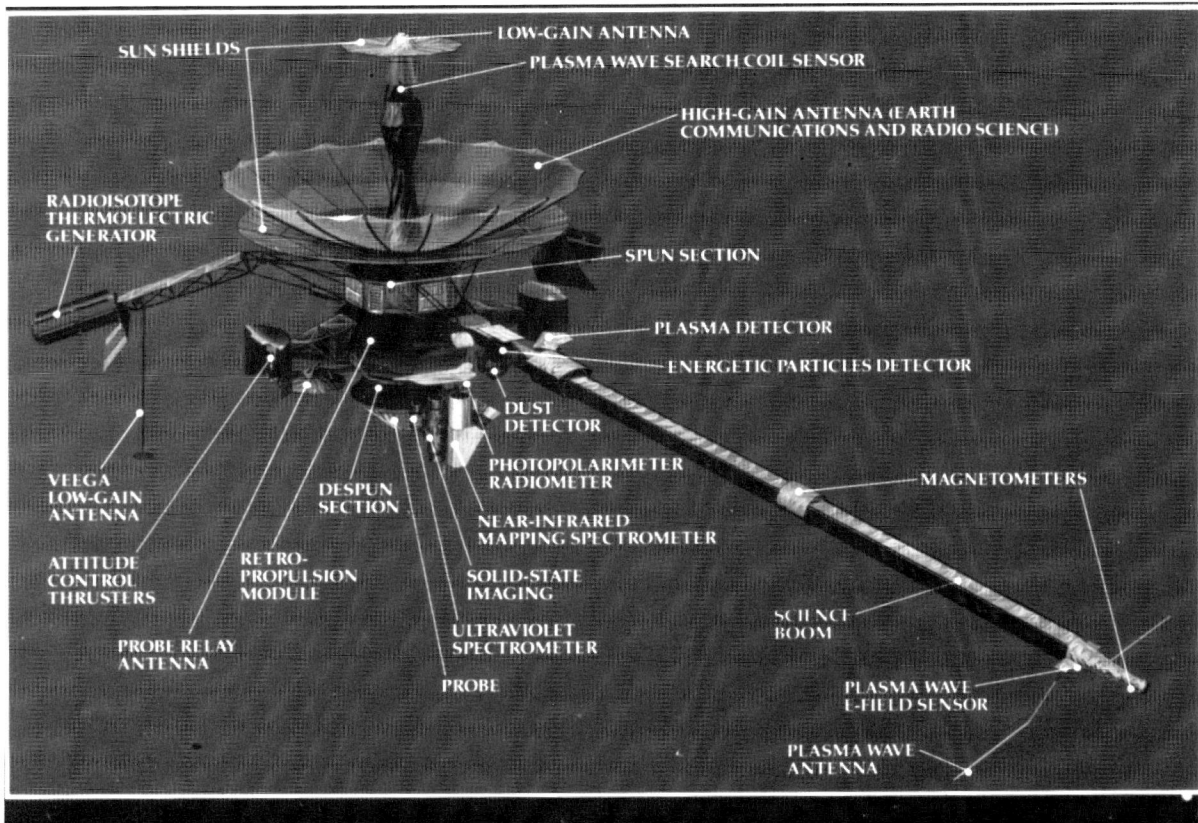

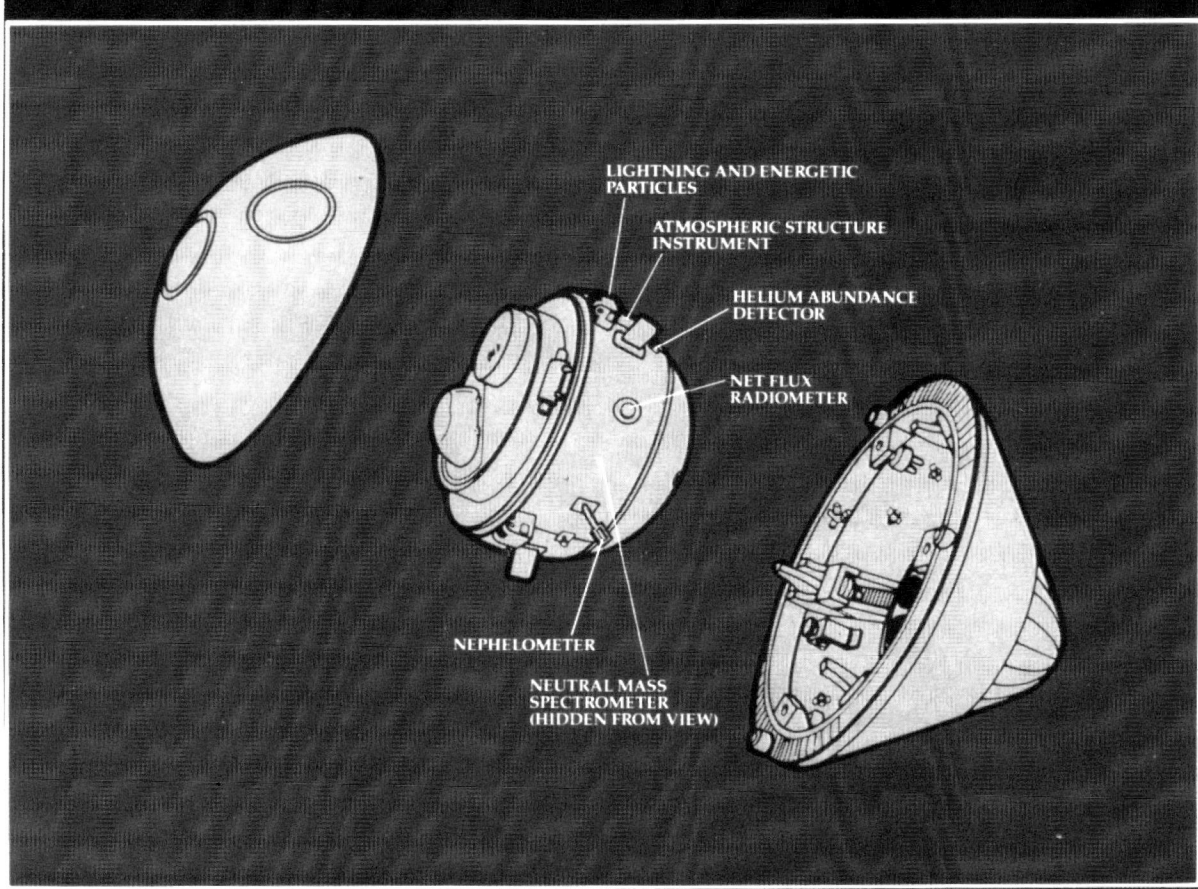

Figure 6.38 Galileo spacecraft with the RTGs located on the booms [23].

Figure 6.39 Solar concentrator design showing the geometry of the mirrors which focus the Sun's rays on to a receiver. *Courtesy of Harris Corporation.*

a liquid alkali metal, either sodium or potassium. The heat pipe walls absorb the heat of the Sun and transfer the energy through the now gaseous alkali metal to the inner tubes. Sufficient heat is provided to push the working fluid through the turbine and melt the PCM. During eclipse periods, the PCM cools, transferring its latent heat to the working fluid and maintaining system power [32].

Selection of the PCM is critical to the performance of the heat pipe. It must have a melting point corresponding to the operating temperature of the engine. In the case of the closed Brayton cycle, temperatures range from 740 to 848°C (1,364 to 1,558°F). Heat of fusion (ΔH_{fus}, J/g) was discussed in Chapter 1 and is the amount of heat absorbed or liberated during a solid-liquid phase change. The greater the value for ΔH_{fus}, the more heat the PCM is able to store during insolation and subsequently transfer in eclipse. Density is important since there is the need to minimize heat pipe weight and volume. Finally, the PCM must have sufficient thermal conductivity to ensure effective heat transfer [31].

Based upon these criteria, a number of inorganic salts were investigated with 80.5% LiF and 19.5% CaF_2 being the most attractive. The LiF-CaF_2 is a eutectic salt mixture consisting of 80.5% mole fraction LiF and 19.5% mole fraction CaF_2. A eutectic composition has the lowest melting point of all mixtures of two or more substances. It also is the mixture which solidifies or melts without changing composition. For all other mixtures, pure solid will separate out from the liquid phase. But at the eutectic composition, the transition from liquid to solid occurs without the premature solid formation [35]. Selected properties for this eutectic are given in Table 6.6.

Spacecraft Power

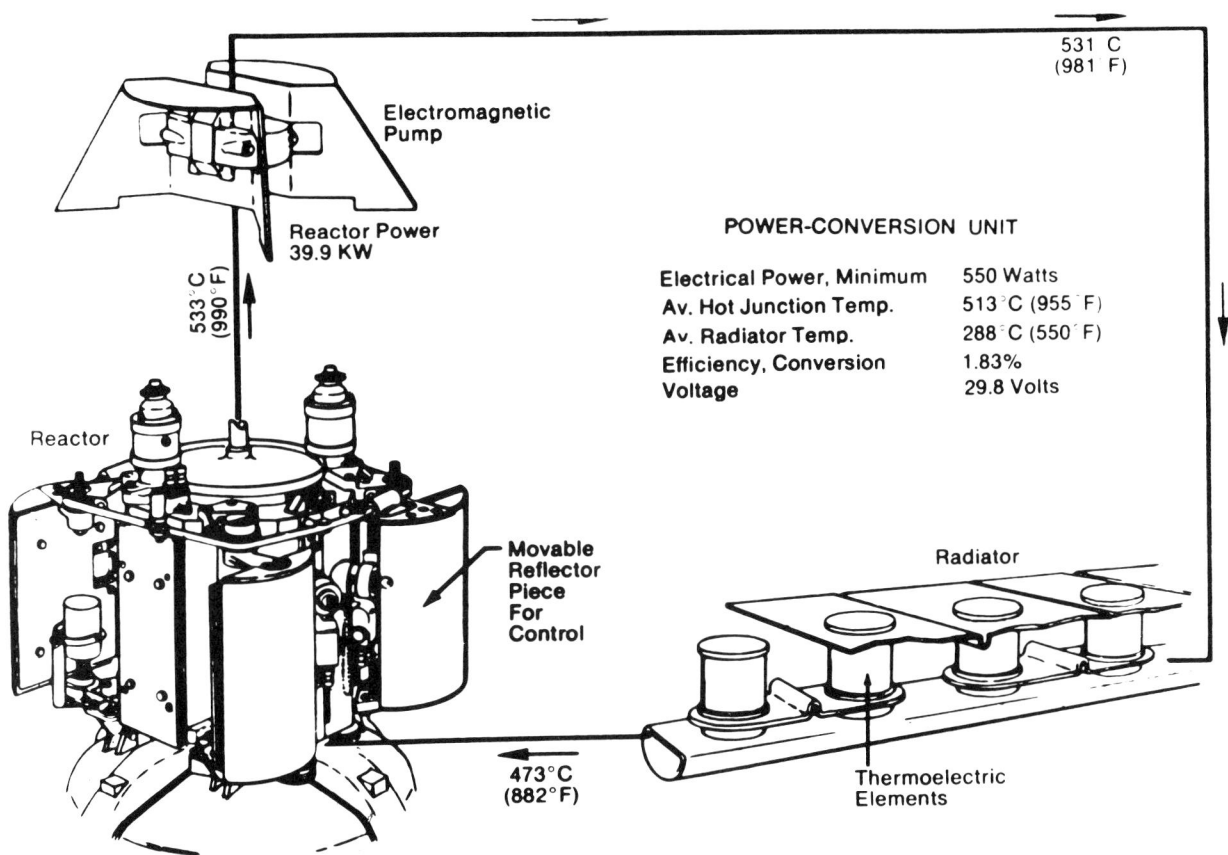

Figure 6.40 SNAP-10A nuclear fission reactor with thermoelectric converters [20].

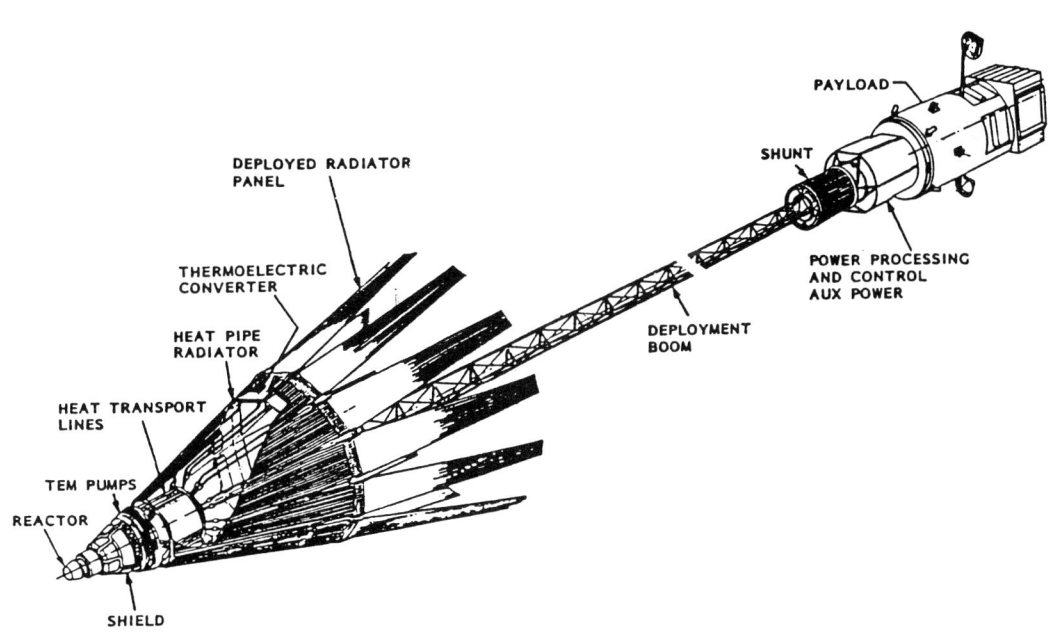

Figure 6.41 The 100 kW$_e$ SP-100 space reactor power system [27].

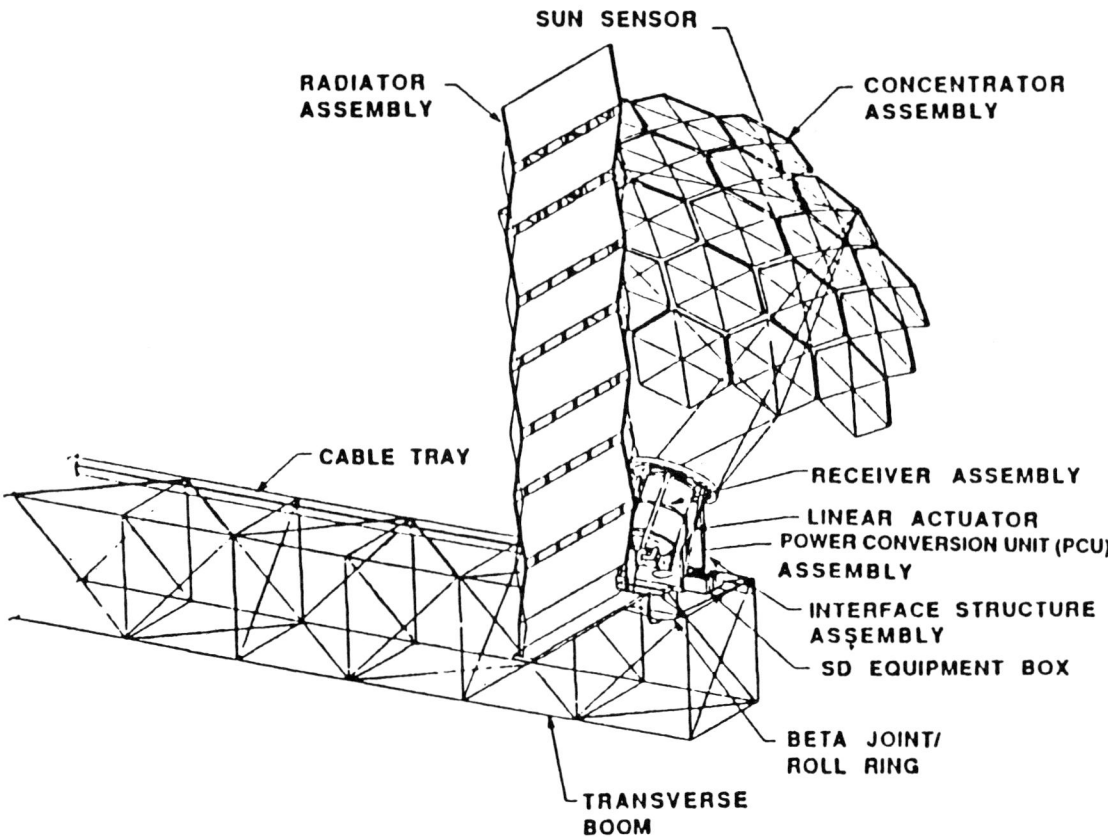

Figure 6.42 Design of the solar dynamic power system for Space Station [31].

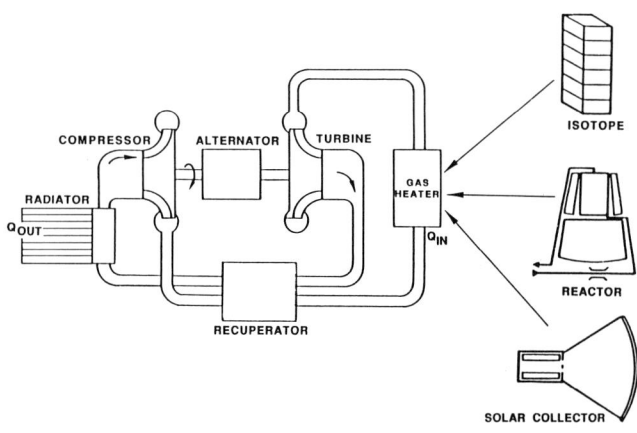

Figure 6.43 Schematic of closed Brayton cycle power system [33].

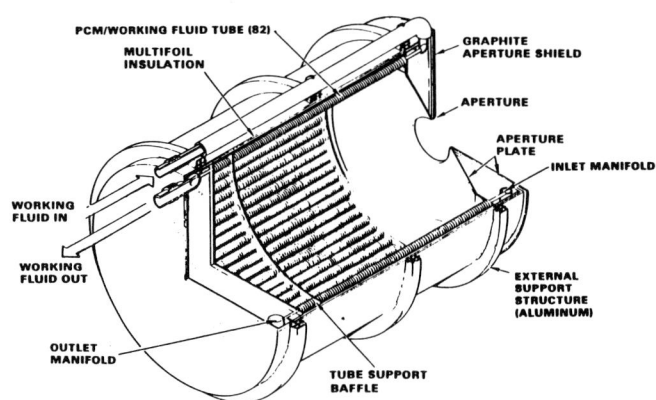

Figure 6.44 Diagram showing the phase change material within the tubes attached to the inner wall of the receiver. The working fluid passes through the receiver exchanging heat with the PCM [34].

Spacecraft Power

Table 6.6 Selected Properties of Eutectic LiF-CaF$_2$ Phase Change Material [36]

Composition	80.5% LiF/ 19.5% CaF$_2$ (by mole)
Melting temperature, K(F)	1,042 (1,416)
Latent heat of fusion, J/g (Btu/lb)	790 (340)
Solid Density*, g/cm^3 (lb/ft^3)	2.68 (167)
Liquid density*, g/cm^3 (lb/ft^3)	2.1 (131)
Liquid coefficient of volumetric expansion, C^{-1} (F^{-1}) °C^{-1} (°F^{-1})	2.7 × 10^{-4} (1.5 × 10^{-4})
Solid heat capacity*, J/kg-K (Btu/lb-F)	1841 (0.440)
Liquid heat capacity*, J/kg-K (Btu/lb-F)	1970 (0.471)
Solid thermal conductivity*, W/m-K (Btu/hr-ft-F)	5.9 (3.4)
Liquid thermal conductivity*, W/m-K (Btu/hr-ft-F)	1.7 (1.0)

*At melting point

Summary

This chapter has discussed a variety of systems in use or being developed for space vehicle power generation and energy. Photovoltaics-batteries are by far the choice for missions of extended length. In those circumstances where solar cells cannot be used, radioisotope thermoelectric generators have often been used to provide power. Short duration manned missions have traditionally used fuel cells for both power and life support needs. Future missions will require much higher power levels than presently available. Systems based upon fission reactors and solar dynamics will meet these increased power needs. Both SDI and Space Station are pushing the frontiers of power generation in space. The future will yield power levels not in kilowatts, but in hundreds and even thousands of kilowatts. In order to accomplish these goals, scientists and engineers will have to overcome many technological hurdles.

Summary of Key Equations

$AgO + H_2O + Cd/Zn \rightleftharpoons$ $Ag + Cd(OH)_2/Zn(OH)_2$	(6.6)	Ag-Cd and Ag-Zn Reactions
$Cd + 2NiOOH + 2H_2O \rightleftharpoons$ $Cd(OH)_2 + 2Ni(OH)_2$	(6.7)	Ni-Cd Reaction
$1/2 H_2 + NiOOH \rightleftharpoons Ni(OH)_2$	(6.12)	Ni-H$_2$ Reaction
$2Na + xS \rightleftharpoons Na_2S_x$ $2 \leq x \leq 5$	(6.18)	Na-S Reaction
$\eta_{ideal} = \Delta G/\Delta H$	(6.22)	Ideal Efficiency of a Fuel Cell
$^{238}_{94}Pu \rightarrow {}^{234}_{92}U + {}^{4}_{2}He$	(6.32)	Plutonium Decay
$\Delta E = mc^2$	(6.34)	Energy-Mass Equation

References

1. Maisel, J., "A Historical Overview of the Electrical Power System; in the U.S. Manned and Some U.S. Unmanned Spacecraft," NASA CR-174806, November 1984.
2. *Basic Photovoltaic Principles and Methods*, Solar Energy Research Institute, Van Nostrand Reinhold Company (New York), 1984.
3. Agrawal, B. *Design of Geosynchronous Spacecraft*, Prentice-Hall Inc. (Englewood Cliffs), 1986.
4. "Space Station Atomic Oxygen Effects," Presentation by LSMC to NASA-LeRC, July 11, 1986.
5. Rauschenbach, H., *Solar Cell Array Design Handbook*, Van Nostrand Reinhold Company, 1980.
6. Bilger, K., Gjerde, H. and B. Sater, "Photovoltaic Array Environmental Protection Program," 24th IECEC, August 6–11, 1989.
7. Leger, L., Visentine, J. and B. Santos-Mason, "Selected Materials Issues Associated with Space Station," *SAMPE Quarterly*, January 1987, 45–54.
8. McCargo, M., Muzzy R., and K. Bilger, "Atomic Oxygen Resistant Coatings for Use in Low Earth Orbit," International Conference on Metallurgical Coatings, 1989.
9. *Handbook of Batteries and Fuel Cells*, ed. by D. Linden, McGraw-Hill Book Company, 1984.
10. McDermott, P., Final Report NASA Grant NSG-5051, CR-14476, GSFC, March 1976.
11. Bauer, P. "Batteries for Space Power Systems," NASA SP-172, 1968.
12. Gutmann, G. "Current Developments in Electrochemical Storage Systems for Satellites," NASA TM-88471, July 1986.
13. Fritts, D. and R. Dueber, "A Discussion of the Mechanism of Cadmium Migration in Sealed Nickel-Cadmium Cells," *Journal of the Electrochemical Society*, 132, 9, pp. 2039–2043.
14. *JANAF Thermochemical Tables*, 3rd ed., American Chemical Society, Vol. 14, 1985.
15. Angrist, S., *Direct Energy Conversion*, 4th ed., Allyn and Bacon Inc. (Boston), 1982.
16. Crowe, B. J., *Fuel Cells: A Survey*, NASA: Washington D.C., 1973, NAS 1.21: 5115.
17. Gregory, D. P., *Fuel Cells*, Millas and Boon Limited (London), 1972.
18. Liebhafsky, H. and E. Cairns, *Fuel Cells and Fuel Batteries*, John Wiley & Sons (New York), 1968.

19. Shelton, W. R., *American Space Exploration: The First Decade*, Little, Brown, and Co. (Boston), 1967.
20. Corliss, W., *Spacecraft Power*, NASA EP-59.
21. Rock, B., "Radioisotope Space Power Programs," US Department of Energy, N85-13887.
22. "Understanding Radiation," Department of Energy/ NE-0074.
23. "Final Safety Analysis Report for the Galileo Mission," Vol. 1, DOE contract DE-ACOI-79 ET 32043, May 1988.
24. "Development of Advanced Thermoelectric Materials," Final Technical Progress Report, NASA JPL No. 9950-1080, N85-24532.
25. Löb, H., "Nuclear Energy In Space," *Atomkernergie Kerntechnik*, Vol. 40, (1982), No. 1, pp. 23–30.
26. Bennett, G. et al, "The General Purpose Heat Source Radioisotope Thermoelectric Generator: Power for the Galileo and Ulysses Missions," 21st IECEC, San Diego, CA, 25–29 August 1986.
27. Terrill, W. and V. Haley, "Thermoelectric Converter for SP-100 Space Reactor Power System," 21st IECEC, August 25–29, 1986, pp. 1950–1955.
28. "Space Power Systems," Part I, AGARD, AGARDograph 123, 2–6 October, 1967.
29. Masterton, W. and C. Hurley, *Chemistry Principles and Reactions*, Saunders College Publishing, 1989.
30. "Solar Dynamic Power Option for the Space Station," Allied-Signal Fluid Systems Division, 41-7961.
31. Faget, N., "Materials Compatability Issues Related to Thermal Energy Storage for a Space Station Dynamic Power System," 21st IECEC, 25–29 August 1986, pp. 811–815.
32. Pidcoke, L., Gay, R., and W. Lee, "Thermal Energy Storage Experiments for Space Applications," 21st IECEC, 25–29 August, 1986, pp. 805–810.
33. "Dynamic Space Power," Allied-Signal Fluid Systems Division, 41-8105-1 A.
34. "Off-Design Modeling and Performance of the Space Station Solar Dynamic Power System," Allied-Signal Fluid Systems Division, 41-8882.
35. Atkins, P., *Physical Chemistry*, Oxford University Press, 1978.
36. "NASA Considers Station Design Changes, Amid Funding and Cancellation Threats," Aviation Week and Space Technology, 15 May 1989, pp. 28–29.

Discussion Questions

1. When light strikes the surface of a silicon solar cell, electron-hole pairs are formed.
   a. What barrier exists in a silicon solar cell that prevents recombination of electron-hole pairs from occurring?
   b. How is this barrier formed?
2. Describe the flow of electrons and holes in an operating solar cell.
3. Solar cells are designed with a thin layer of n-type material placed on top of a thick layer of p-type material.
   a. Why is there a difference in the thickness of the two materials?
   b. Will the cell still operate if the position of the two layers is reversed?
   c. Why aren't solar cells designed with p-type material on top of n-type?
4. a. Draw diagrams showing the electrochemical operation of a Ni-Cd cell during charge and discharge.
   b. Describe the chemical reactions occurring at the electrodes.
   c. Calculate the standard cell potential.
5. Nickel-cadmium batteries have been the standard for energy storage in spacecraft power systems. Now, Ni-H_2 is beginning to replace Ni-Cd in both high and low Earth orbits. Discuss the advantages and disadvantages Ni-H_2 has to offer.
6. Long-duration space missions to the outer planets have used radioisotope thermoelectric generations (RTGs) for power.
   a. Why use RTGs instead of solar arrays and batteries?
   b. What are the two basic components of an RTG?
   c. Briefly explain the basic operating principle of an RTG.
7. This chapter has covered an array of different power systems, each suited for particular missions. Discuss the main factors to consider when designing a power system for a mission.

Problems

1. A spin-stabilized cylindrically shaped satellite 5 m long and 2 m in diameter is in GEO. Solar cells are to be installed along the cylinder's exterior, (excluding the ends).
   a. How many 0.5 cm × 0.5 cm cells are required to completely cover the exterior surface?
   b. If each cell produces 0.1 mW of power how much power is available to the satellite bus initially?
   c. At the end of the satellite's life of 7 to 10 years, what will the power output of the solar array be?
2. Before a new battery cell is developed, scientists and engineers perform a simple set of calculations to determine the cell's theoretical capacity. The calculations are based upon the atomic or molecular masses of the electrodes used as the couples and their oxidation states. Generally speaking, the higher the theoretical capacity the more attractive the cell becomes. Following the example presented in this chapter for Ni-H_2, calculate the theoretical capacity of a silver-zinc cell. You can check your answer against the value given in Table 6.2.
3. Fuel cells are different from batteries in that the chemi-

Spacecraft Power

cals responsible for the reactions are stored externally. By adjusting the temperature and pressure of the reactants and products, the cell voltage can be altered.

a. Calculate the theoretical cell voltage of an H_2-O_2 fuel cell under the following conditions:

Temperature = 150°C P_{O_2} = 4.0 atm
P_{H_2} = 5.0 atm P_{H_2O} = 0.50 atm

b. The actual cell voltage is always less than the theoretical cell voltage. Discuss the three factors which cause this to occur.

4. a. Calculate the theoretical amount of energy released in each of the three reactions given in equation 6.33.
   b. Which of these is the most desirable and why?
   c. Besides producing a tremendous amount of energy, what other advantages do fission reactors possess?

5. The observation that mass and energy are equivalent has revolutionized modern physics. Einstein posed this theory in equation 6.34.
   a. Using this energy-mass relationship, calculate the energy yield in joules if 1 kg of Pu-238 were converted completely into energy.
   b. If this energy were utilized evenly over 3 years time, what constant power output would be available to power a spacecraft?
   c. Why are the calculations accomplished in parts a and b of this problem overly optimistic?

Chapter 7

Propulsion

Perhaps the most significant achievements in our space program have occurred in the area of propulsion. After all, were it not for rockets we would still be earthbound! Beginning with Redstone and continuing through Titan, Saturn, and the SRBs, rocket technology has progressed by leaps and bounds. Higher thrust at lower weight has enabled us to put larger and larger payloads into space. Part of this advancement has been in the development of higher energy propellants for not only boosters, but attitude and stabilization systems as well. This chapter covers the basic thermochemistry of rocket propulsion and the liquid and solid systems used.

Thermochemistry of Rocket Propulsion

Newton's third law states that for every action, there's an opposite and equal reaction. In a rocket combustion chamber, chemicals are burned to produce high pressure, high temperature gases. These rapidly moving gas molecules accelerate and expand as they pass through the nozzle to produce the action. The rocket moving in the opposite direction of the exhausted gas is the reaction (Figure 7.1). Rocket movement is not because of the pushing of escaping gases on the outside air, but rather due to the pressure imbalance between the combustion chamber and the surroundings. Rockets operate best in a vacuum where the pressure imbalance is greatest and escaping gas molecules achieve their greatest kinetic energy.

In order to achieve the necessary continual high speed flow of gas molecules, controlled combustion must occur in the chamber. In a rocket, this occurs when propellants consisting of a fuel and oxidizer are mixed and ignited. Unlike conventional air-breathing engines, rockets carry their own oxygen supply since they operate in the vacuum of space. Propellant selection is a key factor in the design of rockets in order to produce gas molecules with the highest possible velocities. The evaluation is done by studying the thermochemical properties of the propellants. Thermochemistry deals with the energy changes accompanying a chemical process, such as combustion. The three thermochemical properties determining propellant performance are enthalpy (ΔH), entropy (ΔS), and Gibbs's free energy (ΔG), all of which were introduced in Chapter 1.

Enthalpy is simply the heat change associated with a chemical reaction. Recall that an exothermic reaction is one that generates heat while an endothermic one absorbs heat. Obviously, the thrust producing chemical reactions occurring in rockets are exothermic. Figure 7.2 shows the enthalpy of hydrogen combustion with oxygen. For every mole of water vapor formed at standard conditions, 241.93 kJ of energy is released. Using equation 7.2 and Table 7.1, the change in enthalpy for many propellant reactions can be calculated. For example, hydrogen peroxide was once used in rocket applications and can be catalytically decomposed in the following manner:

$$2H_2O_2(l) \xrightarrow{Cat.} 2H_2O(g) + O_2(g) + \text{heat} \quad (7.1)$$

$$\Delta H^o_{rxn} = (2\ \Delta H^o_{H_2O} + \Delta H^o_{O_2}) - (2\ \Delta H^o_{H_2O_2}) \quad (7.2)$$

$$\Delta H^o_{rxn} = (2\ \text{mole})(-241.93\ \text{kJ/mole})$$
$$+ (1\ \text{mole})(0.00\ \text{kJ/mole})$$
$$- (2\ \text{mole})(-187.69\ \text{kJ/mole})$$

$$\Delta H^o_{rxn} = -108.48\ \text{kJ}$$

Evaluation of possible high energy propellants begins with such calculations of enthalpies on the basis of known experimental values. However, enthalpies of formation for many reactants and products are unknown, leaving the rocket designer looking for an alternative to making his own experimental measurements. Estimates of ΔH can be made using bond energies. To do this, return to hydrogen combustion where the reactants have hydrogen-hydrogen and oxygen-oxygen bonds. On the product side two hydrogens are bound to one oxygen. The difference in enthalpy is related to this change in the bond strengths of the product relative to the reactants. By summing average bond energies (Table 7.2) of reactants and subtracting the sum of those for the products, ΔH^o_{rxn} can be estimated.

$$\text{Energy of bonds broken} \quad (7.3)$$
$$= (1\ \text{mole})(432\ \text{kJ/mole})$$
$$+ (1/2\ \text{mole})(494\ \text{kJ/mole})$$
$$= 679\ \text{kJ}$$

$$\text{Energy of bonds formed} \quad (7.4)$$
$$= (1\ \text{mole})(2)(459\ \text{kJ/mole})$$
$$= 918\ \text{kJ}$$

$$\Delta H = 679 - 918 = -239\ \text{kJ} \quad (7.5)$$

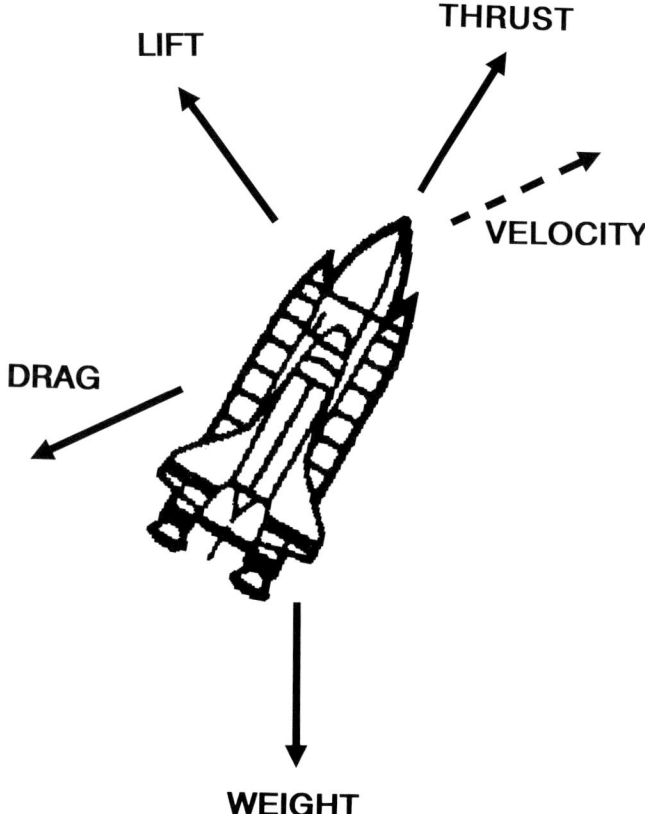

Figure 7.1 A pictorial representation of the forces acting on the Space Shuttle during the latter stages of the launch sequence. (The velocity vector is shown for perspective).

This value compares quite favorably with the -241.93 kJ previously calculated. Differences in the two are due to the use of average bond energies. Actual bond energies vary between molecules with the average calculated from measurements of many different compounds [3].

Reaction enthalpy is important because it is a major factor

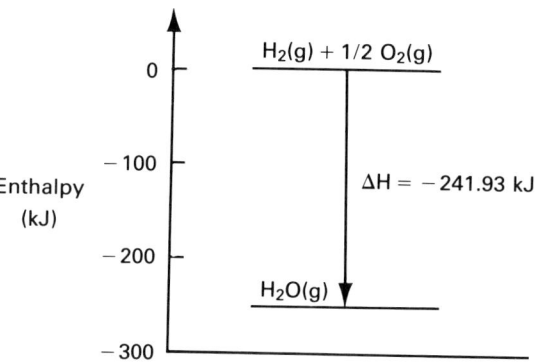

Figure 7.2 Enthalpy diagram for H_2-O_2 reaction in which energy is released as hydrogen and oxygen react to form water vapor.

in the determination of rocket performance. Specific impulse (I_{sp}) is the key measure of rocket performance and is expressed as the ratio of thrust produced (F) to the mass of expelled propellant per unit time (m_p/t).

$$I_{sp} = \frac{F}{(m_p/t)} \qquad (7.6)$$

Theoretical specific impulse can be calculated using thermodynamic values and the following equation:

$$I_{sp} = \left[\frac{2\gamma R}{(\gamma - 1)} \frac{T_c}{\overline{M}} \left(1 - \left(\frac{P_e}{P_c}\right) \frac{\gamma - 1}{\gamma} \right) \right]^{1/2} \qquad (7.7)$$

where

γ = average ratio of specific heats (c_p/c_v) of the combustion products
T_c = combustion chamber temperature
P_e = nozzle exit pressure
P_c = combustion chamber pressure
$\overline{M}$ = average molecular weight of combustion products

Table 7.1 Standard Enthalpies of Formation [1, 2]

Substance	kJ/mol	kcal/mol	Substance	kJ/mol	kcal/mol
Al(s)	0.00	0.00	N_2(g)	0.00	0.00
Al_2O_3(s)	-1670.53	-399.09	N(g)	472.71	112.93
C(s)	0.00	0.00	NH_3(l)	-65.63	-15.68
CO(g)	-110.59	-26.42	NH_3(g)	-46.21	-11.04
CO_2(g)	-393.68	-94.05	N_2H_4(l)	50.48	12.06
C(g)	718.70	171.70	N_2O_4(l)	-20.13	-4.81
CH_4(g)	-74.90	-17.89	NH_4ClO_4(s)	-290.58	-69.42
CH_3OH(l)	-238.76	-57.04	HF(g)	-268.73	-64.20
C_2H_5OH(l)	-277.77	-66.36	HCl(g)	-92.34	-22.06
H_2(g)	0.00	0.00	F_2(l, 85K)	-12.68	-3.03
H_2(l, 20K)	-7.03	-1.68	O_2(g)	0.00	0.00
H(g)	218.04	52.10	O_2(l, 90K)	-9.42	-2.25
H_2O(g)	-241.93	-57.80			
H_2O(l)	-285.96	-68.32			
H_2O_2(l)	-187.69	-44.84			
HNO_3(l)	-173.08	-41.35			

Table 7.2 Average Bond Energies (kJ/mole) [4]

Single Bonds

	H	C	N	O	S	F	Cl	Br	I
H	432								
C	411	346							
N	386	305	167						
O	459	358	201	142					
S	363	272	—	—	226				
F	565	485	283	190	284	155			
Cl	428	327	313	218	255	249	240		
Br	362	285	—	201	217	249	216	190	
I	295	213	—	201	—	278	208	175	149

Multiple Bonds

C=C	602	C=N	615	C=O	799
C≡C	835	C≡N	887	C≡O	1072
N=N	418	N=O	607	S=O(in SO_2)	532
N≡N	942	O_2	494	S=O(in SO_3)	469

The propellant parameter having the greatest impact upon effective exhaust velocity is $T_c/\bar{M}$. Making several assumptions, the value of $T_c/\bar{M}$ can be related to reaction enthalpy of propellants in the following very rough approximation [5]:

$$\frac{\Delta H}{\bar{M}} \approx \left(\frac{\gamma R}{\gamma - 1}\right)\left(\frac{T_c}{\bar{M}}\right). \quad (7.8)$$

Evident from this is the need to maximize enthalpy in selecting propellants which will yield the highest combustion temperatures. In addition, production of low molecular weight exhaust products must also be considered since $\bar{M}$ is in the denominator of the equation. These two factors are of equal importance and their simultaneous achievement is difficult. The conventional approach in chemical rockets is to produce low molecular weight exhaust products at moderate combustion temperatures [5].

A valid question at this point is the role ΔG plays in rocket design and its relationship to ΔH. When propellants react in a combustion chamber they can eventually form a dynamic equilibrium with their products, in which case both forward and reverse reactions occur at the same rate with no net change in species concentrations with time. In a system having reached dynamic equilibrium, ΔG is equal to zero and the following expression can be simplified:

$$\Delta G = \Delta G^o + RT \ln Q.$$
At equilibrium $\Delta G = 0$ and $Q = K_p$; hence,
$$\Delta G^o = -RT \ln K_p, \quad (7.9)$$

where
$$K_p = e^{-\Delta G^o/RT}.$$

K_p is the equilibrium constant for the reaction and is an indicator of the direction in which equilibrium lies. In the H_2-O_2 reaction

$$H_2(g) + 1/2 O_2(g) \rightleftharpoons H_2O(g)$$

the equilibrium constant is given as

$$K_p = \frac{P_{H_2O}}{P_{H_2} P_{O_2}^{1/2}}. \quad (7.10)$$

The position of equilibrium, and therefore K_p, is very temperature dependent; the combustion products at high temperatures are different from those at low temperatures. There exists an interdependence between product composition and temperature, making determination of their performance maximizing values a bit difficult [3].

Essentially, ΔG^o can be used to calculate K_p for a chemical reaction in a rocket. K_p can then be used to find the partial pressures of the gases present in the exhaust (the composition of chemical species at equilibrium) with the ideal gas law being used to relate these partial pressures to mole fractions. If the mole fractions are known, the amount each chemical specie is contributing to the enthalpy of the reaction can be determined. The ΔH, then, is used to find the rocket combustion chamber temperature, which is a major design parameter. Thermodynamics allows us to determine the properties of the reaction products (which include composition, temperature, and pressure) that are subsequently expelled from the rocket at a high velocity, producing the thrust. The calculation of rocket gas properties can become quite complicated. Dozens of interdependent chemical reactions can occur, yielding dozens of products. This makes the calculation of K_p values from ΔG values very difficult. These types of calculations are ideally suited for a computer, which is how it is done by scientists and engineers.

Rocket Propellants

Rocket propellants are divided into two general classes on the basis of their phase. Liquid propellants, like hydrazine, offer the capability for multiple restarting of the engines, a necessity for attitude control. Contrast this with the solid propellants, such as ammonium perchlorate, which can be started only once, but make ideal boosters. Both types have found extensive use in both civilian and military space programs.

Liquid Propellants

A variety of factors contribute to the overall attractiveness of a particular propellant. From an economic standpoint it should be inexpensive, readily available, and easily processed. Its performance is measured in terms of specific impulse, exhaust velocity, and other parameters. Highest performance is attained by propellants with high energy contents which yield high combustion chamber temperatures (T_c). Production of low molecular weight exhaust gases is also desirable, as evident from the inverse relationship between I_{sp} and $\bar{M}$ in equation 7.7. In fact, excess propellant is often used in order to lower the average molecular mass of the exhaust gases. Unfortunately, this also lowers T_c, so a balance exists between maximizing T_c and minimizing $\bar{M}$ values.

As shown in the previous section, calculations can be performed to determine the amount of heat a propellant combination will produce. However, a great deal about how good a propellant will be can be deduced simply from its physical properties. Table 7.3 summarizes the pertinent information.

A low freezing point is normally desirable for a propellant to permit operation in cold temperatures. Therefore, in some cases monomethyl hydrazine and N_2O_4 would need to be heated. Similarly, a high boiling point is normally desired so the propellant will stay liquid at high temperatures. Notice in Table 7.3 that H_2 and O_2 fail miserably in this category, therefore, special pressurized tanks and insulation must be used in rockets using liquid H_2 and O_2 to prevent the propellants from boiling. Although liquid H_2 and O_2 have this shortcoming, they have been used widely as propellants because of their high efficiencies.

Another favorable characteristic of a propellant is a positive heat of formation which yields large, negative values for the enthalpy of reaction. There are very few propellants in this category besides the hydrazine family. Propellants with high specific heats are desirable for rockets that pump propellants in jackets around the thrust chamber to keep it cool. A high specific heat means the propellant can absorb lots of heat. For this reason, liquid hydrogen is an excellent choice for cooling a rocket engine.

A dense propellant means that more weight of propellant can be packed in a given volume. This is a big plus when designing a rocket system. Specific gravity is a measure of the relative density with respect to water (1.0) and the higher the specific gravity, the more dense the propellant. The extremely low specific gravity of liquid H_2 means that hydrogen storage tanks are usually quite large.

Viscosity is a measure of how well the propellant will flow. If it is too viscous, pumping becomes difficult. As a point of reference, water has a viscosity of 1.00 centipoise. All the propellants listed in Table 7.3 have low viscosities.

A low vapor pressure also makes it easier to handle propellants (they don't evaporate as easily) and makes pump designing easier. At room temperature, LH_2 and LO_2 have very high vapor pressures, necessitating the need for cooling to low temperatures. Propellants should also be stable in a wide temperature range and must not react with piping and tank walls. In short, when choosing a propellant, there are many things to consider besides how hot the propellant can burn [6].

Propellants present numerous handling and storage problems. Nitrogen tetroxide (N_2O_4) and hydrogen peroxide (H_2O_2) are quite corrosive and require special materials for pipelines and containers. Hydrogen peroxide is also very unstable and can detonate under certain conditions, such as mixing with liquid oxygen. Flammability is always a problem, especially since many fuels readily ignite when exposed to air and heat. Propellants can also be hazardous to humans, causing injury and even death. Hydrazine, for example, is very toxic, requiring the personnel handling it to wear protective clothing [6].

The ignition system used in a rocket is dependent upon

Table 7.3 Properties of Selected Propellants [6]

	Fuels		Oxidizers	
	Liquid Hydrogen H_2	Monomethyl Hydrazine CH_3NHNH_2	Liquid Oxygen O_2	Dinitrogen Tetroxide N_2O_4
Molecular Weight (g/mol)	2.016	46.08	32.00	92.016
Freezing Point (°C)	−259.2	−52.4	−218.8	−11.6
Boiling Point (°C)	−252.8	87.5	−183.1	21.2
Heat of Formation (kJ/mole)	−7.03	54.8	−9.42	−20.13
Specific Heat (J/g-K)	2.26	0.889	0.5	0.474
Specific Gravity	0.071	0.879	1.23	1.45
Viscosity (centipoise)	0.024	0.855	0.87	0.423
Vapor Pressure (mm Hg)	62 (−259°C)	52 (27°C)	39 (−184°C)	719 (20°C)

the propellant combination. Spontaneously combustible or hypergolic propellants do not require an ignition system, making the propulsion system itself simpler. Unfortunately, the inadvertant combustion of hypergolic fuels may be disastrous. Nine Delta second stage rocket bodies have exploded in orbit for this reason. It is speculated that a common bulkhead between the fuel and oxidizer cracked due to thermal cycling, allowing the two to come in contact. The resulting debris from these events constitute about a third of all trackable objects presently orbiting the Earth [7]. Nonspontaneously ignitable propellants, on the other hand, must be preheated to a certain temperature before ignition can occur. After ignition, combustion is sustained through a steady flow of propellants. The smoothness and uniformity of combustion is again a function of the propellant combination with those giving the greatest stability being most desirable. In many cases the production of smoke should be minimized to avoid the formation of deposits on surrounding instrumentation and equipment [6].

Liquid Oxidizers

Oxygen (LOX)

The most widely used liquid oxidizer today is molecular oxygen. Stored at temperatures below its boiling point of 90 K, cryogenic oxygen has a specific gravity of 1.14 and a ΔH_{vap} equal to 213 kJ/kg. Storing it in the liquid state eliminates the need for thick walled, high pressure gas containers. Unfortunately, a great deal of this weight savings is taken up by the refrigeration system. Still, the attainable performance is relatively high, as seen in Table 7.4. Figure 7.3 shows the heat released in the formation of metal oxides varies periodically with atomic number. Maxima occur for the elements beryllium, aluminum, and scandium [8].

Handling LOX is fairly safe since it does not spontaneously combust with organic materials under ambient temperatures and pressures. Violent explosions can occur between the two during rapid pressurization. LOX supports and accelerates the combustion of other materials, yet is itself noncorrosive and nontoxic. Its biggest drawback is the low temperature it requires to prevent evaporation. Consequently, heavy insulation is necessary as is the need to dispose of condensate formed on exterior walls [6]. Figure 7.4 shows a diagram of the oxygen-burning main engines of the Space Shuttle.

Fluorine (F₂)

Fluorine appeared at one time to be the natural replacement for LOX. It offers higher values of performance coupled

Table 7.4 Theoretical Performance of Liquid Rocket Propellant Combinations [6]

Oxidizer	Fuel	Mixture Ratio By Mass	Mixture Ratio By Volume	Average Specific Gravity, (g/cm³)	Chamber Temp. (K)	C* (m/sec)	M (kg/mole)	I_{sp} (sec)	γ
Oxygen	75% Ethyl Alcohol	1.30	0.98	1.00	2904	1641	23.4	267	1.22
		1.43	1.08	1.01	2957	1670	24.1	279	
	Hydrazine	0.74	0.66	1.06	3027	1871	18.3	301	1.25
		0.90	0.80	1.07	3127	1992	19.3	313	
	Hydrogen	3.40	0.21	0.26	2416	2428	8.9	388	1.26
		4.02	0.25	0.28	2724	2432	10.0	391	
	RP-1	2.24	1.59	1.01	3282	1774	21.9	286	1.24
		2.56	1.82	1.02	3399	1804	23.3	300	
	UDMH	1.39	0.96	0.96	3171	1835	19.8	295	1.25
		1.65	1.14	0.98	3321	1864	21.3	310	
Fluorine	Hydrazine	1.83	1.22	1.29	4218	2128	18.5	334	1.33
		2.30	1.54	1.31	4408	2208	19.4	363	
	Hydrogen	4.54	0.21	0.33	2791	2534	8.9	398	1.33
		7.60	0.35	0.45	3596	2549	11.8	410	
Nitrogen Tetroxide	Hydrazine	1.08	0.75	1.20	2857	1765	19.5	283	1.26
		1.34	0.93	1.22	2977	1782	20.9	292	
	50% UDMH-50% Hydrazine	1.62	1.01	1.18	2957	1731	21.0	278	1.24
		2.00	1.24	1.21	3088	1745	22.6	288	

Combustion chamber pressure-1000 psia (6895 N/m²)
Nozzle exit pressure-14.7 psia (1 atm)
Optimum nozzle expansion ratio
Adiabatic combustion and isentropic expansion of ideal gas
Compositions expressed in mass percent

The density at the boiling point was used for those oxidizers or fuels which boil below 68°F at 1 atm pressure. For every propellant combination, there are two sets of values listed: the upper line refers to frozen equilibrium, the lower line to shifting equilibrium.

140　　Chemical Principles Applied to Spacecraft Operations

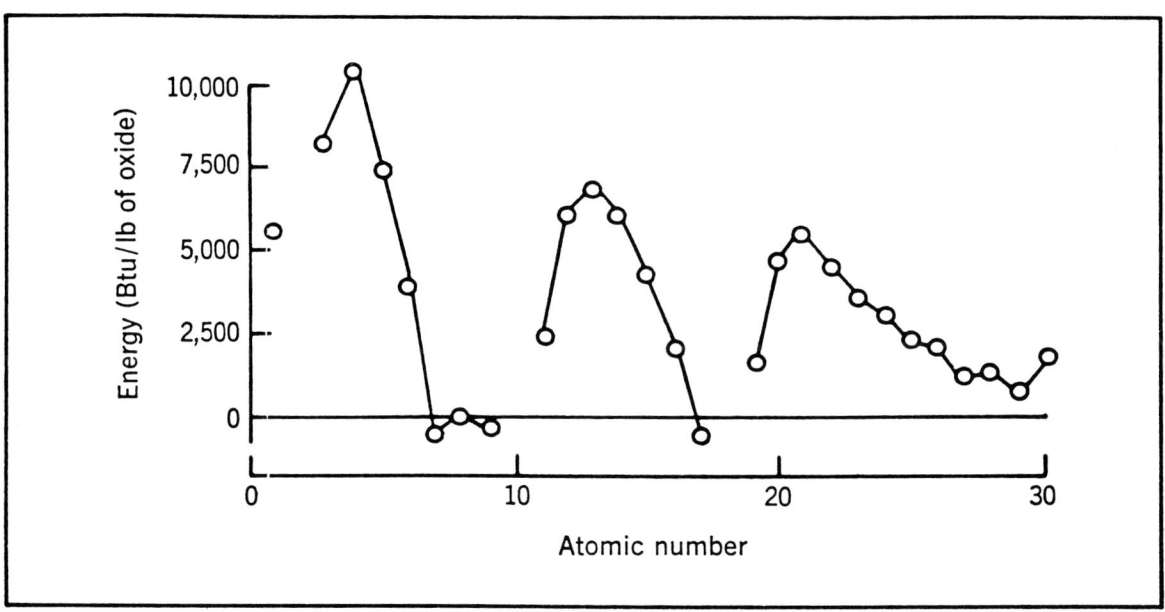

Figure 7.3 Reaction energy with oxygen. The heat released per pound of oxide formed in the reactions of various elements with oxygen is seen to vary periodically with atomic number. Maxima occur for the elements beryllium, aluminum, and scandium [8].

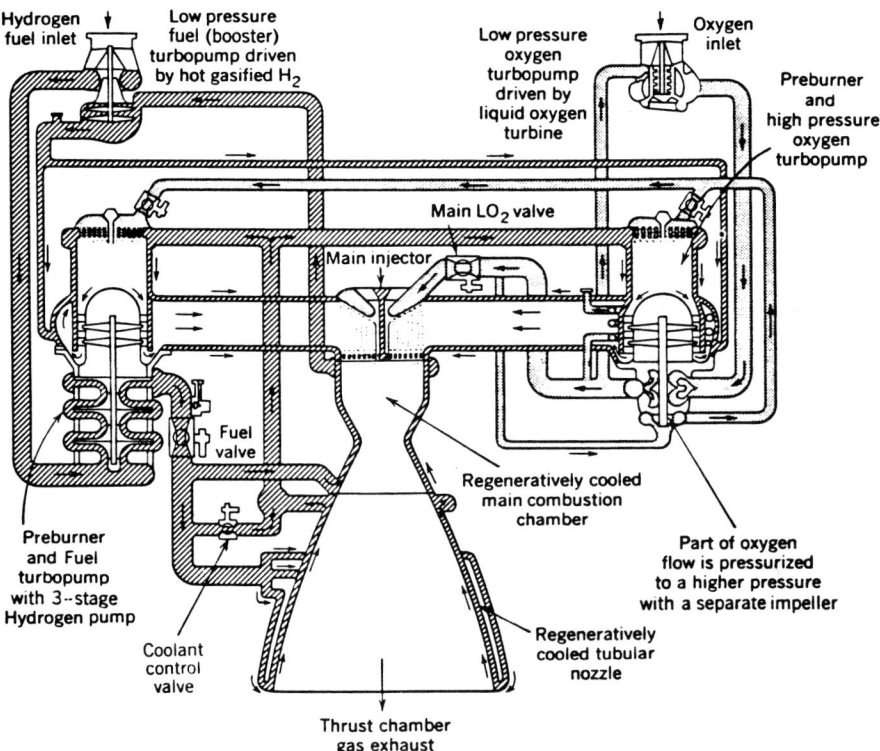

Figure 7.4 Schematic of the Space Shuttle's main engine [6]. *Reprinted by permission of Rocketdyne, division of Rockwell International.*

Propulsion

with a higher specific gravity (1.5), but has several major drawbacks. These include toxicity, corrosiveness, expense, and spontaneous reactivity with a wide variety of materials. By virtue of its position in the periodic table, fluorine is by far the most electronegative element. This makes F_2 a powerful oxidizing agent, as is evident from its half-cell reduction potential of $+2.87$ V (Table 1.7). Reaction products typically contain HF and F_2, both poisonous and toxic. Like LOX, it too requires heavy insulation to prevent evaporation ($T_B = 53.7$ K). Due to these disadvantages, F_2 has found only limited use in experimental rocket engines [6].

Hydrogen Peroxide (H_2O_2)

Hydrogen peroxide is one of those unique compounds which can serve as an oxidizer in bipropellants or be used by itself as a monopropellant. Using catalysts like MnO_2, Pt, and Fe_2O_3, H_2O_2 reacts in the following manner:

$$H_2O_2 \rightarrow H_2O + 1/2 O_2 + \text{heat} . \qquad (7.11)$$

Once used in rocket applications (e.g., the X-1 and X-15) at concentrations as high as 99%, H_2O_2 is spontaneously reactive with many organic materials. It decomposes to form O_2 under even the best of storage conditions. Due mainly to this storage problem, it is no longer used as a propellant [6].

Nitrogen Tetroxide (N_2O_4)

As far as storable oxidizers are concerned, N_2O_4 is by far the most common. It is a high density yellow-brown liquid with a specific gravity of 1.44 and a boiling point of 294 K. The molecules possess weak intermolecular forces resulting in a high vapor pressure. Relatively heavy tanks are therefore necessary for storage. The reddish-brown fumes are quite toxic and can combust with a variety of organic materials. Caution must also be taken to prevent the liquid from freezing ($T_f = 261$ K), which may require the tankage be heated.

Liquid Fuels

Hydrocarbons

Hydrocarbons are a class of chemical compounds containing only covalently bonded hydrogen and carbon. The number of atoms within the molecule varies widely as do the physical and chemical properties. They can consist of all carbon-carbon single bonds (i.e., saturated) or have carbon-carbon multiple bonding (i.e., unsaturated). Tables 7.5 and 7.6 list some of the more common hydrocarbon fuels and their properties.

```
      H   H   H
      |   |   |
  H—C—C—C—H
      |   |   |
      H   H   H
      Saturated
     Hydrocarbon

      H   H   H
      |   |   |
  H—C—C=C—H
      |       |
      H       H
     Unsaturated
     Hydrocarbon
```

Hydrocarbon fuels have been used throughout the life of our space program because they are easy to handle, provide good performance, and are readily available at a low cost. One fuel in particular, RP-1, is the most widely used of the hydrocarbons (Figure 7.5). It is a specially refined kerosene containing both saturated and unsaturated molecules with a narrow range of densities and vapor pressure. Its low vapor pressure (<0.1 psia) minimizes the possibility of accidental ignition while its low freezing point (239 K) means it will remain in the liquid state under most operating conditions [6].

Hydrogen

The greatest performance of a fuel in terms of specific impulse is attained by liquid, molecular hydrogen (LH_2). When combined with LOX, LH_2 supplies a specific impulse of 380–400 sec^{-1} at a T_c of 2,400–2,700 K. Highest performance is not achieved at the stoichiometric fuel/oxidizer ratio, but rather at a 4:1 molar ratio of fuel to oxidizer; namely

$$4H_2 + O_2 \rightarrow 2H_2O + 2H_2 . \qquad (7.12)$$

The reason for using excess fuel is the increased I_{sp} from the reduced average molecular weight of the exhaust.

Due to its low boiling point (20 K) and low specific gravity (0.07), LH_2 requires large, well-insulated tanks. Its

Table 7.5 Common Hydrocarbon Fuels

Hydrocarbon	Common Phase	Composition
Methane	Gas	CH_4
Propane	Gas	C_3H_8
Gasoline	Liquid	C_6H_{14} to $C_{10}H_{22}$
Kerosene	Liquid	$C_{11}H_{24}$, $C_{12}H_{26}$
Diesel	Liquid	$C_{13}H_{28}$ to $C_{17}H_{36}$

Table 7.6 Properties of Some Typical Hydrocarbon Fuels [6]. *Reprinted by permission of John Wiley & Sons, Inc.*

	Jet Fuel	Kerosene	Aviation Gasoline 100/130	Diesel Fuel	RP-1
Specific gravity at 289 K	0.78	0.81	0.73	0.85	0.80 to 0.815
Freezing point, K	213(max)	230	213	250	239(max)
Viscosity at 289 K, centipoise	1.4	1.6	0.5	2.0	16.5(at 239 K)
Flash point, K(TCC)	269	331	244	333	316
ASTM distillation, K					
10% evaporated	347	—	337	—	458 to 483
50% evaporated	444	—	363	—	
90% evaporated	511	—	391	617	
Reid vapor pressure, psia	2 to 3	below 1	7	0.1	—
Specific heat, J/kg-K	2.09	2.05	2.22	1.97	2.09
Average molecular mass kg/mole	130	175	90	—	—

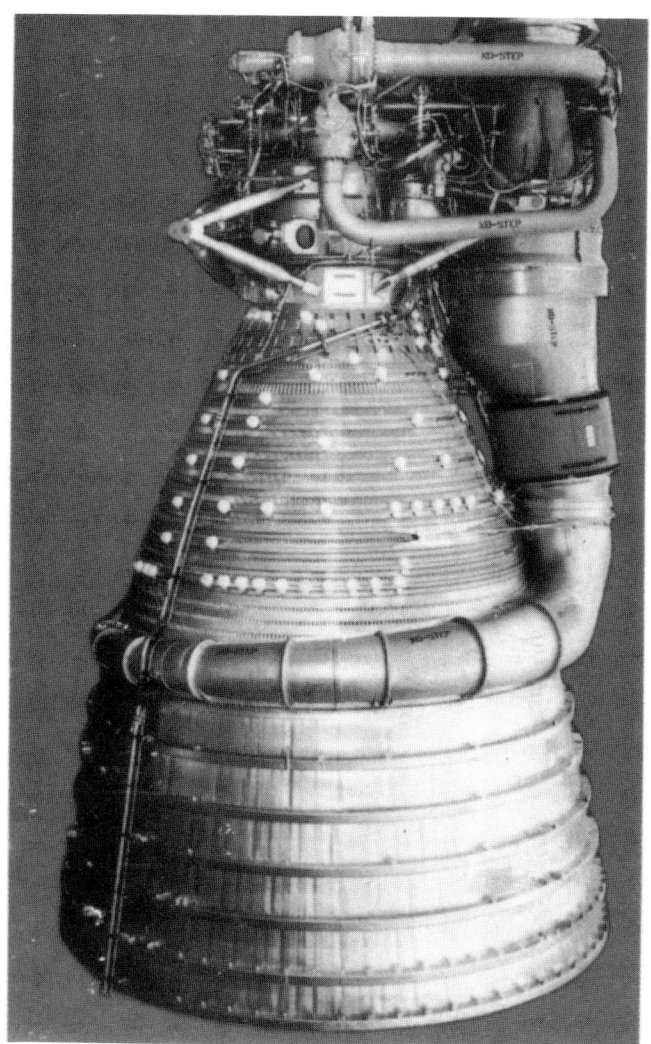

Figure 7.5 The F-1 engine, which used RP-1 and LOX as the propellants, was part of the Saturn V rocket's first stage. *Courtesy of Rocketdyne.*

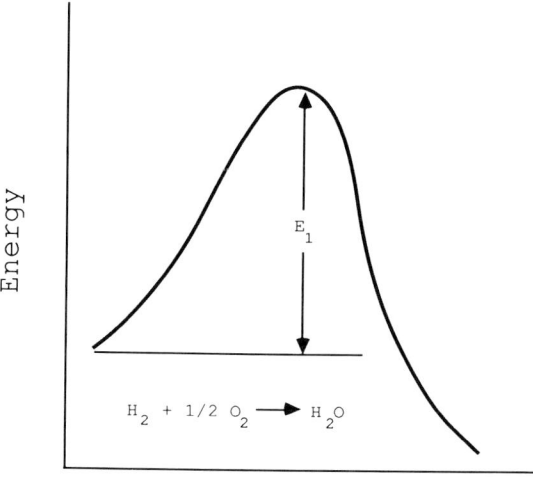

$E_1 > E_2$

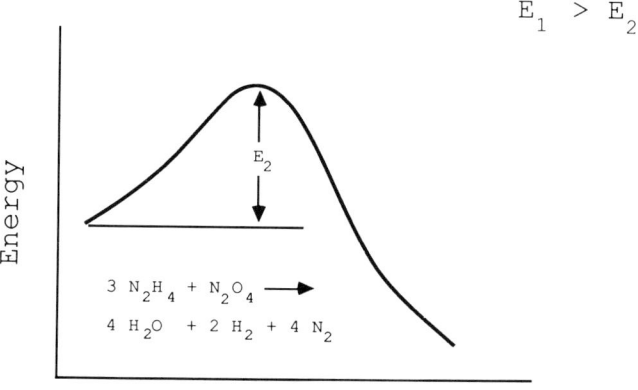

Figure 7.6 Energy diagrams showing the relative difference in activation energy between non-hypergolic (H_2 and O_2) and hypergolic (N_2H_4 and N_2O_4) propellants.

light weight does, however, make it a very efficient regenerative coolant (C_p^o = 14.3 J/g K). Preheating of LH$_2$ before combustion is accomplished by circulating it through cooling tubes attached to the exterior of the nozzle.

Condensation on exterior walls is again a major problem, necessitating the use of a vacuum jacket in many instances. Because all common liquids and gases solidify in LH$_2$, lines and tanks must be completely purged to prevent solids from plugging valves and orifices. The presence of solid air or solid O$_2$ poses an additional hazard since they form explosive mixtures with LH$_2$ [6].

Hydrazine (N$_2$H$_4$)

Hydrazine and its derivatives have been used extensively throughout the space program for propulsion and spacecraft attitude control. The hydrazine compounds are very powerful reducing agents (i.e., easily oxidizable), making them quite reactive. Spontaneous reactions between N$_2$H$_4$ and N$_2$O$_4$ do occur.

Hydrazine itself is a colorless, toxic liquid that freezes at 274.3 K, requiring that it be insulated or heated to prevent solidification. One of its main attributes, however, is its large positive heat of formation (ΔH_f^o); that is,

$$N_2(g) + 2H_2(g) \rightarrow N_2H_4(\ell) \quad \Delta H_f^o = +50.8 \frac{kJ}{mole} . \quad (7.13)$$

The reason for this positive value is the large amount of energy needed to break the strong triple bond of N$_2$. When placed into the enthalpy of combustion equation, the positive ΔH_f^o pushes the ΔH^o_{rxn} toward a negative value.

$$N_2H_4(\ell) + O_2(g) \rightarrow N_2(g) + 2H_2O(g) \quad (7.14)$$

$$\Delta H^o_{rxn} = 2 \text{ mole}\left(-241.8 \frac{kJ}{mole}\right)$$
$$- 1 \text{ mole}\left(+50.8 \frac{kJ}{mole}\right) \quad (7.15)$$
$$= -534.2 \text{ kJ}$$

At this point, it is appropriate to stop and discuss in more detail two very important characteristics of hydrazine. They are its hypergolic and monopropellant aspects.

Hypergolic is a term adopted from German literature to describe self or spontaneous ignition [9]. The act of mixing two hypergolic propellants is sufficient to cause ignition and begin the combustion process. Advantages of such a system include the absence of an ignition system and reliable, repeatable starting capability. Hypergolic propellants can be used not only for propulsion, but also as ignition systems themselves for other non-hypergolic propellants.

The study of spontaneous ignition comes not under the realm of chemical thermodynamics, but rather chemical kinetics. To illustrate this, recall once again the spontaneous reaction between hydrogen and oxygen:

$$H_2(g) + 1/2O_2(g) \rightarrow H_2O(\ell) \quad \Delta G^o = -237 \text{ kJ} . \quad (7.16)$$

At 25°C (standard conditions) this reaction is quite spontaneous, yet such a mixture does not spontaneously burst into flames. As discussed in Chapter 1, there exists a difference between reaction spontaneity and reaction rate. Thermochemistry determines the direction of the reaction, but its speed is a function of chemical kinetics. One of the key governing factors in the rate of any chemical reaction is its activation energy (E_a) or the "energy hump" which reactants must overcome to become products. From Figure 1.13, it is apparent that only a small fraction of the reactant molecules possess sufficient energy to rise above E_a and react at temperature T_1. By increasing the temperature to T_2, the energy distribution of the reactants changes, so now more particles possess energy greater than E_a, and therefore, the rate of the reaction increases. Hypergolic propellant reactions are characterized by low E_a values. Figure 7.6 shows the relative comparison between the activation energies of the following two reactions:

$$H_2 + 1/2O_2 \rightarrow H_2O \quad E_a = E_1 \quad (7.17)$$

and

$$3N_2H_4 + N_2O_4 \rightarrow 4N_2 + 2H_2 + 4H_2O \quad E_a = E_2 . \quad (7.18)$$

Assuming ideal behavior, the energy distributions for both mixtures are Maxwellian. However, many more molecules possess energies greater than E_a in equation 7.18 than in equation 7.17. Upon mixing the hypergolic propellants, the reaction of that proportion of molecules with energies greater than E_a generates sufficient heat to quickly raise the temperature of the mixture to the point where all molecular energies exceed E_a. Once ignition occurs, then the same principles governing any other propellant reaction apply to hypergolics. The key difference, however, is the absence of an initial spark in providing energy above E_a to enough molecules to ignite the mixture.

In addition to being used as a hypergolic propellant, hydrazine is also an excellent monopropellant. The hydrazine is catalytically decomposed to form gaseous nitrogen, hydrogen, and ammonia. This gas generator system consists of liquid hydrazine fed through lines to a catalyst bed made of a noble metal on an inert support material. Upon contact with the catalyst, N$_2$H$_4$ immediately decomposes, with no need of an external heating source, in the following manner:

$$3N_2H_4 \overset{\text{Catalyst}}{\rightarrow} 4(1 - x)NH_3 + (1 + 2x)N_2 + 6xH_2 \quad (7.19)$$

where x is the ammonia dissociation fraction.

Although hydrazine gas generators have lower performance values than other available propellant systems, their simple design, reliable operation, and multiple restart capability

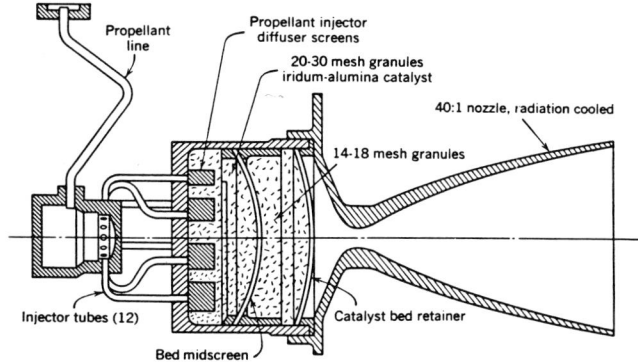

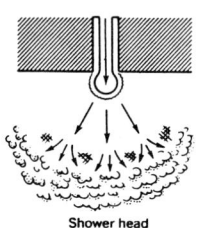
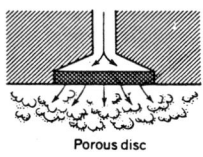
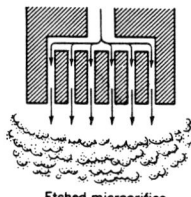

Figure 7.7 General design of a hydrazine monopropellant gas generator used for spacecraft attitude control [6]. *Reprinted by permission of John Wiley and Sons, Inc.*

Table 7.7 Exhaust Gas Species Generated by a Hydrazine Monopropellant Gas Generator [11]

Species	Mole Fraction	Condensation Temperature (K)*
NH_3	0.10–0.77	106
H_2	0.02–0.59	5
N_2	0.20–0.32	25
N_2H_4	0.004–0.05	162
H_2O	less than 0.03	166
C_6H_7N	less than 0.002	178
CO_2	less than 0.0002	83
Other volatiles (M = 100)	less than 0.00006	—
Fe	less than 0.00001	1,362
NVR (M = 500)	less than 0.000003	—
Reaction products of above six	less than parent species	—

*Temperature at which bulk evaporation rate is 10^{-7} g/cm²/sec

make them the standard for spacecraft attitude control (Figure 7.7) [10].

Hydrazine is a stable liquid at ambient temperatures, but slowly decomposes at elevated temperatures to form N_2 and NH_3. Acceleration of decomposition occurs through increasing temperature or by the presence of a metallic catalyst. Virtually all hydrazine monopropellant attitude control rockets use finely divided indium on an alumina (Al_2O_3) support as the catalyst. The liquid hydrazine enters the catalyst bed where it vaporizes and subsequently chemisorbs on the indium surface. Recall from Chapter 1, the purpose of a catalyst is to accelerate the reaction (in this case, decomposition) by lowering E_a. Once chemisorbed on the surface, the bond energies within N_2H_4 are lowered to the point of allowing rapid decomposition to form adsorbed N_2, H_2, and NH_3. These gases then desorb and flow out through the nozzle. Table 7.7 lists all of the exhaust species for hydrazine monopropellant.

Gas generator efficiency is measured by the amount of ammonia remaining in the exit stream. Decomposition of NH_3 has a mixed effect upon performance. First, the dissociation reaction is endothermic which leads to lower reaction chamber temperatures and decreases the specific impulse.

$$2NH_3(g) \rightarrow N_2(g) + 3H_2(g) \quad \Delta H^o = +46.1 \frac{kJ}{mole} \quad (7.20)$$

However, the dissociation does create exhaust products with a lower average molecular mass, thus increasing the value of I_{sp}.

$$2NH_3 \rightarrow N_2 + 3H_2 \quad (7.21)$$

$$\text{Average Reactant Molecular Mass} = 17 \frac{g}{mole}$$

$$\rightarrow \text{Average Product Molecular Mass} = 8.5 \frac{g}{mole}$$

A tradeoff must therefore be made between the two effects in determining the optimum reaction temperature.

Pure, anhydrous hydrazine is a stable liquid at temperatures as high as 530 K. Under sudden shock, it can decompose at temperatures as low as 367 K to produce a violent detonation. Due to its reactivity with a number of materials, care must be taken when selecting compatible materials, which include stainless steel, nickel, and aluminum (1100 and 3003 series). Other materials like iron, copper, monel, magnesium, zinc, and some aluminum alloys are not suitable [6].

Unsymmetrical Dimethylhydrazine

A more stable form of hydrazine at higher temperatures is its derivative, unsymmetrical dimethylhydrazine (UDMH). Two hydrogens on one of the nitrogens are replaced by two methyl groups:

```
    CH₃    H
      \   /
       N—N      UDMH
      /   \
    CH₃    H
```

In addition to being more stable than N_2H_4, UDMH also has a lower freezing point (216 K) and higher boiling point (336 K). When combusted, UDMH has only slightly lower

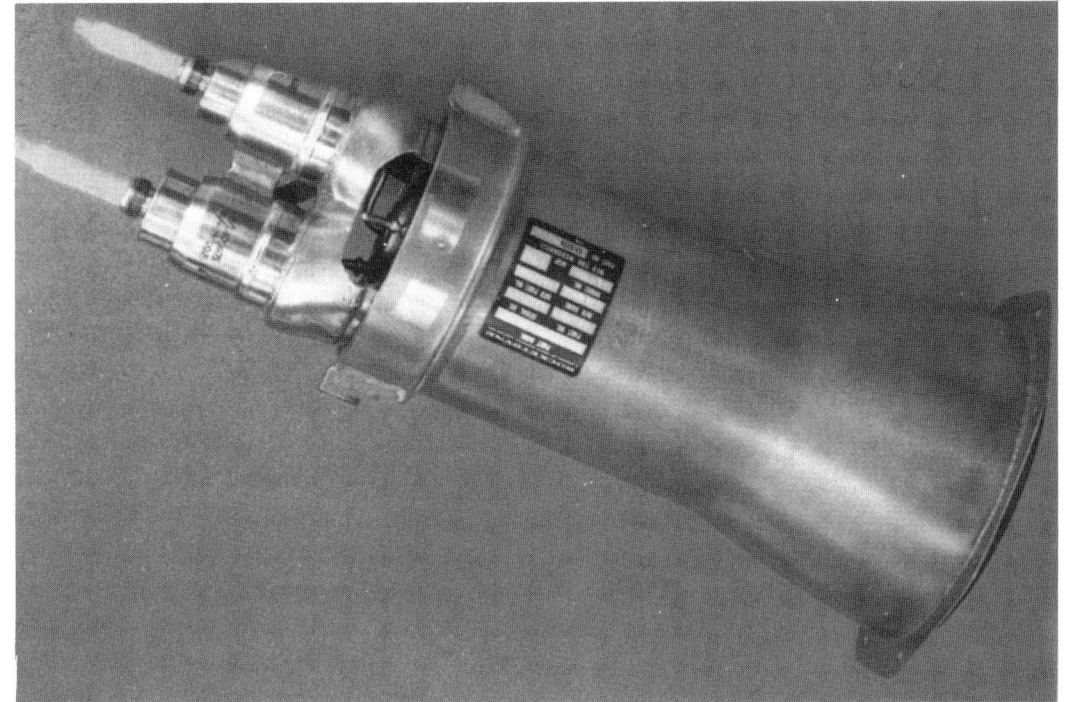

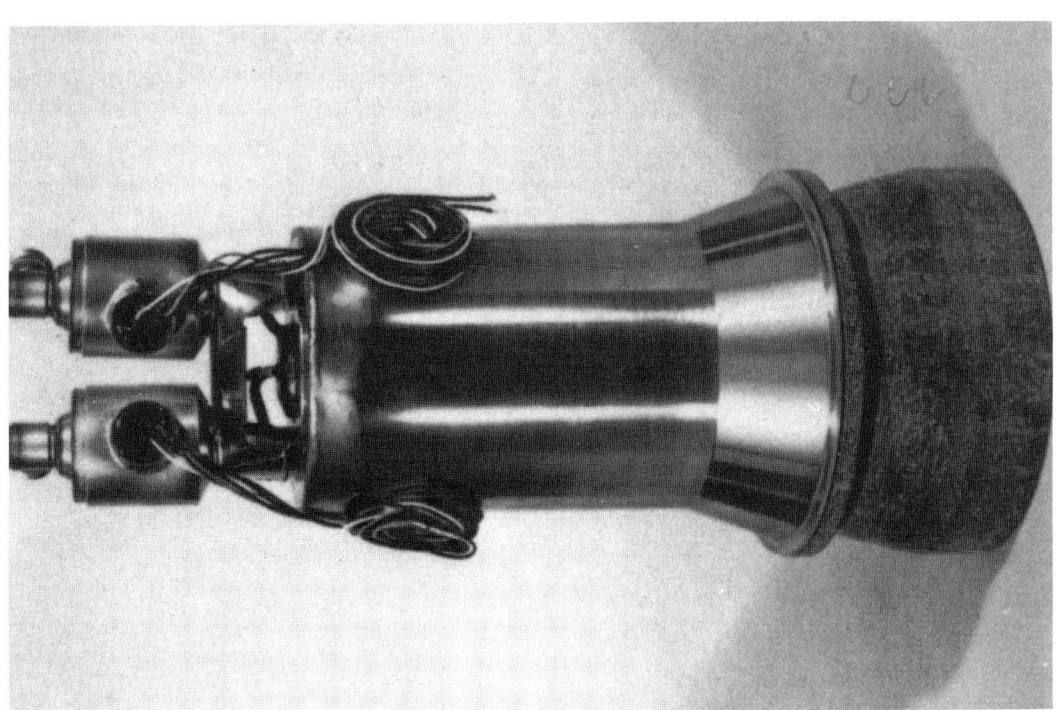

Figure 7.8 The MMH/N$_2$O$_4$ attitude control engines used on (a) Gemini and (b) Apollo spacecraft. *Courtesy of Rocketdyne.*

values for I_{sp} than does hydrazine. Mixing UDMH with up to 50% hydrazine is common. One caution about fuel mixing is the possible separation into distinct layers upon freezing. In the case of Aerozine-50, the mixture must be reblended after thawing [6].

Monomethyl Hydrazine

A second, more stable derivative of hydrazine is monomethyl hydrazine (MMH), which is extensively used in small attitude control engines, like the ones in Figure 7.8, with N_2O_4 as the oxidizer.

$$\begin{array}{c} CH_3 \quad\quad H \\ \diagdown \quad\diagup \\ N\!-\!N \quad\quad MMH \\ \diagup \quad\diagdown \\ H \quad\quad H \end{array}$$

Its advantages over pure hydrazine are (1) better shock resistance, (2) better heat-transfer properties, and (3) better liquid temperature range. Specific impulse is only 1% to 2% lower than that for N_2H_4. When added in small quantities of 3% to 15%, MMH quenches the explosive decomposition of hydrazine [6]. Table 7.8 gives the exhaust gas species for the bipropellant mixture of MMH-N_2O_4.

Solid Propellants

Solid rocket propellants, like the ones used in the Titan IV launch vehicle pictured in Figure 7.9, represent a dramatic simplification over liquid propellants in terms of engine design and operation. Fuel, oxidizer, and binder are stored directly in the combustion chamber in a premixed, solid state. Figure 7.10 shows a typical solid propellant motor with components. Motor performance is dictated by a number of factors, including fuel type, fuel-oxidizer ratio, grain size, and grain configuration. Design for a particular application must therefore be exact because performance cannot be varied once operation begins.

Solid propellants are categorized in the following manner:

1. Double-base (DB): homogeneous propellant grain consisting of a nitrocellulose dissolved in nitroglycerin

Table 7.8 Exhaust Gas Species for MMH-N_2O_4 Thruster [11]

Species	Mole Fraction	Condensation Temperature (K)*
H_2O	0.26 - 0.34	166
N_2	0.30 - 0.32	25
H_2	0.16 - 0.22	5
CO_2	0.04 - 0.15	83
CO	0.003 - 0.140	27
CH_6N_2	0.02 - 0.05	162
N_2O_4	0.02 - 0.05	181
NO	0.02 - 0.04	55
NH_3	less than 0.024	106
H	less than 0.02	—
"MMH-nitrate"	less than 0.02	300
CH_4	less than 0.017	37
OH	less than 0.015	—
CH_5N	less than 0.014	104
O_2	less than 0.002	29
Cl	less than 0.0005	—
NO_2	fraction of N_2O_4	152
N_2O	fraction of N_2O_4	75
N_2O_3	fraction of N_2O_4	129

*at which bulk evaporation rate is 10^{-7} g/cm^2/sec

Propulsion

Figure 7.9 The Titan IV launch vehicle powered by two solid rocket boosters attached to the main body. *Courtesy of Martin Marietta.*

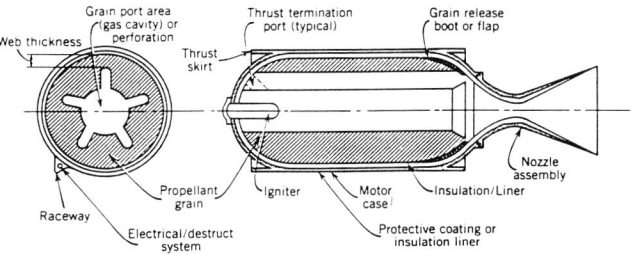

Figure 7.10 General design of a solid propellant rocket motor [6].

2. Composite: heterogeneous propellant grain formed using a polymeric binder to hold together oxidizer crystals and powdered fuel
3. Composite modified double-base (CMDB): polymeric binder in composite is replaced by a nitrocellulose-nitroglycerin matrix

Table 7.9 lists the chemical composition of each of these three types of solid propellants.

Oxidizers

Crystalline oxidizers supply the oxygen necessary for sustained combustion. Table 7.10 is a comparison of various crystalline oxidizers. Apparent from these compounds is the presence of a relatively large number of oxygen atoms. The choice of an oxidizer is based upon it having as positive an enthalpy of formation as stability permits, a high density, and a large oxygen content. With these criteria in mind, the oxidizer most often chosen is ammonium perchlorate. In general, all of the perchlorates are exceptional oxidizers which makes them suitable for high specific impulse propellants. Both ammonium and potassium perchlorate are insoluble in water, a desirable characteristic, but produce toxic, corrosive chlorine compounds during combustion. Ammonium perchlorate comes in the form of small, white crystals. Particle size is carefully controlled due to its impact upon propellant burning rate [6].

Fuels

Just as NH_4ClO_4 is the standard for oxidizers, aluminum metal is the standard solid propellant fuel. When reacted with ammonium perchlorate in the Space Shuttle's SRBs, the following occurs:

$$3Al(s) + 3NH_4ClO_4(s) \rightarrow Al_2O_3(s) + AlCl_3(s) + 3NO(g) + 6H_2O(g) . \quad (7.22)$$

This reaction is very exothermic, producing -2674 kJ of energy. Unfortunately, the exhaust products are rather heavy, thereby decreasing I_{sp}. Alumina presents an additional problem because it can solidify within the nozzle and

Table 7.9 Chemical Composition of Each of the Three General Types of Solid Propellants [6]

Double-Base		Composite		Modified Composite Double-Base	
Ingredients	Wt%	Ingredients	Wt%	Ingredients	Wt%
Nitrocellulose	51.5	Ammonium perchlorate	70.0	Ammonium perchlorate	20.4
Nitroglycerin	43.0	Aluminum powder	16.0	Aluminum powder	21.1
Diethyl phthalate	3.2	Polybutadiene-acrylic acid-acrylonitrile	11.78	Nitrocellulose	21.9
Ethyl centralite	1.0	Epoxy curative	2.22	Nitroglycerin	29.0
Potassium sulfate	1.2			Triacetin	5.1
Carbon black	< 1%			Stabilizers	2.5
Candelilla wax	< 1%				

Table 7.10 Properties of Common Crystalline Oxidizers [12]

Oxidizer	Formula	Molecular Weight (g/mole)	Density (g/cm^3)	Percent Oxygen Available	Heat of Formation at 25°C (kJ/mole)	Remarks
Ammonium perchlorate	NH_4ClO_4	117.49	1.95	34.04	−290	Low n, low cost, readily available
Ammonium nitrate	NH_4NO_3	80.05	1.725	20.0	−365	Smokeless, medium performance
Hydrazine nitrate	$N_2H_5NO_3$	95.07	1.685	8.4	−247	
Lithium nitrate	$LiNO_3$	68.95	2.38	58.02	−402.4	
Lithium perchlorate	$LiClO_4$	106.40	2.429	60.15	−410	
Nitronium perchlorate	NO_2ClO_4	145.46	2.22*	66	+33*	Very reactive, hygroscopic, high performance, unstable
Potassium perchlorate	$KClO_4$	138.55	2.524	46.19	−433.5	Low burning rate, medium performance
Hydrazine perchlorate	$N_2H_5ClO_4$	132.49	1.939**	24.2	−178**	

*Siegel and Schieler, *Energetics of Propellant Chemistry*, Wiley p. 175, 1964.
**Levy, von Elbe, et al. *Research on the Deflagration of High-Energy Oxidizers*, Atlantic Research Corp., Final Tech. Rept., Contract AF 49 (638)-1169, 1965.

on the combustion chamber walls if they are improperly designed.

Higher energy, lighter weight fuels like boron and beryllium appeared at one time to be attractive alternatives to aluminium. Boron, however, has a high melting point of 2,304°C and is difficult to combust efficiently within a reaction chamber of reasonable length. Beryllium burns more easily than boron and has a higher I_{sp} than aluminium by about 15 seconds. Its drawback is its extreme toxicity when inhaled by both humans and animals, making its application virtually impossible [6].

Binders

An organic binding material is used to hold the fuel and oxidizer together within the cast. The binder is also a fuel because it too is oxidized during the combustion process. Polymers are routinely used as binders because of their primary effect on motor reliability, mechanical properties, storability, and costs [6].

The first polymer used in early solid rocket motors was from the polysulfide family. Liquid polysulfide replaced the asphalt-pitch combination which was commonly used

Propulsion

Table 7.11 Properties of Three Common Solid Propellant Formulations [6]

Propellant Formulation	Oxidizer	Fuel	ρ (g/cm³)	T_1(°K)	C^* (m/sec)	M_c (kg/mole)	I_{sp} (sec)	γ
1	Ammonium nitrate	11% binder and 7% additives	1.51	1282	1209	20.1	192	1.26
2	Ammonium perchlorate 78–66%	18% organic polymer binder and 4–20% aluminum	1.69	2816	1590	25.0	262	1.21
3	Ammonium perchlorate 84 to 68%	12% polymer binder and 4 to 20% aluminum	1.74	3371	1577	29.3	266	1.17

Conditions for calculating I_{sp} and C^*:
Combustion chamber pressure–1000 psia
Nozzle exit pressure–14.7 psia
Optimum nozzle expansion ratio and frozen equilibrium assumed

ρ–average mass density of propellant
γ–specific heat ratio (c_p/c_v)

in composite propellants in the 1940s. A typical example is the polysulfide polymer $H(SCH_2CH_2-O-CH_2-OCH_2CH_2S)_n$. By controlling the polymerization process, the molecular weight was decreased to around 1,000 g/mole which allowed for very uniform mixing with the oxidant, NH_4ClO_4. The propellant mixture was then subjected to elevated temperatures which cured the polymer through crosslinking of sulfide linkages (i.e., vulcanization). The final product was a fairly homogeneous propellant mixture.

Research in the mid-1950s began on the replacement for polysulfide binders with polybutadiene polymers offering the advantage of a higher hydrogen content. A typical example is the polybutadiene-acrylonitrile (PDAN) polymer

$$-[CH_2-CH_2-CH_2-CH-CH_2-CH]_n-$$
$$|||$$
$$CH=CH_2 \quad C\equiv N \quad COOH$$

At the same time aluminum powder was added in order that the heat evolved as it reacted with oxygen would generate large amounts of hydrogen in the exhaust. The much higher hydrogen content would significantly reduce the average molecular weight of the exhaust gases, resulting in an increase in the specific impulse I_{sp}. A variation of the initial polybutadiene polymers, PBAN, is used today in the SRBs for the Space Shuttle. Rapid gains in solid rocket motor performance have been mainly due to advances in binder technology [12]. Table 7.11 lists the key properties of three common solid propellant formulations and Table 7.12 lists the exhaust gas species.

Table 7.12 Solid Rocket Motor Exhaust Gas Species [11]

Specie	Mass Fraction	Condensation Temperature (K)*
H_2	0.3576	5
CO	0.2477	27
HCl	0.11999	73
H_2O	0.09227	166
Al_2O_3	0.081518	1453
N_2	0.07769	25
CO_2	0.01688	83
$AlCl_3$	0.013378	277
NaCl	4.918×10^{-4}	699
$AlCl_2$	1.923×10^{-4}	—
Na	6.046×10^{-5}	425
KCl	4.384×10^{-5}	666
$FeCl_2$	3.466×10^{-5}	—
Fe	2.7734×10^{-5}	1362
AlO_2H	1.5997×10^{-5}	—
Al_2O	1.338×10^{-5}	—
NH_3	1.251×10^{-5}	106
NH_2	1.131×10^{-5}	—
AlCl	1.1164×10^{-5}	—
HCN	1.09775×10^{-5}	118
AlOH	9.745×10^{-6}	—
AlOCl	7.1285×10^{-6}	—
H	5.627×10^{-6}	—
NH	1.543×10^{-6}	—
Cl	9.651×10^{-7}	—
Na^+	1.1290×10^{-9}	—
K^+	9.2437×10^{-9}	—
Cl^-	9.244×10^{-9}	—

*Temperature at which bulk evaporation rate is 10^{-7} g/cm²/sec

Applications of Thermochemistry

Now that some of the fundamentals of propellant thermochemistry have been introduced, we will apply these principles to calculate the H_2-O_2 combustion chamber temperature using the available heat method. This problem is included as part of the introductory course taught to chemistry students at the U.S. Air Force Academy [3].

How well a propellant will perform depends upon the total energy released during the burning of the reactants to form products. It is desirable to choose propellants which will create a high combustion temperature and low molecular weight products. If the flame temperature is too high (greater than about 2,000 K), dissociation of the combustion products can occur. Dissociation, as has been noted, is the breakup of molecular species, which consumes energy released during the burning of propellants.

The combustion chamber temperature is a key variable in calculating rocket engine performance. To do this, we will use the available heat method. The following is an adaptation of the methods described in references 1 and 2.

1. Assume a combustion chamber temperature $T_c = T_c'$ at a particular chamber pressure.
2. Obtain a solution to the equilibrium constant equations for a value of the chemical composition of the combustion gases.
3. Use the calculated composition at the assumed temperature T_c' to determine the available heat, Q.
4. Calculate the heat absorbed by the products at T_c'. This is called Q'.
5. If $Q < Q'$, then the assumed value of T_c' is too large. If $Q > Q'$, then the assumed T_c' is too small. If $Q = Q'$, then the correct combustion temperature has been found: $(T_c = T_c')$.
6. Using the criteria in step 5, iterate until the correct T_c' is determined.

To illustrate this process, we will use a slightly more complicated version of the oxygen-hydrogen reaction:

$$5H_2(l) + O_2(l) \rightleftarrows 2H_2O(g) + 3H_2(g) \quad (7.23)$$

This example is more realistic than the one used previously. Liquid hydrogen and oxygen are used and the rocket runs fuel-rich from a mole-standpoint and oxygen-rich from a weight standpoint. It has an oxidizer-to-fuel ratio of 3.2 from a mass standpoint. The $H_2(g)$ in the exhaust reduces the average molecular weight of the products, which increases the rocket performance by increasing the exhaust velocity. The combustion chamber temperature is lower than it would be for a 2:1 (stoichiometric) H_2/O_2 mole mixture, but the extra H_2 in the exhaust helps make up some of the performance loss.

Before beginning, one must first understand what's going on thermodynamically in the reaction. At some equilibrium temperature T_c, there will be a thermal balance between the enthalpy released by the reactants and a rise in the enthalpy of the resulting products [7]. In other words, enthalpy released by the reactants equals enthalpy absorbed by the products.

Using the concepts discussed earlier, the amount of heat available from the reaction (ΔH_{rxn} or Q or Q_{rxn} or $Q_{available}$) is found. The step-by-step procedure is as follows.

1. Assume a combustion chamber temperature $T_c = T_c'$ *at a particular chamber pressure.* Any temperature can be selected, however, it's best to choose a temperature close to the correct value. $T_c' = 2500$ K will be used for this reaction.

2. Obtain a solution to the equilibrium constant equations for a value of the chemical compositions of the combustion gases. This is by far the most complicated step in the procedure, and both stoichiometry and the law of mass action come into play. Most net chemical reactions in a rocket problem are combinations of simpler reactions, all of which have a certain value for their equilibrium constants. At a particular temperature, the equilibrium constant for each reaction has a certain value (K is temperature dependent). In what is often a very complicated series of equations, the equilibrium constants can be used to determine the stoichiometric coefficients in the overall chemical equation at equilibrium. This method is known as the equilibrium constant approach and essentially allows one to do a mass balance on the chemical equation.

Another approach toward solving the mass balance is by the minimization of free energy. Recall, for a system at equilibrium, the change in free energy is equal to zero. In this approach free energy is minimized for all of the chemical reactions describing the combustion process. Although the approach is very different, the end result is the same as for the equilibrium constant approach. Both of these methods rely on an iterative process with a convergence on a solution involving many equations and many unknowns! The calculations are best performed by a computer. However, though they may seem complicated, the methods are an extension of simple chemical and thermodynamic principles.

For the sake of simplicity, assume the example reaction goes completely to products; that is, the combustion gases are only hydrogen gas and water vapor, and there are no dissociation reactions.

3. Use the calculated composition at the assumed temperature T_c' *to determine the available heat,* Q. At the estimated temperature $T_c' = 2,500$ K, it is assumed that the composition is completely products, in exactly the mole ratios shown in the balanced equation

$$5H_2(l) + O_2(l) \rightarrow 2H_2O(g) + 3H_2(g). \quad (7.24)$$

Now the available heat, Q (or Q_{rxn} or $Q_{available}$) at standard conditions is found using values from Table 1.4 thus,

$$\Delta H°_{rxn} = [2 \text{ mol}(-241.93 \text{ kJ/mol}) + 3 \text{ mol}(0.00 \text{ kJ/mol})] \\ - [5 \text{ mol}(-7.03 \text{ kJ/mol}) + 1 \text{ mol}(-9.42 \text{ kJ/mol})]$$
(7.25)

$$\Delta H°_{rxn} = -439.29 \text{ kJ/5 mol products}$$
$$Q = Q_{rxn} = Q_{available} = 439.29 \text{ kJ/5 mol product of released energy}$$

This Q is now the benchmark energy. This is the amount

Propulsion

Table 7.13 Enthalpy Values for Selected Compounds at Various Temperatures [1]

T(K)	H_2	H_2O	CO	CO_2	O_2	N_2
300	0.0	0.0	0.0	0.0	0.0	0.0
400	2.909	3.395	2.926	3.943	3.018	2.918
500	5.831	6.869	5.877	8.246	6.057	5.856
600	8.761	10.488	8.895	12.859	9.222	8.841
700	11.704	14.149	11.980	17.715	11.851	11.888
800	14.659	17.966	15.132	22.776	15.815	14.994
900	17.631	21.910	18.360	28.013	19.230	18.180
1000	20.641	25.987	21.646	33.388	22.692	21.424
1100	23.680	30.185	25.003	38.892	26.280	24.727
1200	26.761	34.522	28.402	44.493	29.867	28.084
1300	29.880	38.980	31.839	50.186	33.446	31.479
1400	33.048	43.530	35.317	55.954	37.034	34.920
1500	36.268	48.219	38.825	61.785	40.621	38.398
1600	39.512	52.995	42.362	67.679	44.338	41.898
1700	42.806	57.858	45.929	73.619	48.059	45.435
1800	46.142	62.815	49.516	79.605	51.777	48.989
1900	49.512	67.838	53.116	85.625	55.494	52.568
2000	52.928	72.937	56.737	91.682	59.215	56.164
2100	56.369	78.106	60.375	97.776	63.045	59.772
2200	59.839	83.339	64.021	103.896	66.867	63.338
2300	63.343	88.622	67.683	110.050	70.697	67.030
2400	66.871	93.959	71.350	116.203	74.519	70.672
2500	70.438	99.346	75.026	122.407	78.349	74.335
2600	74.021	104.708	78.726	128.632	82.293	78.010
2700	77.617	110.251	82.426	134.864	86.236	81.694
2800	81.250	115.768	86.131	141.118	90.179	85.390
2900	84.900	121.310	89.848	147.393	94.126	89.095
3000	88.576	126.899	93.570	153.676	98.065	92.804
3100	92.264	132.520	97.291	159.997	102.071	96.521
3200	95.972	138.163	101.033	166.326	106.090	100.242
3300	99.702	143.839	104.767	172.647	110.125	103.968
3400	103.444	149.536	108.514	179.010	114.177	107.702
3500	107.208	155.267	112.260	185.360	118.242	111.440
3600	110.979	161.019	116.015	191.752	122.311	115.182
3700	114.763	166.791	119.774	198.132	126.409	118.933
3800	118.573	172.593	123.537	204.536	130.511	122.679
3900	122.759	178.403	127.305	210.949	134.638	126.438
4000	126.220	184.226	131.076	217.379	138.778	130.201

of energy that is available to be consumed in order to raise the temperature of 5 moles of product from the standard temperature up to T_c.

4. *Calculate the heat absorbed by the products at T_c'.* This is called Q'. It could also be called Q_{required}. To calculate Q', an enthalpy table is needed. An enthalpy table is a compilation of ΔH values of various chemical species at different temperatures relative to a standard temperature; see Table 7.13.

The enthalpy table indicates how much heat it takes to raise one mole of the compound of interest from the standard temperature to the guessed temperature T_c'. Remember, only the products, H_2 and H_2O, are of interest in this example.

T_c' (guessed temp)	H_2	H_2O
2,500 K	70.438 kJ/mol	99.346 kJ/mol

Since there are 3 moles of H_2 and 2 moles of H_2O as products, at 2,500 K:

$$Q' = (3 \text{ mol})(70.438 \text{ kJ/mol}) + (2 \text{ mol})(99.346 \text{ kJ/mol})$$
$$Q' = 410.006 \text{ kJ/5 mol products} .$$

(7.26)

The Q' just found using Table 7.10 is the amount of heat absorbed by the products at T_c' (2,500 K).

5. *If $Q < Q'$, then the value assumed for T_c' is too large, if $Q > Q'$, then the assumed T_c' is too small. When $Q = Q'$,*

the correct temperature has been found. Since 439.29 kJ > 410.006 kJ, T'_c is too small.

6. *Using the criteria in step 5, iterate until the correct T'_c is determined.* Using a new T'_c of 2,700 K:

T'_c (guessed temp)	H_2	H_2O
2,700 K	77.617 kJ/mole	110.251 kJ/mole

$$Q' = (3 \text{ mol})(77.617 \text{ kJ/mol})$$
$$+ (2 \text{ mol})(110.251 \text{ kJ/mol}) \quad (7.27)$$
$$Q' = \underline{453.353 \text{ kJ/5 mol products}}.$$

Since 439.29 kJ < 453.353 kJ, the new T'_c is too large; therefore, the actual T_c is somewhere between 2,500 K and 2,700 K. At this point, 2,600 K is used to see how Q' compares to Q. However, if T_c is narrowed to a 200 K temperature range, the enthalpy function is fairly linear and linear interpolation to find T_c is acceptable. There is a 43.347 kJ difference between 2,700 K and 2,500 K. That amounts to 0.217 kJ/K. The difference between Q and Q'_{2500} is 439.29 − 410.006 = 29.28 kJ. So, 29.28 kJ/(0.217 kJ/K) = 135 K. Therefore T_c = 2,500 K + 135 K = 2,635 K.

With T_c in hand, a myriad of performance information can be found. These performance parameters include ratio of specific heats, exhaust velocity, specific impulse, characteristic velocity, and others. All are critical to the rocket designer.

Summary

Current spacecraft propulsion technology is based upon conversion of chemical energy into motion. Highly exothermic chemical reactions produce high velocity exhaust which produces the thrust to propel the vehicle. Propellants consist of a fuel and an oxidizer, both of which can be in a solid or liquid phase. Solid propellants have low I_{sp} values, but are relatively simple systems with high reliability. Liquid propellant engines are more complex than solid motors, yet yield higher I_{sp}. In addition, they have multiple restart capability. Applications vary for both systems with solid motors finding their greatest use in boosters. Liquid propellant engines are more flexible and can accomplish duties ranging from boosting to attitude control. Selection of the appropriate design is driven solely by the mission.

References

1. Thermochemistry Handout, Astro 495, U.S. Air Force Academy, Spring 1985.
2. Wilkins, R., *Theoretical Evaluation of Chemical Propellants*, Prentice-Hall, Inc. (Englewood Cliffs, NJ), 1963.
3. "The Thermochemistry and Thermodynamics of Rockets," Department of Chemistry, U.S. Air Force Academy, Spring 1987.
4. Ebbing, D., *General Chemistry*, Houghton Mifflin Co. (Boston), 1984.
5. *Chemistry in Space Research*, ed. by R. Laudeland and A. Rembaum, American Elsevier Publishing Co. (New York), 1972.
6. Sutton, G., *Rocket Propulsion Elements*, John Wiley & Sons (New York), 1986.
7. Johnson, N. L., and McKnight, D. S., *Artificial Space Debris*, Krieger Publishing (Malabar, FL), 1986.
8. "Space Resources for Teachers: Chemistry," NASA, NAS 1.19:87, February 1971.
9. Altman, D. et al., *Liquid Rocket Propellants*, Princeton University Press (Princeton N.J.), 1960.
10. Curran, F. and M. Whalen, "In-situ Analysis of Hydrazine Decomposition Products," NASA TM-89916, 23rd Joint Propulsion Conference, June 29–July 2, 1987.
11. Shaw, C. G. "Surface Effects Evaluation Study," Final Report for Air Force Contract F04611-84-C-0007, AFRPL TR-86-093, February 1987.
12. Williams, F., Barrere, M., and N. Huang, *Fundamental Aspects of Solid Propellant Rockets*, Technivision Services (Slough, England) AGARDograph #116, October 1969.
13. Sutton, E., "From Polysulphides to CTPB Binders—A Major Transition in Solid Propellant Binder Chemistry," AIAA, Paper 84-1236, June 1984.

Discussion Questions

1. Aircraft engines and rockets are similar in that they both use fuel and oxidizer. However, the form of the fuel and oxidizer can differ dramatically between the two. Discuss these differences as they pertain to operation.
2. The yardstick for measuring propellant performance is specific impulse.
    a. Explain what specific impulse means.
    b. What two parameters make the greatest contribution to maximizing I_{sp}?
3. Discuss the desirable characteristics of a liquid rocket propellant.
4. Fluorine offers higher values of performance as an oxidizer than does oxygen. Why don't we use F_2 propelled rockets?
5. Monopropellant gas generators are used extensively as attitude control thrusters aboard spacecraft.
    a. Describe the operation of a hydrazine gas generator.
    b. Why are they so widely used?
6. The word *hypergolic* is used to describe certain propellants.

a. What does hypergolic mean?
b. Discuss in terms of chemical kinetics what happens when two hypergolic propellants are mixed. How does a hypergolic reaction differ from a non-hypergolic one?

7. Hydrazine has been replaced over the years by UDMH and MMH. Why?

8. Solid propellants are much more complicated than liquid propellants due to the larger number of individual components required to make them. Describe the function of each of the following components within a composite propellant.
   a. ammonium perchlorate
   b. polybutadiene
   c. epoxy
   d. aluminum

9. The combustion of H_2 and O_2 is thermochemically spontaneous, yet an ignition is needed to start the reaction. Explain the fundamental difference between thermochemistry and chemical kinetics in the context of this problem.

Problems

1. Rather than perform actual experiments to measure reaction enthalpy, scientists often estimate the value using bond energies.
   a. Estimate the enthalpy of reaction for the combustion of hydrazine.

   $$N_2H_4(g) + O_2(g) \rightarrow N_2(g) + 2H_2O(g)$$

   b. Why is this calculated value different from the experimental value?

2. A new rocket has a specific impulse of 900 sec but will provide a thrust of only 500 lbs.
   a. What is the mass flow rate of propellant?
   b. If this thrust is required for a 100 minute burn, how much propellant is needed?
   c. What volume tank would be required if the propellant is hydrogen? Is this a realistic size?

3. Equation 7.7 describes the theoretical specific impulse.
   a. If the combustion chamber temperature is doubled, how is specific impulse changed?
   b. If the average ratio of specific heats is 2.4 and the pressure ratio is tripled, how is the specific impulse changed?
   c. If the nozzle exit pressure is equal to the combustion chamber pressure, what is the specific impulse?

4. Table 7.13 lists enthalpy values for various compounds over a range of temperatures.
   a. What is the percentage change in enthalpy for hydrogen going from 1,500 K to 3,000 K? for water?
   b. Does the difference between these derived values make sense? Explain.

Chapter 8

Thermal Control and Protection Systems

The control of heat into and through a satellite and its components is one of the most important aspects of spacecraft design. Yet, it receives proportionally little attention. The dark black expanse of space appears to be a consistently cool environment for operations. It is not. A satellite in Earth orbit receives energy from a variety of sources. It may come directly from the Sun, as reflected sunlight off the Earth, or as Earth-radiated energy. The amount of energy absorbed is dependent on the satellites's orientation, location, and design. In addition, numerous internal thermal loads, such as electronic devices and energy-conversion hardware, complicate the thermal control process. The design problem entails maintaining the temperature of many different components within their respective limits. These temperature extremes must not be exceeded during any phase of the mission: prelaunch, launch, deployment, or operation [1, 2].

A subset of thermal control is thermal protection. Vehicles which return to Earth must be designed to withstand the extremely high temperatures generated during reentry. Thermal control systems distribute modest amounts of energy from sources to sinks while thermal protection devices dissipate large quantities of heat to maintain the environmental and structural integrity of the spacecraft. Thermal protection systems usually are used only during a short period of a satellite's mission, while in contrast, thermal control generally spans a craft's entire useful lifetime.

The pioneering work in thermal protection materials was initiated during the development of intercontinental ballistic missiles (ICBM). An ablative was used to dissipate the great heating experienced in flight. The early manned space missions used similar technology, but the Space Shuttle has ushered in a new era in thermal protection. The Shuttle has the most complex and efficient series of heat absorbing and transferring materials ever engineered. This unique design will be highlighted to explain thermal protection methodology.

The design process for thermal control and protection systems is summarized in Figure 8.1. First the temperature limits of all satellite subsystems are compiled and compared to the expected temperatures. This establishes the thermal boundary conditions [1]. Next, the internal loads are factored in. Last, the necessary thermal control and protection devices are incorporated into the design.

As will be discussed later, both passive and active devices may be used to control the thermal balance of a satellite. The emphasis, however, will be on the chemical processes which allow thermal control and protection systems to function. Also covered will be the chemical effects which degrade their effectiveness as well as the basics of heat transfer.

Nomenclature from thermal control and protection designs is not always consistent so a standardized set of terminology is proposed. The following terms and units will be used throughout this chapter.

$$Q = \text{heat, J}$$
$$q = \text{heat per unit area, J/m}^2$$
$$\dot{Q} = \text{heat per unit time, J/s or watts}$$
$$\dot{q} = \text{heat per unit time and area}$$
$$\phantom{\dot{q}} = \text{heat flux, W/m}^2$$

Basics of Heat Transfer

Energy may be transmitted via conduction, convection, or radiation. Heat, as a form of energy, may be transferred via conduction or convection. Conduction is the transfer of energy through a stationary material while convection transmits energy via the motion of fluids. Thus, both conduction and convection require some medium for the transfer of energy. The last mode of energy exchange, radiation, needs no such medium. Radiation transmits energy through photons or electromagnetic waves.

Conduction and radiation are the two mechanisms of the greatest importance for thermal control design of spacecraft, while convection is important during the reentry phase which requires thermal protection.

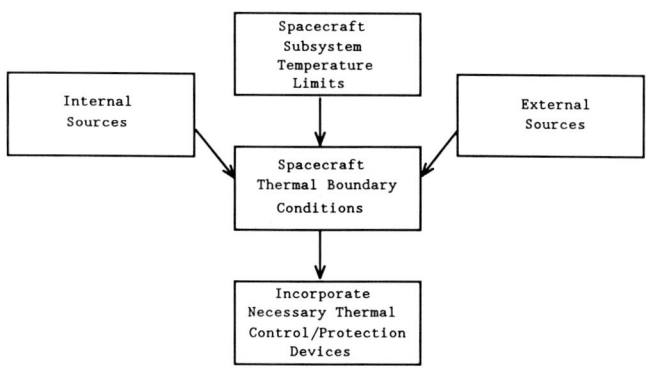

Figure 8.1 An outline of the design process for thermal control and protection systems. The central aspect of the design process is the determination of spacecraft thermal boundary conditions.

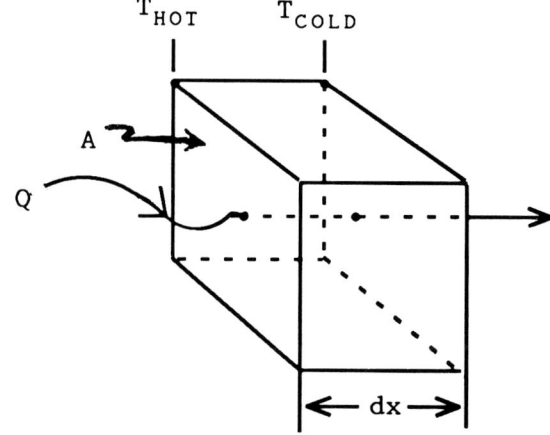

Figure 8.2 Conduction is a function of thermal conductivity, k, and the temperature gradient.

Conduction

Heat will naturally flow from areas of high temperature to low temperature. A material's temperature increase is manifested by an increase in the random motion within the molecular structure. This excitation is dissipated by a transfer of kinetic energy to less energetic atoms, thus increasing their temperature. In this way, heat is conducted through a medium and can be described by the following equation [3]:

$$\dot{q} = -k \frac{dT}{dx} \quad (8.1)$$

where

$\dot{q}$ = heat flux, W/m^2
k = thermal conductivity, W/(m · K)
T = temperature, K
x = path of heat flow, m
$\frac{dT}{dx}$ = temperature gradient, K/m

The above equation, called the Fourier law, may be modified by noting that [3]

$$\dot{q} = \dot{Q}/A \quad (8.2)$$

where $\dot{Q}$ is the rate of heat transfer in watts or joules/second, and A is the cross-sectional area of heat path normal to the x-axis in square meters. Combining equations 8.1 and 8.2 produces the most widely used version of the conduction equation which is graphically described in Figure 8.2 and is given as

$$\dot{Q} = -kA \frac{dT}{dx}. \quad (8.3)$$

The thermal conductivity, k, is a property of a given material at a given temperature. Table 8.1 lists the k values for a variety of substances [1, 3].

Table 8.1 Thermal Conductivity, k, at Different Temperatures

Material	k(W/m · K)	T(°C/K)
Aluminum	210	25/298
	249	400/673
Aluminum alloys	117–175	25/298
Copper	386	20/293
	369	300/573
Magnesium	157	25/298
Silver	407	20/293
Titanium	21	25/298
Stainless Steel	16.2	25/298
Teflon	0.25	25/298
Tungsten	163	20/293

A larger value of k means a greater capacity to transfer heat by conduction. The heat flow rate and the cross-sectional area through which the heat passes are also directly proportional to one another. The negative sign in equation 8.3 denotes that the heat flow is from high to low temperature regions (i.e., $dT/dx < 0$).

As mentioned earlier, conduction will occur as heat is transmitted through a solid medium. This is likely to occur between components of a satellite. Often, heat balance is calculated over more than one material or component and a simplified method of representing conduction through a series of dissimilar subsystems is required.

Thermal conduction can be modeled as current flow through a resistor where the rate of heat flow is $\dot{Q} = (T_{HOT} - T_{COLD})/R$ as shown in Figure 8.3 [3].

The resistance to the flow of heat, Q, is given by $\Delta X/kA$ which will be referred to simply as R. Figure 8.4 shows that just as resistors in series add to represent an equivalent resistance, the thermal resistance of two conducting materials may be added to provide an equivalent thermal resistance.

Thermal Control and Protection Systems

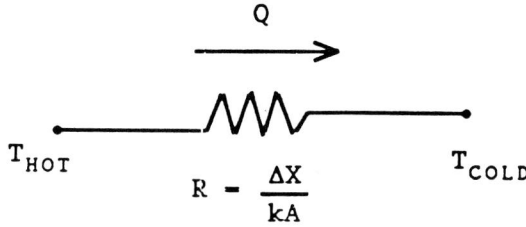

Figure 8.3 The resistive analogy to conduction simplifies analysis.

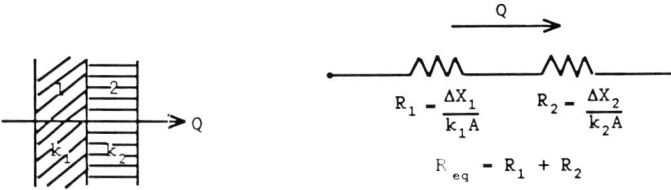

Figure 8.4 Series resistance to heat flow combines like electrical resistors.

The same relationship holds true for parallel conducting solids; see Figure 8.5.

Using these basic identities, very complex heat transfer conditions may be addressed. Take a spacecraft wall made of aluminum with an external coating of paint and an internal blanket of Kapton. A steel bolt penetrates all three materials as shown in Figure 8.6. The equivalent resistive diagram is shown next to it [3].

The equivalent resistance values for two geometries commonly found in satellite structures are given in Figure 8.7 [3].

Example—Conductor Problem [4]

The outer skin of a 1.5m long cylindrical rocket section is subjected to heating which conducts a $\dot{q} = 315$ J/(s · m²)

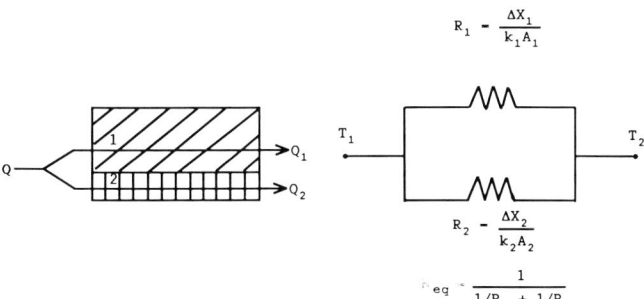

Figure 8.5 The parallel resistance analogy holds for heat that may conduct through parallel paths.

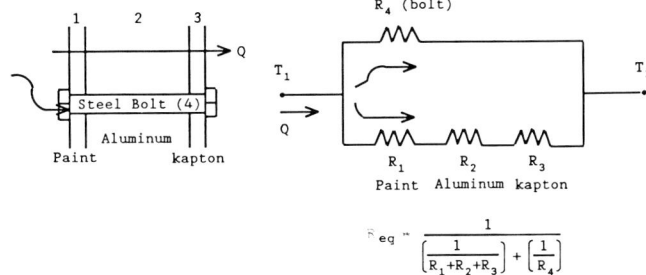

Figure 8.6 Example of spacecraft wall heat conduction is simplified by use of resistance analogies.

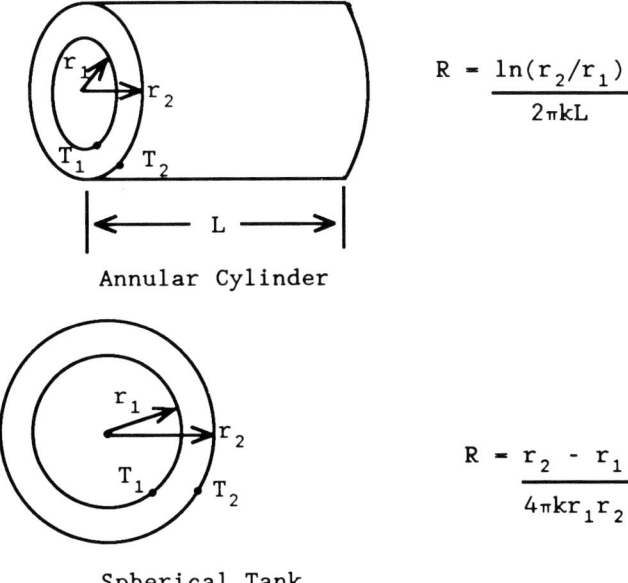

Figure 8.7 Cylinders and spheres are commonly used geometries in satellite design.

into the rocket's interior. The inner radius is 30cm while the rocket's insulation is 5 cm thick. The inner surface of the rocket's wall/insulation is at 10°C (50°F). The conductivity of the insulation is 0.18 J/(s·m ·°C). Calculate the temperature of the outer skin.

1. Determine the surface area of rocket section:
$$A = 2\pi r L = 2\pi(0.35 \times 1.5) = 3.30 \text{m}^2$$

2. Define temperature and radius terms.
   inner $T_1 = 50°F$, $r_2 = 30$cm
   outer $T_2 = ?$, $r_2 = 35$cm

3. Determine $\dot{Q}$:
$$\dot{Q} = \left(315 \frac{\text{J}}{\text{s} \cdot \text{m}^2}\right)(3.30\text{m}^2) = 1041 \text{ J/s}$$

4. Determine equivalent resistance:

$$R = \frac{\ln(r_2/r_1)}{2\pi k L} = \frac{\ln(35/30)}{2\pi \left(0.18 \dfrac{J}{s \cdot m \cdot °C}\right)(1.5 \text{ m})}$$

$$R = 0.164 \frac{s \cdot °C}{J}$$

5. Solve the conduction equation for outer temperature T_2:

$$\dot{Q} = \frac{T_1 - T_2}{R} \quad \text{or} \quad T_2 = T_1 + \dot{Q}R$$

$$T_2 = 10°C + \left(1041 \frac{J}{s}\right)\left(0.091 \frac{s \cdot °C}{J}\right)$$

$$T_2 = 104.8°C\ (220.7°F)$$

Radiation

Any object whose temperature is higher than the ambient temperature will emit electromagnetic radiation. This energy is generally radiated nondirectionally with random phase and frequency. Thermal radiation is grouped in the 0.1 to 100 μm wavelength range. The upper limit of this range includes ultraviolet (UV) radiation—1 to 100 nm. When the radiation has a wavelength in the 0.38 to 0.76 μm range, the energy is visible. An example of this phenomenon is the glowing red electric burner on a stove. If the wavelength of the radiation is slightly longer than visible (1 μm–1 mm), the infrared (IR) region is encountered. Radiation of this type is less energetic and is emitted by an object unless it is very cold. IR radiation can be detected with the aid of night-vision observation devices.

The total energy radiated per unit surface can be expressed as [1]

$$\dot{q} = \sigma \epsilon T^4 \tag{8.4}$$

where

$\dot{q}$ = energy (or heat) flux, W/m^2
σ = Stephan-Boltzmann constant
  = 5.669 × 10^{-8} W/(m^2 · K^4)
ϵ = emissivity
T = absolute temperature, K

The emissivity is a material constant which may range from zero to one. When $\epsilon = 1$ the body is called a "black body" and emits the maximum amount of energy. The emissivity may be thought of as the ratio between the actual energy emitted by a body, at some temperature T, and the radiation produced by a perfectly emissive object (black body) at that same temperature.

Once radiant energy encounters another body, it may be absorbed, reflected, or transmitted. These three phenomena are quantified by the following three dimensionless constants [3]

α = absorptivity (fraction of incident radiation absorbed)
ρ = reflectivity (fraction of incident radiation reflected)
τ = transmissivity (fraction of incident radiation transmitted).

By the conservation of energy it must be true that [3]

$$\alpha + \rho + \tau = 1. \tag{8.5}$$

Just as a "black body" is considered a perfect emitter, it is also assumed to be a perfect absorber. Some black paints have emissivity values near unity while polished gold or silver have values as low as $\epsilon = 0.1$–0.3.

Thermal equilibrium, a goal for thermal control systems, is attained when the incident radiation is emitted from a body at the identical temperature as the original emitting body. Thermal equilibrium occurs when $\epsilon = \alpha$.

The energy radiated in each wavelength interval, $d\lambda$, contributes to the total energy emitted by a body. The total energy produced is [3]

$$E = \int_0^\infty E_\lambda \, d\lambda \tag{8.6}$$

where E_λ is the energy produced with wavelengths between λ and $\lambda + d\lambda$. A plot of the energy radiated versus wave-

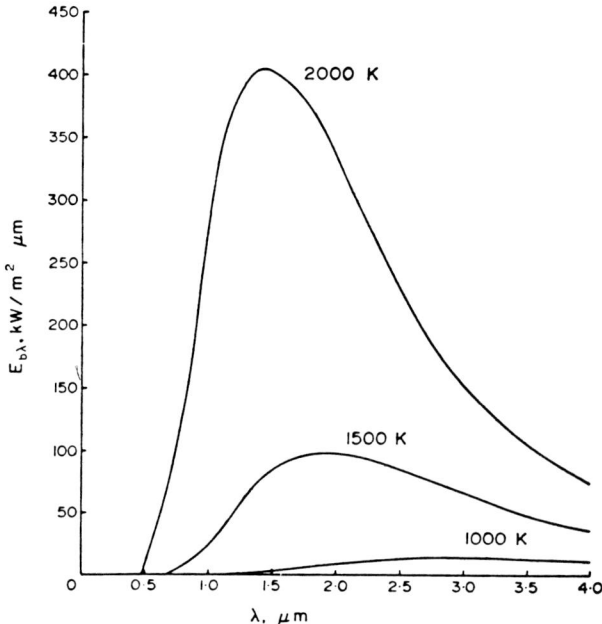

Figure 8.8 Black body radiation peaks at lower wavelengths for higher temperatures [2]. *Reprinted by permission of Prentice-Hall.*

Thermal Control and Protection Systems

Table 8.2 Thermal Properties of Surfaces [6]

Surface	Typical Application	Absorptance, α		Emittance, ϵ	
		Beginning of Life	End of Life (7 years)	Beginning of Life	End of Life (7 years)
Silicon solar Cell (1 mm thick)		0.94	—	0.32	—
Glass, 3 mils Silicon solar cell		0.93	—	0.84	—
Black paint	Interior structure	0.9	0.9	0.9	0.9
White paint	Antenna reflector	0.2	0.6	0.9	0.9
Optical solar reflector	North and South panel radiators	0.08	0.21	0.8	0.8
Graphite/epoxy	Solar panel and antenna structure	0.84	0.84	0.85	0.85
Chromium plate (0.1 mil thick)		0.78	—	0.15	—
Stainless steel (Type 410)		0.76	—	0.13	—
Aluminized Kapton	Thermal insulation	0.35	0.50	0.6	0.6
Aluminum foil with 10 mil silicon		0.52	—	0.12	—
Anodized titanium	Apogee motor thermal shield	0.6	0.6	0.6	0.6
Aluminum, aluminum tape, deposited aluminum	Propellant insulation	0.12	0.18	0.06	0.06
Anodized aluminum	Interior structure	0.2	0.6	0.8	0.8
Solar cells	Solar panels	0.65–0.75	0.65–0.75	0.82	0.82
Gold		0.2–0.3	—	0.03–0.06	—

length, for a single temperature, is shown in Figure 8.8. This typical radiative energy transfer plot will flatten and the peak move down the x-axis (lower λ) as the temperature increases.

From Figure 8.8, it can be seen that as temperature decreases less of the radiated energy is in the visible spectrum. The high temperature curve is representative of direct solar radiation with great amounts of energy in the visible region. If a body needs to be protected from direct solar radiation, then a low α value in the visible wavelength range is desirable. On the other hand, the value of ϵ should be large so that any excess energy absorbed by the body will be radiated.

Table 8.2 lists values of α and ϵ for a variety of surfaces. The changes in these parameters result from interaction with the environment via particulate impacts and chemical reactions.

Example—Radiation Problem [3]

Assume that a deep space probe is a 1 m diameter sphere covered in aluminized Kapton. What is the equilibrium temperature of this probe if the incident solar energy rate is 300 W/m^2? (The solar energy flux density just outside Earth's atmosphere is about 1,350 W/m^2 while the high end value on Earth is about 946 W/m^2).

To reach thermal equilibrium the rate of energy absorption and radiation must be equal. The energy absorbed is the product of the absorptance, solar energy flux density (E), and exposed area to the Sun (A_α). Similarly, the energy radiated per unit area, from equation 8.4, times the total area (A_ϵ) times the total area represents the energy radiated.

$$\text{Energy Absorbed} = \text{Energy Radiated}$$
$$(\alpha E)A_\alpha = (\sigma \epsilon T^4)A_\epsilon$$

The absorptivity for aluminized Kapton from Table 8.2 is taken to be $\alpha = 0.35$ while the projected area of the sphere (cross-section) to the Sun is $A_\alpha = \pi d^2/4 = 0.785$ m^2.

The emissivity of the same type of material is $\epsilon = 0.6$ from Table 8.2. The area for radiation is the surface area of the sphere, $A_\epsilon = 4\pi r^2 = 3.142$ m^2. Making substitutions into the above equation yields

$$0.35 \times 300 \text{ W/m}^2 \times 0.785 \text{ m}^2$$
$$= 0.6 \times \left[5.669 \times 10^{-8} \frac{\text{W}}{\text{m}^2 \cdot \text{K}^4}\right] \times T^4 \times 3.142 \text{ m}^2$$
$$82.425 \text{ W} = \left[1.069 \times 10^{-7} \frac{\text{W}}{\text{K}^4}\right] \times T^4$$

Solving for the temperature yields

$$T^4 = 7.725 \times 10^8 \text{ K}^4$$
$$T = 167 \text{ K}$$

Since some of the material properties may vary by temperature and time in orbit, this equilibrium temperature is an approximate value. If a tight temperature range is necessary for mission operations, active heat transfer devices may be used to control the energy balance.

Convection

Convection is the transfer of energy through a moving fluid medium. Energy is transferred from a higher temperature solid surface to the fluid via conduction. The moving fluid then transports, or convects, the energy downstream. The energy is further diffused through the fluid by conduction. This form of energy transport is termed convective heat transfer.

Convective heat transfer theory was developed in the 1950s. Empirical work up to that point had shown the dependency of convection on fluid velocity and properties as well as the geometry of the solid interfaces with the fluid. The material properties of the solid do not influence the problem [1].

The science of convection involves numerous parameters but basically the amount of heat transferred is proportional to the temperature difference ΔT between the solid surface T_s and fluid T_f.

$$\dot{q} = h \, \Delta T . \quad (8.7)$$

The term h is the convective conductance and has units of W/(m$^2 \cdot$ K). The circuit resistance analogy may be used for convection as it was for conduction. The convective thermal resistance is

$$R = \frac{1}{hA} \quad (8.8)$$

where A is the area of the fluid-solid interface. Now equation 8.7 may be written as

$$\dot{q} = \frac{\Delta T}{R} . \quad (8.9)$$

This relationship may now be teamed with earlier equations to analyze heat transfer problems with convection and conduction. Yet, due to the numerous varying parameters possible in convective problems, a series of dimensionless groupings are used in solving for heat transfer characteristics [4]. Two important parameters, the Prandtl and Reynolds numbers, are defined as follows:

$$\text{Prandtl number: Pr} = \frac{\mu C_p}{k} \quad (8.10)$$

and

$$\text{Reynolds number: Re} = \frac{VL\rho}{\mu} \quad (8.11)$$

where

μ = viscosity, kg/(hr $\cdot$ m)
C_p = specific heat at constant pressure, W/(kg $\cdot$ K)
k = thermal conductivity, W/(m $\cdot$ K)
V = velocity, m/s
L = length, m
ρ = fluid density, kg/m^3

The Prandtl number quantifies the ratio of a fluid's ability to dissipate momentum versus internal energy. Thick, viscous fluids have large Pr values (100 to 10,000) so momentum is diffused quickly in comparison to internal energy. Conversely, gases and thin liquids will dissipate their internal energy quickly. Table 8.3 lists the values for Pr for various gases and fluids [3].

The Reynolds number quantifies the relative effects of inertia and viscosity for a moving fluid [1]. Viscous characteristics dominate inertial qualities for low Re. This means that the flow remains organized and is termed laminar. High Re values indicate a greater dependence on inertial effects thus leading to turbulent flow.

A third parameter is used to define fluid convective flow: the Grashof number. It determines the relation between buoyancy and viscous forces and is expressed as [1]

$$\text{Gr} = \frac{g\beta(\Delta T)L^3\rho^2}{\mu} \quad (8.12)$$

where

β = isobaric compressibility, $\beta = \frac{1}{v}\left(\frac{\partial v}{\partial T}\right)_p$
v = volume, m^3
g = acceleration due to gravity, m/s^2.

Two final parameters are directly related to the Prandtl and Reynolds numbers and are given as the

$$\text{Nusselt number: Nu} = \frac{hL}{k} \quad (8.13)$$

and

$$\text{Stanton number: St} = \frac{h}{V\rho C_p} . \quad (8.14)$$

Thermal Control and Protection Systems

Table 8.3 Prandtl Number

Fluid	Prandtl Number	Temperature
H_2O	13.6	$T = 273$ K ($0°C$)
	4.3	$T = 313$ K ($40°C$)
	1.7	$T = 373$ K ($100°C$)
Air	0.770	$T = 100$ K ($-173°C$)
	0.680	$T = 500$ K ($227°C$)
	0.702	$T = 1000$ K ($727°C$)
	0.705	$T = 1500$ K ($1227°C$)
H_2	0.664	$T = 600$ K ($327°C$)
O_2	0.704	$T = 600$ K ($327°C$)
N_2	0.691	$T = 700$ K ($427°C$)

These last two terms are used in tandem, since after one of these is found, all other terms may also be calculated.

In summary, natural convection is defined through the Nusselt number which is a function of Grashof and Prandtl numbers. More general convection problems are solved using the values for the Nusselt and Stanton numbers.

Convection calculations for reentry vehicles are beyond the scope of this text. It is, however, important to understand effects of fluid characteristics and vehicle velocity on spacecraft heating. Thermal protection systems must respond to heat flow by conduction and convection. Complex analytical and computational tools are necessary for the design and construction of vehicles such as the Space Shuttle, whose thermal protection system will be described later in the chapter.

Table 8.4 Typical Equipment Temperature Limits [2]

	Thermal Design Temperature Limits (°C), Min/Max	
Subsystem/Equipment	Nonoperating/Turn-on	Operating
Communications		
Receiver	$-30/+55$	$+10/+45$
Input multiplex	$-30/+55$	$-10/+30$
Output multiplex	$-30/+55$	$-10/+40$
TWTA	$-30/+55$	$-10/+55$
Antenna	$-170/+90$	$-170/+90$
Electric power		
Solar array wing	$-160/+80$	$-160/+80$
Battery	$-10/+25$	$0/+25$
Shunt assembly	$-45/+65$	$-45/+65$
Attitude control		
Earth/sun, sensor	$-30/+55$	$-30/+50$
Angular rate assembly	$-30/+55$	$+1/+55$
Momentum wheel	$-15/+55$	$+1/+45$
Propulsion		
Solid apogee, motor	$+5/+35$	
Propellant tank	$+10/+50$	$+10/+50$
Thruster catalyst bed	$+10/+120$	$+10/+120$
Structure		
Pyrotechnic mechanism	$-170/+55$	$-115/+55$
Separation clamp	$-40/+40$	$-15/+40$

Thermal Control Systems

Subsystems of a satellite may include communications, electric power, attitude control, propulsion, and structures hardware. Each of the components have thermal design limits for both nonoperational/turn-on status and operational mode as listed in Table 8.4. Usually, the acceptable temperature limits are lower for systems while in use.

The energy deposited on a satellite may come directly from several sources. The Sun can directly illuminate a spacecraft with radiation spanning a wide range of wavelengths. The incident solar radiant energy flux is around 1,350 W/m², as discussed earlier [4]. More of the Sun's energy may be reflected off the Earth toward the satellite. A measure of the Earth's ability to reflect solar energy is called its albedo. The Earth may actually reflect up to 75% of the energy hitting it, although the average is about 50%. Lastly, the fairly stable temperature of the Earth, normally around 240 K, creates radiant energy which must be considered. This nearly steady-state radiation source is present even when a satellite is above the unlit side of the Earth [4]. This Earth-radiated energy is fueled partially by solar energy and by chemical reactions on the Earth and in its atmosphere.

The iterative design process starts with a conceptual design where structural and thermal requirements are compiled. Preliminary analyses are conducted to quantify the performance of the original thermal control design to anticipated external and internal thermal loads. Compromises may be required since several features such as antenna and solar panel locations are often not flexible. Electrical power sources contribute greatly to the internal thermal loading problem which changes each time the system configuration is altered.

The spacecraft configuration now includes thermal control devices which serve to provide sources and sinks of thermal energy to keep all subsystems within their temperature limits. Coatings, insulation/structure, heat pipes, and radiators may also be used [2].

After the design has been determined to be acceptable in a thermal computer model, a spacecraft thermal test model is constructed. Thermal loads are simulated during testing to measure the adequacy of the thermal design. Eventually, after the fine tuning of the thermal control design, the flight spacecraft will even undergo thermal vacuum qualification and acceptance tests. The environment which is anticipated for the satellite is reconstructed as accurately as possible. Usually, during qualification and acceptance testing the measured temperatures cannot vary by more than 10°C and 5°C from the accepted values, respectively [2].

Thermal control devices will be outlined in later sections, followed by an example of a thermal control design for a geosynchronous spacecraft.

Coatings

The radiative properties of various surfaces were outlined in Table 8.2. These coatings are usually used to radiate internal heat and to prevent external heat from being absorbed. Thus, low absorptance, α, and high emittance, ϵ, values are best.

In their simplest form, paints are a two phase mixture consisting of a binder and a pigment. Some common binders include epoxies, acrylics, silicones, silicates, and polyurethanes. Optical properties are controlled by the type and amount of pigment used. White pigments used in low absorptance–high emittance applications include zinc oxide (ZnO), titanium dioxide (TiO_2), zinc orthotitanate ($ZnTiO_4$), and zirconium dioxide (ZrO_2). For medium absorptance and emittance, aluminum flakes work well. Carbon black is the most common pigment for high α and high ϵ paints [5].

Table 8.2 shows that over time, in orbit, the absorptance increases while the emittance value is fairly stable. The degradation in absorptance is due to particulate impact, charged particles, UV radiation and effects of high vacuum [2]. Volatile solvents used in the manufacturing process outgas appreciably in space. The organic-based paints are particularly harmful because their condensing volatiles contaminate optical and electrical components in the vicinity. Particle and radiation impact embrittle paint layers and cause loss of adhesion. However, the most damaging effect of paint contamination is degradation of optical properties. Black paints are generally immune, except for some minor bleaching. White paints, on the other hand, are quite susceptible to drastic darkening. Color centers form in the pigment and the optical absorption edge of the binder shifts to a longer wavelength. Inorganic based white paints are generally more stable than their organic based counterparts.

Atomic oxygen again is a problem for organic-base paints, but those with a silicone or perfluorinated base stand up best to its effects [5]. Due to this loss of effectiveness in coatings, especially for white paints, optical mirrored reflectors of vapor deposited silver are becoming more popular [2]. A mirror substrate of this type is transparent to sunlight but emits efficiently in the infrared [4].

The effect of atomic oxygen attack on paint will not only adversely change the paint's thermal control properties but also cause it to flake or peel off. These paint flakes are in the 1–100 μm size range. Yet, these small particles may still present significant hazards to spacecraft due to the high relative impact velocities encountered in low Earth orbit. Metals and coatings will generally be eroded by these impacts while some thin insulation might actually be penetrated. The greatest concern is the impact of paint flakes on brittle surfaces such as lenses and windows. In 1983, a side window of the Space Shuttle *Challenger* was found to have a 0.5 mm diameter chip. The window was removed and replaced at a cost of $50,000. Upon analysis by a scanning electron microscope the culprit was found to be a 50 μm titanium-based paint chip striking the window at approximately 3–4 km/s. The average relative collision velocity between low Earth orbit objects is 9–11 km/s. Since 1983 over 25 more windows have been replaced due to similar-sized pits. Analysis has shown that over 80 percent of the impacts were probably caused by high velocity paint flakes.

Insulation

The next level of passive thermal control systems is insulation. The properties of selected insulators are also included in Table 8.2. A single material may be used as insulation but multilayer insulation (MLI) of a variety of materials has proven more effective. MLI is used to (1) minimize heat flow between components, (2) control temperature fluctuations caused by changing external heat sources, and (3) reduce temperature gradients within subsystems [2].

The key to MLI is to use insulation between the layers that reflect radiation. The gas between the layers is removed to prevent the conduction of heat through the gas. The use of MLI has been found to be tens to hundreds of times more effective than foam and fiberglass batting [2]. A common MLI configuration for a geosynchronous satellite is shown in Figure 8.9.

The satellite structure may also be used to conduct heat away from high temperature regions and also perform as an insulator. The double use of the structure is a part of the fine tuning of efficient satellite design.

Another positive side effect of using MLI is its ability to protect a structure from space debris (e.g., paint flakes) or micrometeoroid impacts. The outermost layer serves to fragment the incoming impactor so that the debris is spread over a larger area on the next layer. The original Whipple bumper design, using only one bumper, was developed for this express purpose. MLI conveniently performs a similar function.

Phase-Change Materials (PCM)

Materials that change phase (solid to liquid or liquid to solid) in the temperature ranges experienced by spacecraft may be used to control the transfer of heat energy. The use of PCM devices are most appropriate for control of large, rapid incident fluxes over short periods of time [6].

A PCM system contains a material which melts fairly easily. The extra heat incident on the satellite will be used to melt the PCM. As the temperature drops, the PCM solidifies and slowly releases the heat previously absorbed. The optimum PCM material will have a melting point similar to the operating temperature of the subsystems of the satellite and have a high heat of fusion (ΔH_f). The use of a PCM for thermal control, however, does have limitations. Once all of the PCM is melted, it cannot absorb any ad-

Thermal Control and Protection Systems

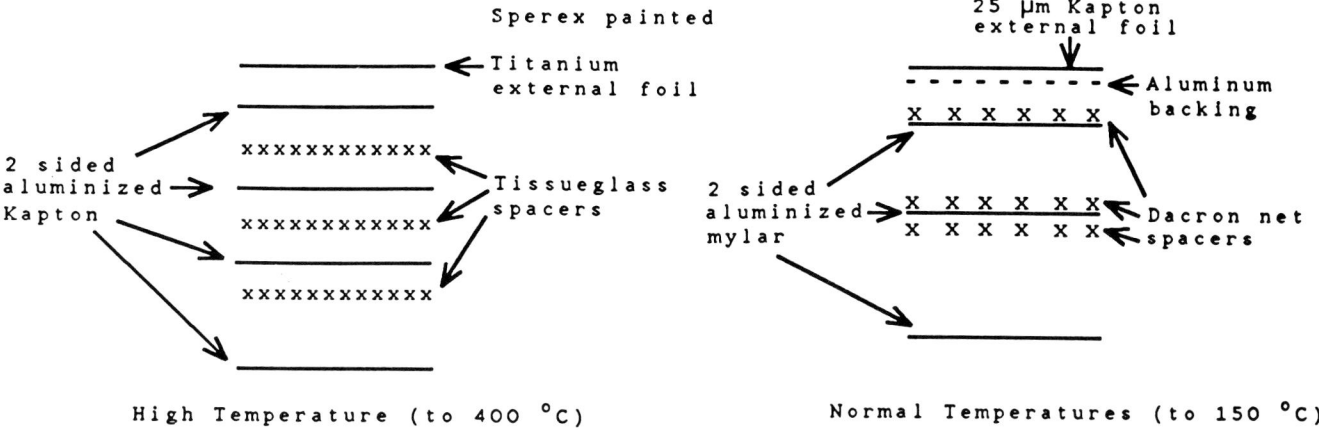

Figure 8.9 Two examples of typical geosynchronous multilayer insulation (MLI) [2].

ditional thermal energy [6]. A common PCM system is wax in an aluminum container [1]. PCM systems have been used on a variety of interplanetary probes to Venus, for example, the Venera and Pioneer-Venus crafts used this form of thermal control for their visits to that incredibly hot environment.

Louvers

A louver system may be used to vary the absorptance and emittance of the surface of a satellite or satellite component. The louver blades are opened to decrease the satellite's temperature by inhibiting the absorption of external heat sources (i.e., low absorptance, high emittance). Closed louvers will create a lower emittance, allowing less heat to radiate away, so the satellite temperature will increase [6].

This method is reliable and used often in satellite design but it requires a capability to sense the temperature and move the louvers to maintain the desired temperature.

Radiators

When passive devices cannot control the thermal state of the satellite sufficiently, active systems must be used to eliminate excess heat. A convective system uses moving fluid/gas to transport the heat away from high temperature areas. A common type of convector is the heat pipe which routinely works between regions which have small temperature differences. Figure 8.10 depicts the structure and functions of a heat pipe.

The surface tension forces of the fluids inside a heat pipe allow the two phase loop to remove heat effectively and efficiently [4]. The fluid/gas (often ammonia) changes state as heat is absorbed. That is, the liquid evaporates. Heat is then released via the condensation of the gas. When condensation occurs, the energy produced is simply the latent heat of vaporization. The heating of the fluid creates a pressure difference which forces the fluid/gas to flow to a region of lower temperature. No pump is necessary. A solid metal pipe will conduct heat from high temperature to low temperature, but a heat pipe is much more efficient and lighter. A "heat pipe will transfer 200 to 300 times as much energy as a solid copper bar of the same diameter" [4].

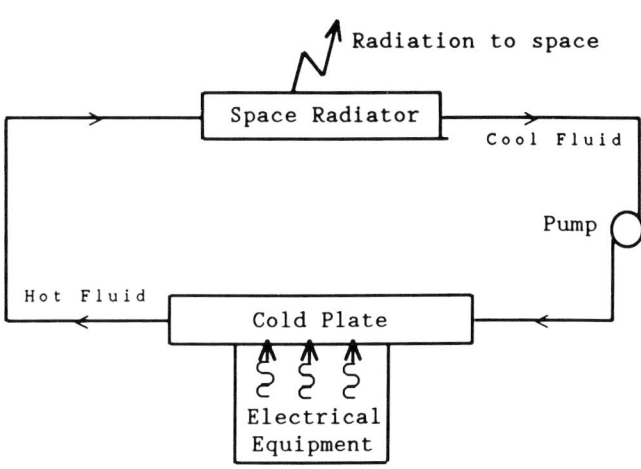

Figure 8.10 Illustration of the operation of a heat pipe [1].

Figure 8.11 Schematic of a space radiator system [1].

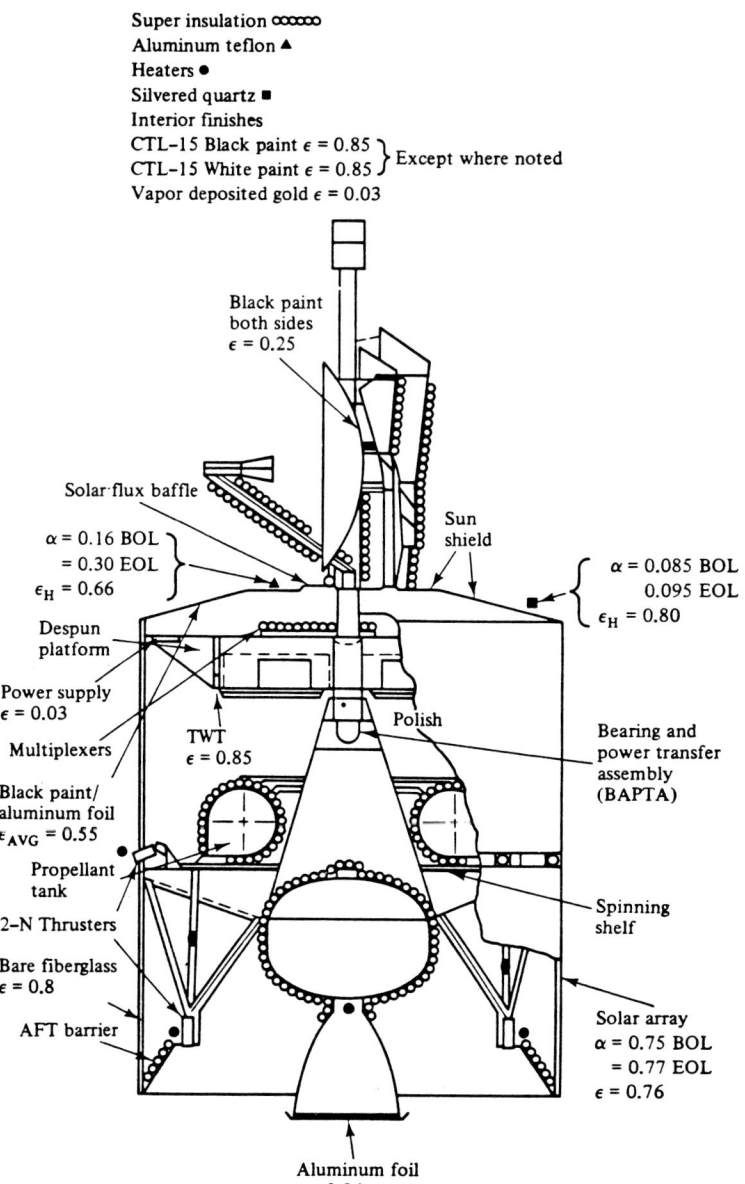

Figure 8.12 The dual-spin stabilized Intelsat IV has a thermal control design which utilizes the satellite's rotational motion to its benefit [2]. *Reprinted by permission of COMSAT Technical Review.*

A space radiator will convect excess heat to the space environment. Any heat-generating system near the exterior of a satellite may actually serve as a radiator itself. However, for electrical equipment contained well within the satellite structure a working fluid may be used to transfer the heat to the radiator. A pump routes the fluid past the hot electrical equipment thereby heating up the working fluid. The space radiator then removes the energy from the system returning the fluid to a lower temperature. This process is shown in Figure 8.11.

Thermal Control System Design

The mission and structure of a satellite drive its thermal control design. The demands on thermal control for geosynchronous satellites are also greatly dependent on the attitude stabilization system [2]. A spin-stabilized spacecraft naturally spreads out the incident solar radiation. A three-axis-stabilized satellite maintains a constant orientation with respect to the Earth. Thus, parts of the satellite have significant daily variations in solar flux (east, west, zenith, and nadir) while other regions (north-south) experience only small seasonal variations in incident energy [2].

Figures 8.12 and 8.13 depict the thermal control design for a spin-stabilized and a three-axis-stabilized spacecraft, respectively. The Intelsat IV is dual-spin-stabilized so the nonspinning part of the satellite requires different types of thermal control than the spinning section does. The Intelsat V is a three-axis-stabilized spacecraft as shown in Figure

Thermal Control and Protection Systems

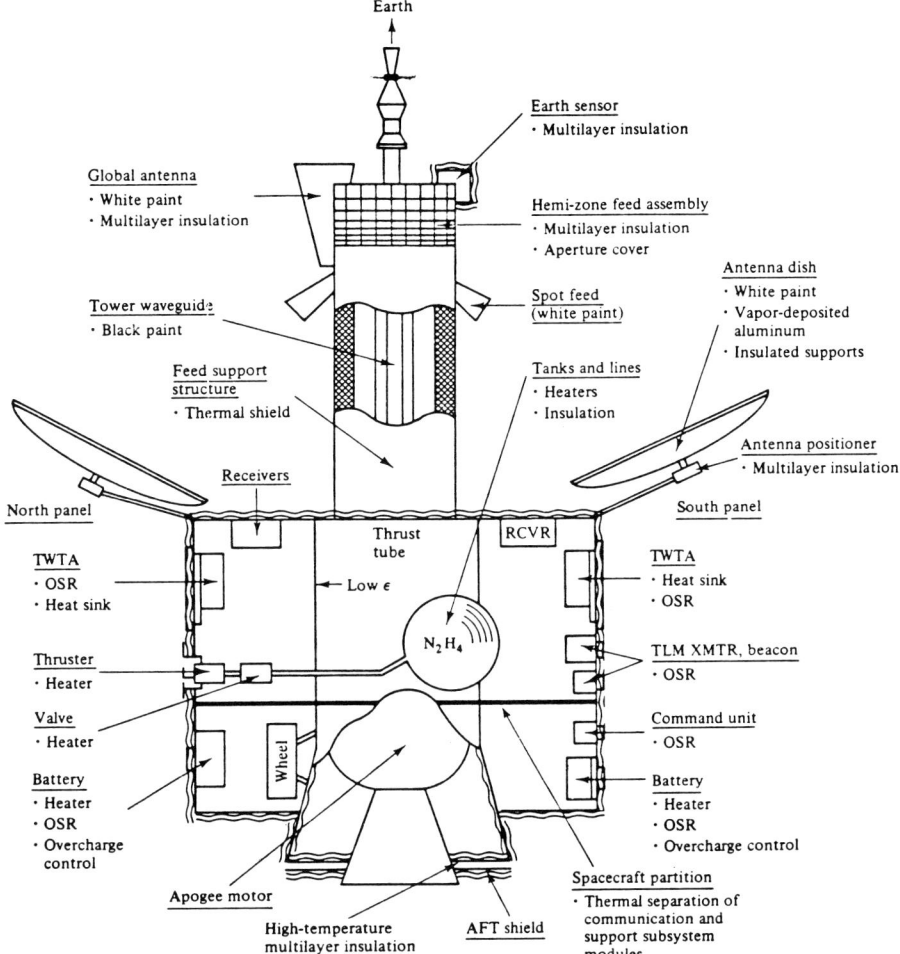

Figure 8.13 A three-axis stabilized spacecraft, such as the Intelsat V, may use active means to cool equipment in the interior of the craft while MLI is often added to protect external hardware [2]. *Reprinted by permission of INTELSAT/Ford Aerospace & Communication Corporation.*

8.13. Heat generating subsystems must be cooled actively while the circumference of the satellite exposed to the drastic daily fluctuations from the Sun are protected by MLI [2].

Thermal Protection Systems

Thermal protection systems differ from thermal control systems in the time scale considered and the amount of heat encountered. These systems are only needed for space structures which are to be recovered after their return to Earth. Manned missions and recoverable film canisters from reconnaissance satellites are typical examples. Ablative materials were the first attempts to dissipate large amounts of heat by absorbing the energy. This heated material is shed before the heat is conducted to the spacecraft structure. After the physics of atmospheric heating are outlined, ablatives and the Shuttle thermal protection system will be discussed.

Atmospheric Heating

For a spacecraft reentering the atmosphere, the considerable heat generated by friction must be absorbed in an orderly fashion or radiated away from the spacecraft. The amount of heat which must be dissipated is a function of the returning vehicle's structure, orientation, and trajectory. The goal of any thermal protection system is to maximize the amount of heat absorbed per unit of weight of material. In general, the total heat input to the vehicle must be kept as low as possible [6].

During reentry, the strength of the spacecraft's structure is dependent on the magnitude and location of stresses. Temperature gradients will induce significant stresses and thus must also be minimized. The rate at which heat is absorbed drives these temperature gradients. The external structure of a spacecraft must be able to withstand the overall force of the atmosphere particularly over its frontal area. Therefore, an important parameter when determining the effectiveness of a thermal protection system is the dynamic pressure, P, which can be written as

$$P = \frac{1}{2}\rho V^2 \qquad (8.15)$$

where

ρ = atmospheric density, kg/m^3
V = velocity relative to the atmosphere, m/s.
P = dynamic pressure, N/m^2 or J/m^3.

(Note that in most reentry texts the dynamic pressure is given as q but here in order not to conflict with the thermal nomenclature used in this chapter, it is referred to as P.)

Several other quantities are also important in determining the thermal protection required. First, the total heat input to a satellite is a major design factor. The total heat input is represented by Q and is measured in units of Joules. Second, the rate of the heat input per unit area, $\dot{q}$, is needed to determine the protection needed. This quantity has units of J/m^2 · sec or W/m^2. The time rate at which heat is introduced to a spacecraft area may be categorized and analyzed in a variety of forms. The average and maximum values for $\dot{q}$ describe the bounds of the design. The ability to identify where $\dot{q}_{max}$ will occur further refines the thermal protection devices required.

As the reentering spacecraft encounters the atmosphere, a shock wave is created. The gas between the shock wave and the vehicle attains a high temperature. If this region has a higher temperature than the vehicle's surface, heat will be absorbed by the spacecraft. Thus, heat is exchanged via convection from the gas behind the shock wave to the satellite while the body also radiates heat away [6].

Assumptions of this complex heat transfer problem by Allen and Eggers allow for a quick explanation of the key heating parameters described earlier [8]. The difference in temperature between the wall and the air behind the shock wave drives the heat transfer process. This quantity is given by [6]

$$\Delta T = (T_a - T_w) = \frac{V^2}{2C_p} \qquad (8.16)$$

where

T_a = temperature of air, K
T_w = temperature of wall, K
C_p = specific heat at constant pressure, J/kg · K
V = velocity of wall, m/s.

The rate of heat exchanged by convection from the air to the spacecraft surface $\dot{q}$ is proportional to this temperature difference [6] and the two can be equated as follows:

$$\dot{q} = h(T_a - T_w) \qquad (8.17)$$

where

h = heat transfer coefficient, J/(m^2 · sec · K)
$= \frac{1}{2}C_F C_p \rho V$

C_F = equivalent skin-friction coefficient, unitless

Equation 8.17 was derived by substituting the heat transfer coefficient h as the convective conductance in equation 8.7. The h term shows that the ability of a temperature difference to drive a transfer of energy is a function of the spacecraft's surface (C_F), the type of gas in contact with spacecraft wall (C_p and ρ), and the relative velocity between the surface and the gas (V). Substitution of equation 8.16 into equation 8.17 yields

$$\dot{q} = \frac{V^2}{2c_p}(C_F C_p \rho V) = \frac{V^3}{2} C_F \rho . \qquad (8.18)$$

This equation shows that the rate of heat transfer is proportional to the reentering vehicle's velocity, skin friction, and atmospheric density. Intuitively, these parameters would cause increased heating with C_F and ρ contributing to the spacecraft/environment interaction. Since the velocity is cubed it clearly has the greatest effect on the rate at which this heat is transferred.

The rate at which total heat is added to the system is found by summing up the convective heat transfer across the body's surface area as represented in equation 8.18. The integral form of this equation is

$$\dot{Q} = \int_A \dot{q}\, dA = \frac{1}{4} C_F A \rho V^3 \qquad (8.19)$$

where A is the total surface area of the vehicle. Again, despite the complexity of the derivation, the final result yields a physically understandable relationship. dQ/dt is proportional to the skin friction, surface area, and atmospheric density. Yet, velocity has a much greater effect with dQ/dt being proportional to V^3, as stated earlier.

For a point on the surface, the stagnation point, where the relative velocity between the air and surface has gone to zero, a special relationship is developed. The heating rate per unit area at the stagnation point, $\dot{q}_s$, is [6]

$$\dot{q}_s = \frac{b}{\sqrt{R}}\left(\frac{\rho}{\rho_o}\right)^{1/2}\left(\frac{V}{V_o}\right)^3 \qquad (8.20)$$

where

b = a constant, J/m$^{3/2}$ sec
R = radius of curvature of body surface at stagnation point, m
V_o = initial velocity, m/s
ρ_o = initial atmospheric density, kg/m^3.

The most important observation from equation 8.20 is that $\dot{q}_s$ is inversely related to the radius of curvature. For example, as R increases the $\dot{q}_s$ decreases, so as the body becomes more blunt the heating rate is reduced. This characteristic of heating was incorporated in the early space vehicles such as Gemini with the very blunt reentry surfaces. Shapes with smaller R values (more pointed) will produce larger heating rates. This holds true for the leading edges of the Space Shuttle and thus new materials tech-

Thermal Control and Protection Systems

nology had to be developed to allow the use of the sharp leading edges of the Shuttle which were required so that it could perform like an airplane in addition to a spacecraft.

The total heat absorbed during reentry, Q, is found by simply integrating the heating rate over a vehicle's entire reentry trajectory. Trajectories may vary greatly, but the key parameters affecting Q are common for all types of missions. The physical characteristics of a reentering object important to its resulting trajectory can be lumped into its ballistic coefficient, B, which is simply the object's cross-sectional area-to-mass ratio times its coefficient of drag. Integrating equation 8.19 over a reentering body's trajectory using B produces the relationship

$$Q = \frac{AC_F}{4}\left[\frac{m}{SC_D}\right]V^2 \quad (8.21)$$

where
- A = exposed surface area, m²
- C_D = coefficient of drag, unitless (1.5 to 2.5)
- m = mass of reentry vehicle, kg
- S = cross-sectional area, m²
- V = velocity of reentry vehicle, m/s.

This relationship shows that the total heat transferred is minimized by reducing the skin-friction coefficient and body's area. Additionally, the term m/SC_D must be kept as small as possible. This ratio basically represents a vehicle's mass-to-area ratio which is the inverse of the ballistic coefficient. A large mass-to-area ratio denotes a dense object and the total heat transferred to this structure is not a function of C_D since it will be affected so little by atmospheric drag. On the other hand, a vehicle with a low mass-to-area ratio (large B) has its velocity reduced significantly by aerodynamic drag forces.

By rearranging equation 8.21 a kinetic energy term may be factored out of the right hand side resulting in the equation

$$\frac{Q}{1/2\ mV^2} = 1/2\left[\frac{AC_F}{SC_D}\right]. \quad (8.22)$$

Equation 8.22 represents the percentage of the initial kinetic energy absorbed by the space vehicle during its reentry trajectory. A goal of thermal protection design is to keep this relationship as small as possible. Minimization of the ratio C_F/C_D keeps this fraction low, while C_F/C_D is small for a blunt body. Thus, again a blunt body leads to less heat transfer than a sharp vehicle [6]. (Note that A is the total surface area while S is the cross-sectional area.)

The time scale of the effect of aerodynamic forces is also of importance. During reentry the maximum stagnation heat transfer will be reached first. A high lift-to-drag ratio, which is proportional to C_L/C_D, will help to minimize the maximum stagnation heating rate. Next, along a reentry vehicle's trajectory, the maximum average heat transfer occurs. Finally, the structure will encounter a region of maximum deceleration. The deceleration is often measured with respect to the acceleration due to gravity, g. Thus, if it is said that a spacecraft feels a deceleration of $3g$'s the deceleration is three times the acceleration due to gravity ($g = 9.81$ m/s²). The use of a high C_L/C_D ratio also decreases maximum deceleration values. The use of a high C_L/C_D design will, however, increase the time of flight and thus the total heat absorbed by the space vehicle.

The design of spacecraft, which must withstand the rigors of reentry, is a very complicated process. Trade-offs must be made between vehicle trajectories, materials, and geometry. The design and construction of the Space Shuttle's thermal protection/control systems have required the integration of classical aerodynamic heating theory with new material technology.

Ablation

Before the advent of the Space Shuttle and its passive thermal protection system, manned vehicles relied upon ablative cooling to provide thermal protection from the intense heat of reentry. Unlike the Shuttle thermal tiles, which act as passive super insulators, ablator material will dissipate the heat generated during reentry via endothermic thermal decomposition (burning) followed by removal of the charred ablator [11]. The ablator must be designed so that the heated material is removed as quickly as possible to minimize conduction of heat into the spacecraft's structure.

Figure 8.14 shows the shock wave and the gas envelope for a sphere moving at hypersonic velocity ($M \gg 1$). A sphere is a good model for the Mercury, Gemini, and Apollo spacecraft which utilized ablation shields. During reentry, these spacecraft reached speeds of Mach 25, generating shield temperatures of 2,760°C (5,000°F) [7].

The basic principle behind ablative cooling is dissipating thermal energy reaching the body by making the heat do work at the surface of a sacrificial protective layer. Endothermic processes occurring at the surface include melting, vaporization, and thermal decomposition which create volatile products that decrease heat transfer. Heat is thus transported away by the heat capacity of the ablating mass. The high surface temperatures also mean heat is rejected through the radiation process. In the end, the heat shield is expended, but the payload survives without damage [7].

In summary, energy is introduced to a spacecraft by atmospheric heating via radiation and conduction. This heat is dissipated through surface radiation, conduction into the structure, vaporization/melting of material, and blocking of the heat (transpiration). The balance of heat transfer is shown in Figure 8.15.

The cooling process utilizes endothermic chemical reactions and phase changes to dissipate heat. A typical ablator is a composite material made of a silica (quartz) cloth filler impregnated with a phenolic or epoxy resin. Upon exposure to a stream of hot gas, the resin undergoes en-

168 Chemical Principles Applied to Spacecraft Operations

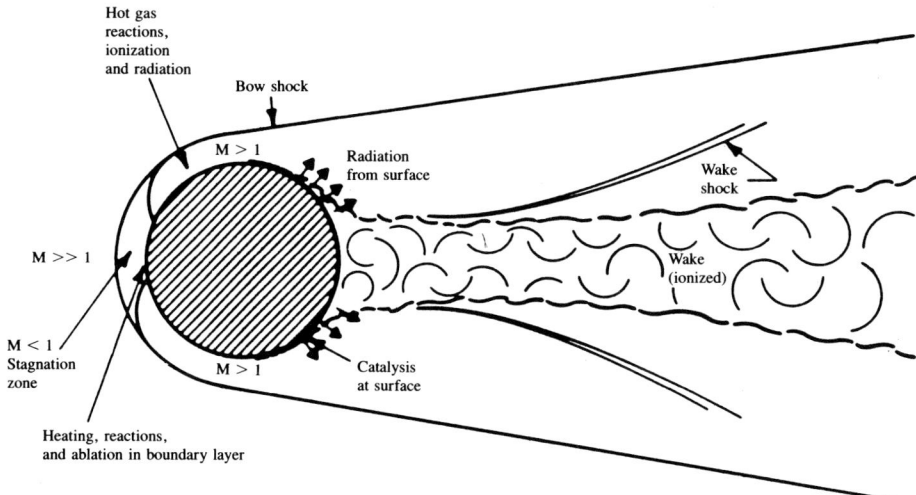

Figure 8.14 The bow shock wave about a hypervelocity sphere [7].

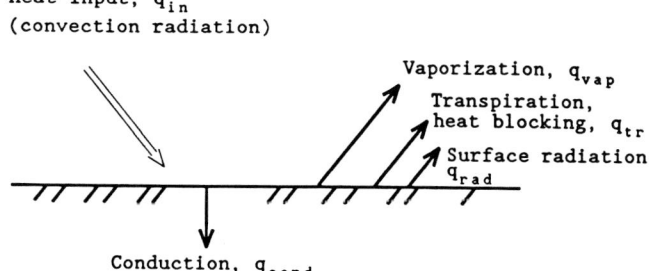

Figure 8.15 The heat balance at the surface of a simple ablator [7].

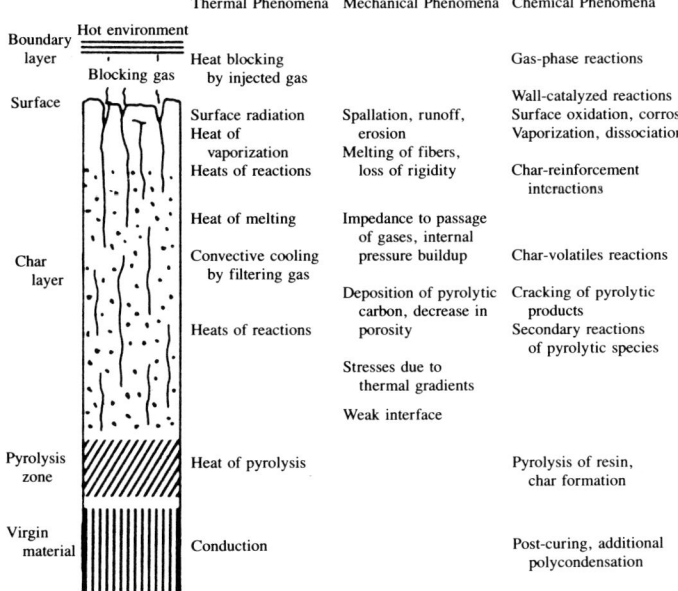

Figure 8.16 Mechanisms of charring ablation involve thermal, mechanical, and chemical processes [7].

dothermic decomposition (i.e., pyrolysis) to create volatile products and an adhering char. As the pyrolysis zone increases in thickness into the virgin material, the volatile gases diffuse through the char layer to the surfaces absorbing heat along the way. Here they absorb more heat by fragmenting into still smaller molecules. Secondary reactions with the volatile gases provide additional cooling for the char. These gases also block incoming convective heat. The combined effect of the blocking action of gases and the heat exchange within the porous char is termed transpiration cooling [7].

Eventually, the pyrolysis reaction slows down as less heat reaches the char-virgin material interface. Temperatures within the char do, however, rise above the melting point of the silica. As SiO_2 vaporizes, the glowing shield radiates heat and begins to recede or ablate. Material is lost not only by physical and chemical changes, but also by runoff, erosion, and spallation. A point is reached where the rate of surface recession equals the recession rate of the pristine material. When this quasi-steady state exists, external heating and heat dissipation are nearly in equilibrium. This state will continue to exist as long as the ablator is of adequate thickness. Figure 8.16 shows the mechanism of charring ablation. Table 8.5 is a list of the various chemical reactions and phase changes occurring in ablation [7].

Ablators must possess the following thermal and mechanical properties:

- High heat-protective performance
    Low thermal conductivity
    High specific heat
    High emissivity
    High yield of low molecular weight gases
    High endothermicity of chemical reactions and phase changes

Thermal Control and Protection Systems

- Good mechanical properties
    - Thermal stability
    - High strength and ability to absorb elastic stresses
    - Low mechanical erosion and dimensional changes
    - Low density [7].

With this rather stringent list of requirements, the materials available for ablators are rather limited. Of the ablative plastics, heat-resistant polymers (i.e., thermosets) like phenolics and epoxy-novolacs are most often used. A characteristic of these polymers is the presence of aromatic rings which lend oxidative stability to the char which is important in preventing combustion.

Table 8.5 Chemical Reactions and Phase Changes Occurring in Ablation [7]

Type	Example
Depolymerization	$(C_2F_4)_n \rightarrow nC_2F_4$
Pyrolysis (resin)	Phenolic $\rightarrow$ Products
"Cracking" (volatiles)	$CH_4 \rightarrow C + 2H_2$
Char-volatiles reactions	$C + H_2O \rightarrow CO + H_2$
Char-reinforcement reactions	$3C + SiO_2 \rightarrow SiC + 2CO$
Phase transitions	α-Quartz $\rightarrow \beta$-Quartz
Melting and vaporization	$SiO_{2(s)} \rightarrow SiO_{2(l)} \rightarrow SiO_{2(g)}$
Sublimation	$C_{(s)} \rightarrow C_{(g)}$
Dissociative vaporization	$SiO_{2(l)} \rightarrow SiO_{(g)} + \frac{1}{2}O_{2(g)}$
Combustion	$C + O_2 \rightarrow CO_2$
"Chemical corrosion"	$SiO_2 + 4HF \rightarrow SiF_4 + 2H_2O$
Molecular dissociation	$N_2 + M \rightleftharpoons N + N + M$
Wall-catalyzed recombination	$H + H + M \rightleftharpoons H_2 + M$
Reactions of atomic species	$C + N + M \rightleftharpoons CN + M$
Electron-producing reactions	$N + O \rightleftharpoons NO^+ + e$
Charged particle reactions	$CHO^+ + H_2O \rightleftharpoons CO + H_3O^+$

Biphenol-formaldehye   Epoxy-Novolac

The primary purpose of the filler is to provide structural integrity and mechanical strength to the plastic. Materials receiving the most use are those made of glass, quartz, graphite, and carbon. These materials can be in the form of fibers, tape, and rigid honeycomb form. Project Mercury used a phenolic-Fiberglas ablator while the one for Apollo was a honeycomb matrix filled with phenol-epoxy [7].

Shuttle Thermal Protection System

This section on the Shuttle thermal protection system is taken in its entirety from the Space Shuttle News Reference, "Orbiter Systems," published by NASA and is reproduced with permission of NASA.

The overall economic feasibility of a reusable Space Transportation System hinges on protecting the Space Shuttle orbiter—which will experience widely varying thermal and aerodynamic environments typical of both aircraft and spacecraft—in a way that does not require significant refurbishment between trips. The thermal protection system (TPS) is designed to limit the temperature of the Orbiter's aluminum and graphite epoxy structures to a nominal value of 450 K (177°C or 350°F) during ascent and entry.

Maximum surface temperatures during entry vary from 1,783 K (1,510°C or 2,750°F) on the wing leading edge to less than 589 K (316°C or 600°F) on the upper fuselage. The thermal protection system must also endure exposure to nonheating environments during prelaunch, launch, onorbit, landing, and turnaround operations similar to those encountered by conventional aircraft. It must also sustain the mechanical forces induced by deflections of the airframe as it responds to the same external environment. The system is designed to withstand 100 ascents and entries with a minimum of refurbishment and maintenance.

The thermal protection system (Figure 8.17) consists of materials applied externally to the orbiter that maintain the airframe outer skin within acceptable temperature limits. Internal insulation, heaters, and purging facilities are used to control interior compartment temperatures. The orbiter thermal protection system is a passive system consisting of the following four materials, selected for weight efficiency and stability at high temperatures.

1. Coated reinforced carbon-carbon (RCC) for the nose cap and wing leading edges where temperatures exceed 1,533°K (1,260°C or 2300°F)
2. High-temperature reusable surface insulation (HRSI) for areas where maximum surface temperatures reach 922 to 978 K (649 to 704°C or 1,200 to 1,300°F)
3. Low-temperature reusable surface insulation (LRSI) for areas where surface temperatures reach 644 to 922 K (371 to 649°C or 700 to 1,200°F)
4. Flexible reusable surface insulation (FRSI) (coated Nomex felt) for areas where the surface temperature does not exceed 644 K (371°C or 700°F).

The reinforced carbon-carbon is an all-carbon composite of layers of graphite cloth contained in a carbon matrix formed by pyrolysis. To prevent oxidation at elevated temperatures, the outer graphite-cloth layers are chemically converted to silicon carbide. RCC covers the nose cap and wing leading edge areas of the Orbiter, which receive the highest heating.

The high-temperature reusable surface insulation consists of approximately 20,000 tiles located predominantly on the lower surface of the vehicle. The tiles nominally measure

Chemical Principles Applied to Spacecraft Operations

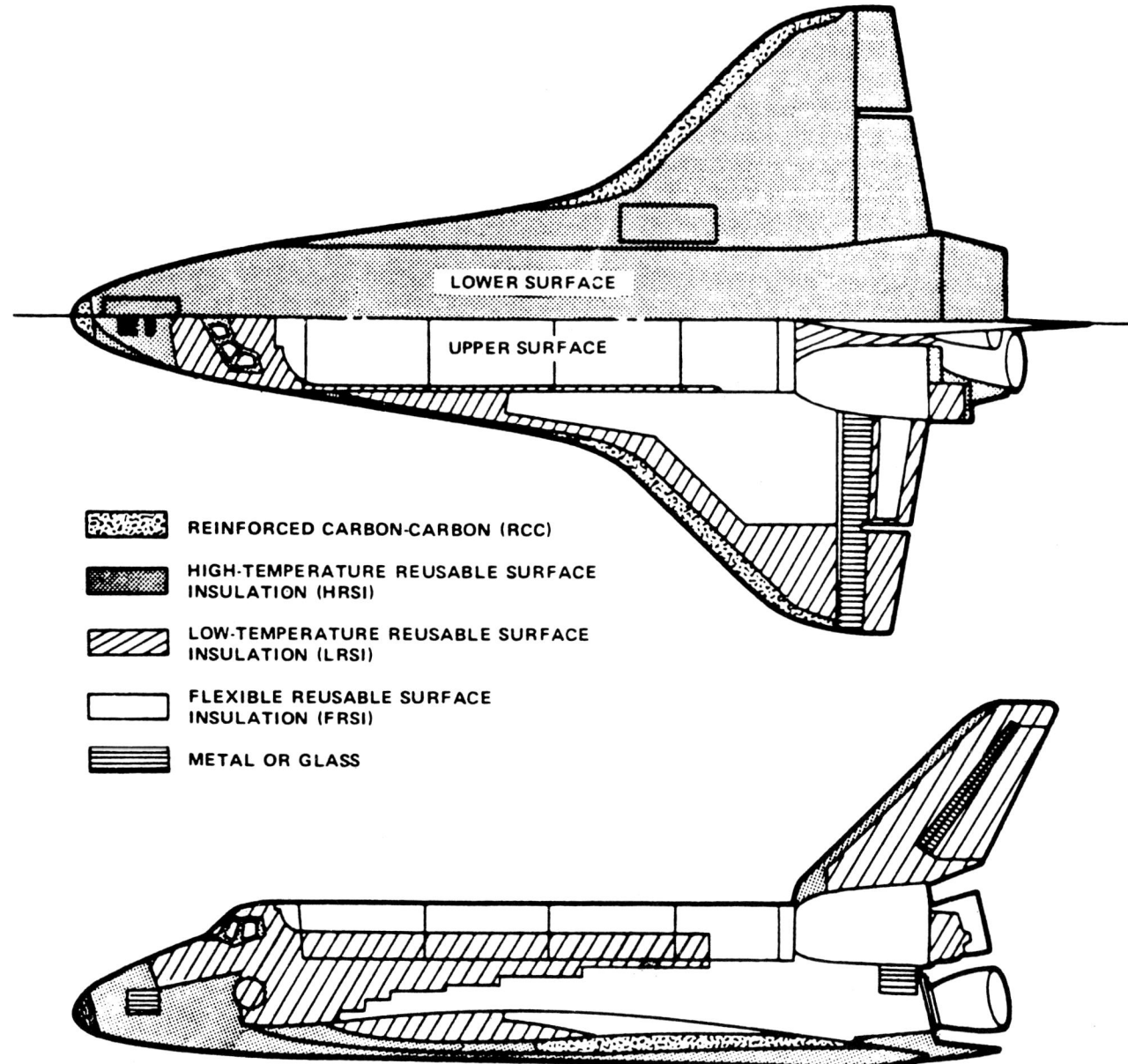

Figure 8.17 The Orbiter thermal protection system (TPS) is tailored to withstand the peak temperatures felt at certain locations on the Shuttle.

15 by 15 centimeters (6 by 6 inches) and vary in thickness from 1.3 to 8.9 centimeters (0.5 to 3.5 inches), depending on local heating. The high-temperature material is composed of a low-density high-purity silica (SiO_2) fiber insulator made rigid by ceramic bonding. Each tile is bonded to a strain isolator pad made of Nomex fiber felt and the total composite is directly bonded to the vehicle.

The low-temperature reusable surface insulation consists of approximately 7,000 tiles applied to the upper wing and fuselage side surfaces of the Orbiter in the same manner as the HRSI. These tiles nominally measure 20 by 20 centimeters (8 by 8 inches) and vary in thickness from 0.5 to 2.5 centimeters (0.2 to 1 inch). The low-temperature tiles are the same as the high-temperature tiles except that the coating has a different optical pigment for obtaining low solar absorbance and high solar emittance. These two sets of tiles cover approximately 70 percent of the Orbiter.

The basic raw material for the tiles is a high-purity amorphous short-staple silica fiber that was selected for its low thermal conductivity, low thermal expansion, and high-temperature stability. The reusable tiles can easily be repaired. Coating scratches can be repaired in place by spray techniques and torch firing and small gouges or punctures can be cored out and replaced with standard size plugs. A complete set of tiles can be removed and replaced in 45 hours.

Thermal Control and Protection Systems

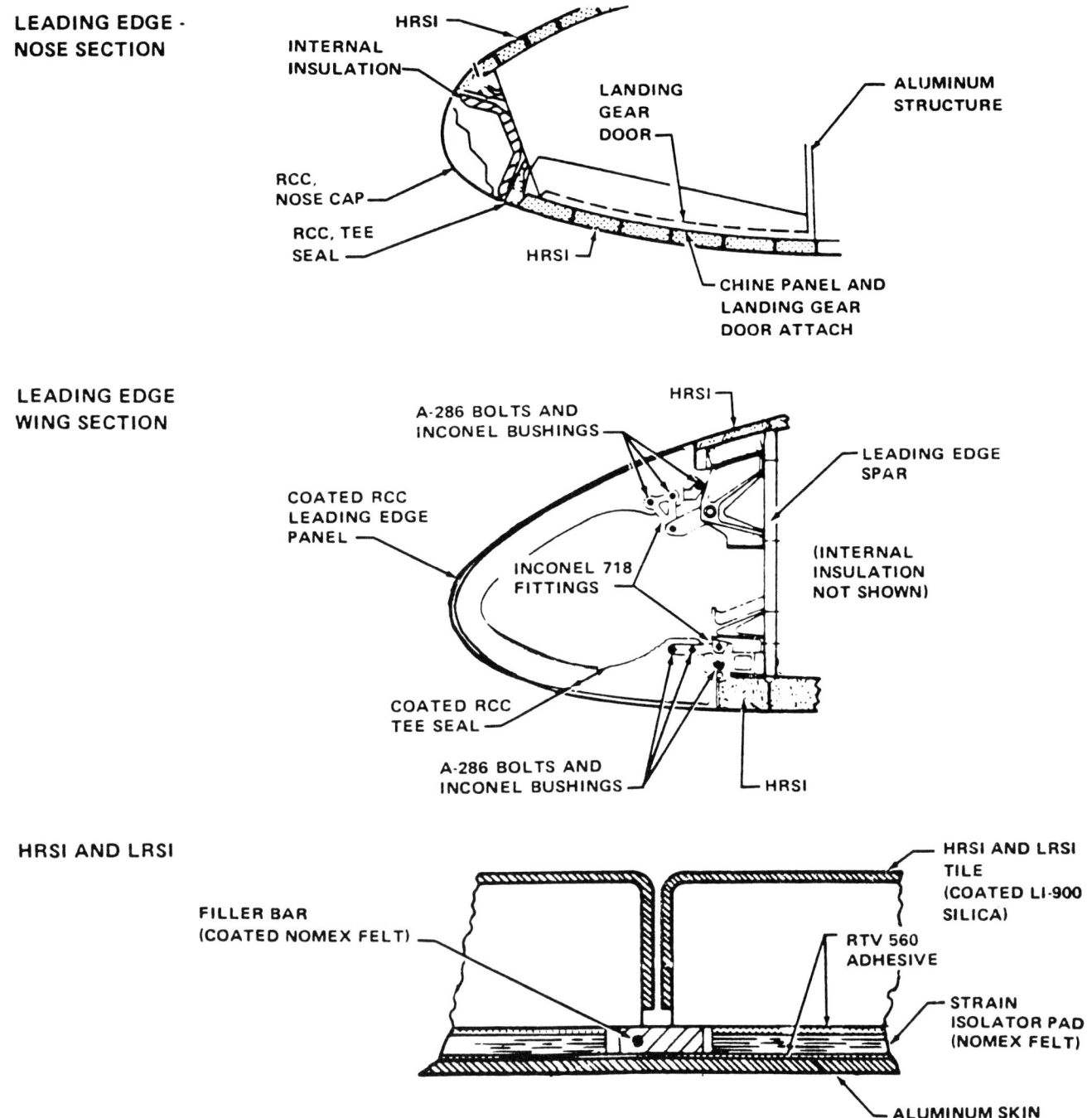

Figure 8.18 Thermal protection system interfaces are crucial for a maintainable, durable system.

The flexible reusable surface insulation consists of 0.9 by 1.2 meter (3 by 4 foot) sheets of Nomex felt that are directly bonded to the structure. Before installation, the Nomex felt, which varies in thickness from 0.41 to 1.62 centimeters (0.16 to 0.64 inch), is coated with a silicone elastomeric film to waterproof it and to give the surface the desired optical properties. The Nomex felt is applied to the upper parts of the payload bay doors, the sides of the fuselage, and the upper wing.

Typical interfaces with the thermal protection system are shown in Figure 8.18. The leading edge subsystem to HRSI transition is shown for the nose and outer wing leading edge. A typical joint for the two tile systems is also shown.

The tiles are bonded to the aluminum skin with RTV 560 (a silicone resin cement) with a strain isolator pad in between. Filler bars are installed under the tiles at the intertile gaps. In areas where high surface pressure gradients would

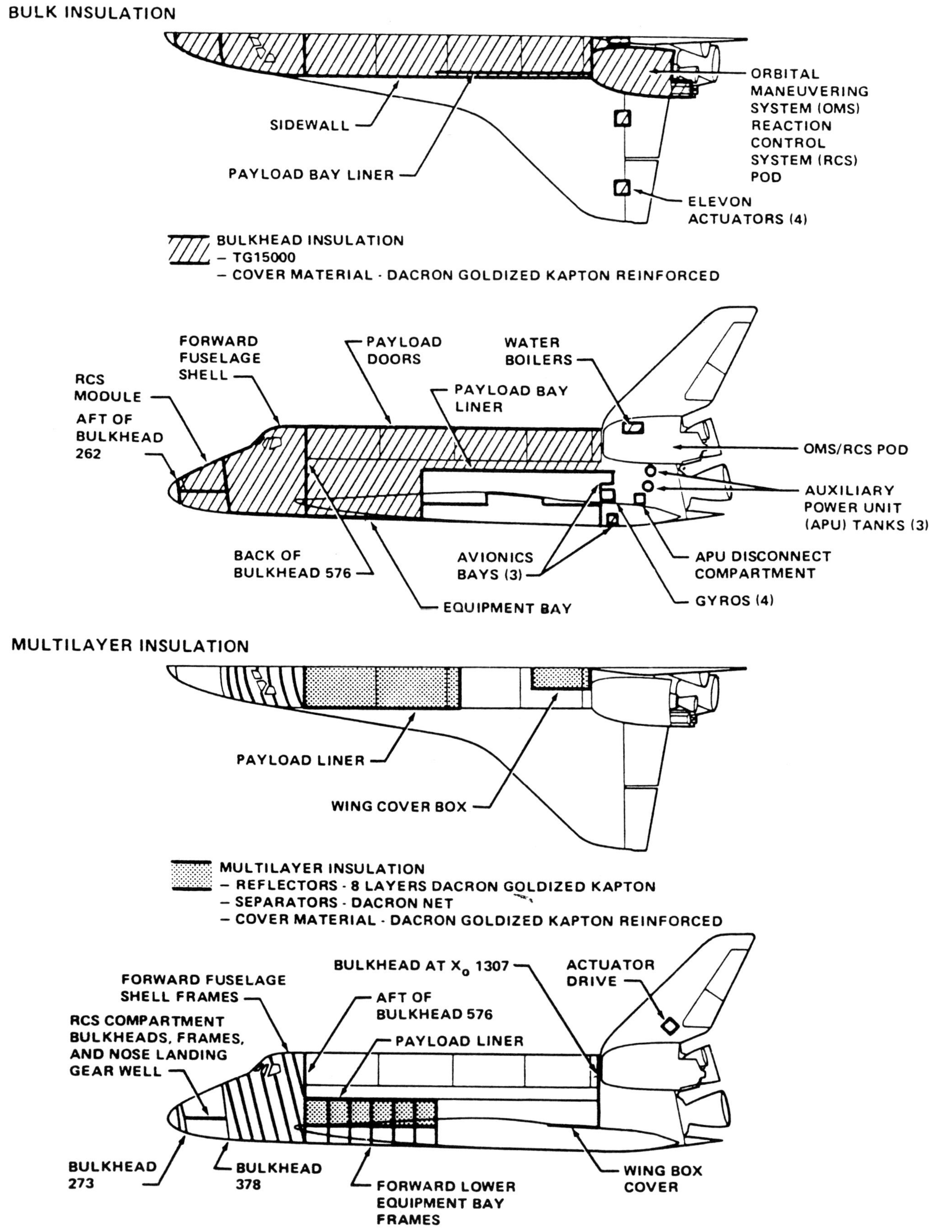

Figure 8.19 Orbiter passive thermal control system.

cause crossflow of boundary-layer air within the intertile gaps, tile gap fillers are provided to minimize increased heating within the gaps. The gap fillers are fabricated using a silica fiber cloth cover with an alumina fiber filler.

A passive (nonactive) thermal control system helps maintain spacecraft systems and components at specified temperature limits. This system uses available spacecraft heat sources and heat sinks supplemented by insulation blanket, thermal coatings, and thermal isolation methods. Heaters are provided on components and systems where passive thermal control techniques are not adequate to maintain required temperatures.

The insulation blankets are of two basic types: fibrous bulk and multilayer (Figure 8.19). The bulk blankets are made of 32 kg/m³ (2 lb/ft³) density fibrous material with a sewn cover of reinforced double-goldized Kapton. The cover materials have numerous small holes for venting purposes. Goldized tape is used for cutouts, patching, and reinforcements. Tufts are used throughout the blankets to minimize billowing during venting.

The multilayer blankets are constructed of alternate layers of perforated double-goldized Kapton reflectors and Dacron net separators for a total of 20 reflector layers, with the two cover halves counting as two layers. The covers, tufting, and goldized tape are similar to those of the bulk blankets.

Aerospace Plane Thermal Protection

A number of aerospace planes are being studied worldwide: the HOTOL (Horizontal Takeoff and Landing) in Great Britain, the Sänger II in Germany, and the NASP (National Aero-Space Plane) in the United States. The technical challenge is to create a vehicle that can take off from a conventional runway, deploy satellites in orbit, withstand the heat and stress of reentry, and land on a conventional runway. Peak temperatures felt by these new vehicles are comparable to those of the Shuttle, 1,200–1,800 K.

The high temperature reusable surface insulation (HRSI) presently used by the Shuttle can withstand temperatures above 1260°C (2,300°F). The HRSI, developed in the 1970s, is composed of silica 160–352 kg/m³ (10–22 lb/ft³). The new breed of spacecraft will have to maintain low density while improving the strength and durability of the heat shield system [9].

Fibrous refractory composite insulation (FRCI) was developed in response to this need at Ames Research Center. The FRCI is one fifth aluminoborosilicate fiber (11 nm diameter with 62% Al_2O_3, 14% B_2O_3, and 24% SiO_2) and four-fifths silica fibers (1.5 nm in diameter). This material was immediately incorporated into the design of the Shuttle due to its low density, similar thermal properties, and better mechanical properties. The FRCI, however, does not have a high enough temperature capability for use on future aerospace planes [9].

In the early 1980s a much smaller diameter (2 nm) aluminoborosilicate fiber became available and combined with alumina (3 nm) fibers to improve the temperature capability of FRCI to 1482°C (2,700°F). This new heat shield material was named the alumina enhanced thermal barrier (AETB). The AETB has exhibited a 20% increase in tensile strength and a 15% increase in temperature capability, with no change in the thermal conductivity and only a slight increase in the thermal expansion coefficient. Its development is a major step toward reliable, lightweight, and durable thermal protection systems for advanced aerospace planes like NASP, HOTOL, and Sänger [9].

Summary

Both thermal control and protection systems are vital to the operation of all space systems. The use of new materials has revolutionized this science. These materials have chemical properties engineered to provide efficient heat transfer, yet they remain durable against other environmental effects. The epitome of this type of engineering is the thermal protection/control system of the Space Shuttle. Yet, more difficult tasks are presently being addressed in the design of the new breed of spacecraft—the aerospace plane. only through the melding of chemistry and aerospace design will the unique technology be developed that will make this type of spacecraft a reality.

Summary of Key Equations

Conduction	(8.3)	$\dot{Q} = -kA \dfrac{dT}{dx}$
Radiation	(8.4)	$\dot{q} = \sigma \epsilon T^4$
Conservation of Radiant Energy	(8.5)	$\alpha + \rho + \tau = 1$
Convection	(8.7)	$\dot{q} = h \, \Delta T$
Aerodynamic Heating	(8.18)	$\dot{q} = \dfrac{V^2}{2C_p} (C_F C_p \rho V)$
	(8.19)	$\dot{Q} = \dfrac{1}{4} C_F A \rho V^3$

References

1. Brown, C. and McMordie, R., *Spacecraft Design: Aero 556*, Professor Publishing, Boulder, CO, 1985.
2. Agrawal, B. N., *Design of Geosynchronous Spacecraft*, Prentice-Hall, Inc., Englewood Cliffs, NY, 1986.
3. Reynolds, W. C. and Perkins, H. C., *Engineering Thermodynamics*, McGraw-Hill Book Company, New York, 1977.
4. Woodcock, G. R., *Space Stations and Platforms*, Krieger Publishing Company, Malabar, FL, 1986.
5. "Data for Selection of Space Materials," ESA pss-01-701, Issue 1, November 1985.

6. Vinh, N. X., Buseman, A. and Culp, R. D., *Hypersonic and Planetary Entry Flight Mechanics*, Ann Arber, The University of Michigan Press, 1989.
7. *Chemistry in Space Research*, ed. by R. Landel and A. Rembaum, American Elsevier Publishing Co. (New York), 1972.
8. "Orbiter Systems," Space Shuttle News References, NASA.
9. "Alumina-Enhanced Thermal Barrier," NASA Tech Briefs, ARC-12135.
10. Sutton, G., Wagner, W. and J. Scader, *Astronautica Aeronautica*, 60–71 (January 1966).

Discussion Questions

1. Proper thermal control and thermal protection are essential to effective spacecraft operations.
    a. How is thermal control important to a satellite mission?
    b. What are the major differences between thermal control and thermal protection?
2. The earliest stages of research into thermal protection accompanied the development of intercontinental ballistic missiles (ICBMs).
    a. What specific thermal protection device was used on ICBMs?
    b. How are thermal protection requirements affected by the maximum altitude of an orbiting craft?
3. The three basic types of heat transfer are conduction, radiation, and convection.
    a. Which are most important for thermal control design?
    b. Which are most important for thermal protection design?
4. Conduction is the transfer of heat via a solid medium.
    a. What parameters determine the rate of heat conduction?
    b. What material is the best conductor of heat? The worst?
5. The transmittal of energy by photons or electromagnetic waves is called radiation.
    a. What is a "black body" radiator and why is it also called an ideal radiator?
    b. What is the key parameter in the radiation process?
6. Convection is a heat transfer mechanism which uses a moving fluid to transfer energy.
    a. Why is convection not important in thermal control design?
    b. Describe what the Reynolds number represents physically.
    c. Describe what the Prandtl number represents physically.
7. Thermal control systems must be designed to maintain satellite equipment within stringent limits for its entire lifetime.
    a. Do acceptable temperature limits differ within a spacecraft?
    b. Explain how a passive thermal control system, such as a coating, can be degraded over time.
8. The earliest interplanetary probes to the extremely hot environment of Venus used thermal control components.
    a. What was the reason for the use of a phase-change material (PCM) in these probes?
    b. What are the limitations of the use of a PCM?
9. Thermal control designs depend on the mission and structure of satellites.
    a. Explain how these designs might differ for a three-axis stabilized satellite compared to a spin-stabilized craft.
    b. How might the thermal control design be different for an Earth-orbiting satellite versus an interplanetary probe to Jupiter?
10. The quantities that drive thermal protection system design are the total heat and time rate of heat transferred to a reentering object.
    a. Why is the rate of change of heat transfer critical?
    b. What is the most important variable in the determination of the heat created during reentry?
    c. What basic forms of heat transfer are exhibited during reentry?
11. The Space Shuttle thermal protection is a unique aerospace engineering feat.
    a. How does it differ from the ablative-type systems used on Gemini and Apollo spacecraft?
    b. Why must the thermal protection system of the Shuttle be configured in tiles?
    c. It has been shown that blunt leading edges (like Gemini craft) reduce the heat generated at the surface of the vehicle. Why then does the Shuttle have sharp leading edges?

Problems

1. A small scientific payload is contained within a 2 foot insulated cylinder. The outer radius is 15cm while the insulation is 0.65cm thick. The outer skin is subjected to heating of $\dot{q} = 160$ J/(s · m^2) into the cylinder's interior. The payload must be kept at 10°C (50°F). The conductivity of the insulation is 0.22 J/(s·m·°C).
    a. What is the surface area of the cylinder (neglect the ends)?
    b. What is the equivalent resistance of the cylindrical capsule?
    c. What is the heat per unit time being absorbed ($\dot{Q}$)?
    d. What is the temperature of the skin?
2. An exploratory device is landed on the sunlit side of the Moon. It is a 5m diameter spherical structure covered

Thermal Control and Protection Systems

in aluminum foil with a 10 mil thick coating of silicon (see Table 8.2). The probe is suspended above the surface. To maintain a temperature of 100 K, twice as much energy must be radiated as absorbed.

   a. What is the average incident solar energy rate for this to occur?
   b. Is this device in thermal equilibrium?

3. Black body radiation curves are shown in Figure 8.8.
   a. For a 1,500 K object what is the most likely wavelength of energy produced?
   b. What is the frequency of this energy?

4. Atmospheric heating is a function of atmospheric density, velocity, and the surface of the structure reentering.
   a. If the velocity is halved, what is the effect on the total heat produced?
   b. What is the effect on the rate at which heat is produced by a similar halving of reentry velocity?
   c. Why do these two differ?

Chapter 9

Manned Space Flight

The presence of people in space introduces a whole new set of requirements for spacecraft designers. Unmanned probes, such as Pioneer and Voyager, are relatively autonomous once launched and need only occasional course adjustments and signals for equipment operation. Manned spacecraft, on the other hand, must provide food, water, atmosphere, and waste disposal for its crew throughout the duration of the mission. Chemistry plays a critical role in the design of these life support systems. This chapter looks at the environmental requirements for manned systems with specific emphasis on atmospheric control, water and waste management, thermal control, toxic hazards, and radiation protection.

Cabin Atmospheres

Manned explorations have always been hampered by human physiological limitations. In choosing to explore space, we are placing even greater demands on our knowledge of our own physiological needs and on our technological capability of meeting them. We are adapted only to the environmental conditions prevalent near the surface of the Earth. These conditions include an atmosphere composed principally of nitrogen and oxygen gases (Table 9.1) at a total pressure of about 760 mm Hg (or 1 atm). This atmosphere (troposphere) extends to an altitude of about 16 km (10 miles) with little change in composition [2]. Yet, with only a few exceptions, human communities are located at altitudes of less than 1.6 km (1 mile) [3]. If we are exposed to altitudes in excess of 3.2 km (2 miles) for extended periods of time, we begin to develop symptoms of oxygen deficiency [4].

We have an absolute need for oxygen; an insufficient supply quickly results in unconsciousness and eventually in death. The partial pressure of oxygen in the atmosphere at sea level is 159 mm Hg. Although this is sufficient to supply our oxygen requirements, the total pressure and hence the partial pressure of oxygen decrease with increasing altitude (Figure 9.1). At an altitude of about 3.7 km (12,000 feet), for example, the ambient pressure drops to 480 mm Hg and the partial pressure of oxygen is only 100 mm Hg. At this oxygen partial pressure, borderline hypoxia (oxygen deficiency) occurs in humans [4]. From 3.7 to 12.2 km (12,000 to 40,000 feet), we can use a supplementary oxygen supply at ambient pressure to increase the percent of oxygen of inhaled air; but beyond this altitude, compression of the ambient atmosphere or a sealed cabin is required [4].

In the near vacuum of outer space, a sealed cabin is the only feasible means of survival. As the name implies, a sealed cabin is completely isolated from its surroundings. Within the cabin, the total pressure, the partial pressures of oxygen and carbon dioxide, the temperature, the relative humidity, and the levels of microcontaminants are controlled to meet the necessary requirements. Ideally, this atmosphere should duplicate that found near sea level on Earth. Nevertheless, constructing a spacecraft capable of maintaining a duplicate of a "normal" atmosphere and placing it into the near vacuum of outer space presents many formidable technological problems, some of which are discussed later. In resolving the dilemma of our physiological needs versus our technological capability to fulfill them, NASA scientists arrived at a compromise solution; thus, a cabin atmosphere of pure oxygen gas at a pressure of 1/3 atm was selected for America's early manned missions in space. Let us examine further some of the relevant physiological and technological aspects of this problem.

Oxygen Partial Pressure

Humans require a constant supply of energy to live. This energy is released through oxidation of the food we eat. The ultimate oxidizing agent for this process is molecular oxygen that is obtained from inhaled air and transported to our body tissues by the blood. A brief description of the process of breathing may be helpful in understanding the importance of controlling the partial pressure of this gas in the atmospheres of manned spacecraft.

When we breathe, air flows alternately into and out of our lungs because the volume of the lungs is alternately increased and decreased by the respective downward and upward movement of the diaphragm and the expansion and contraction of the rib cage. The process of breathing is represented schematically in Figure 9.2. Following exhal-

Table 9.1 Composition of Air at Sea Level (Dry Basis) [1]

Gas	Percent Composition by Volume	Partial Pressure, mm Hg at sea level
O_2	20.95	159
CO_2	0.04	0.3
N_2	79.0	600

Note: The majority of this chapter is taken directly from references 1 and 8. We would like to express our gratitude to NASA for supplying us with these documents for their inclusion here.

ation, the volume of gas contained in the lungs is at the normal minimum, and the pressure in the lungs is equal to that of the atmosphere. During inhalation, the volume of the lungs increases, and the pressure in the lungs decreases slightly below that of the atmosphere causing air to be inhaled.

The total pressure in the lungs, however, is not entirely caused by the inhaled air. Carbon dioxide and water vapor in sufficient quantities to maintain relatively constant partial pressure of 40 and 47 mm Hg, respectively, are also found in the lungs. Carbon dioxide is an end product of the oxidation of foodstuffs in the body and is eliminated through the lungs. The water vapor arises from the evaporation of the water serving as the dispersion medium of the body cells. Considering the relatively constant vapor pressures of carbon dioxide and water vapor and, assuming that atmospheric pressure is 760 mm Hg, the inhaled air exerts a partial pressure of $760 - 40 - 47 = 673$ mm Hg.

Because air is 21% oxygen (by volume), the partial pressure of oxygen in the lungs should be 141 mm Hg. Oxygen, however, dissolves in the blood; and its partial pressure is less than this calculated value. The relatively constant partial pressure produced by oxygen in the lungs is about 108 mm Hg.

From the lungs, oxygen is physically dissolved in the blood in accordance with Henry's law, which states that the amount of dissolved gas is directly proportional to the vapor

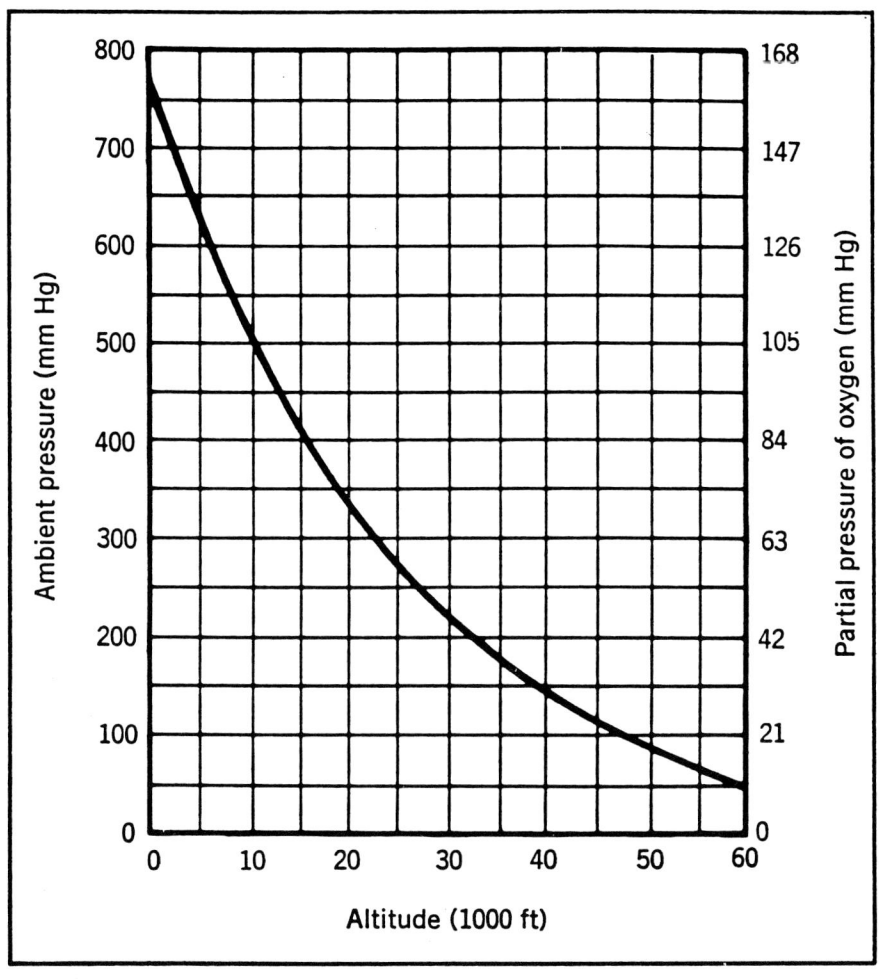

Figure 9.1 Variation of atmospheric pressure and oxygen partial pressure with altitude [1].

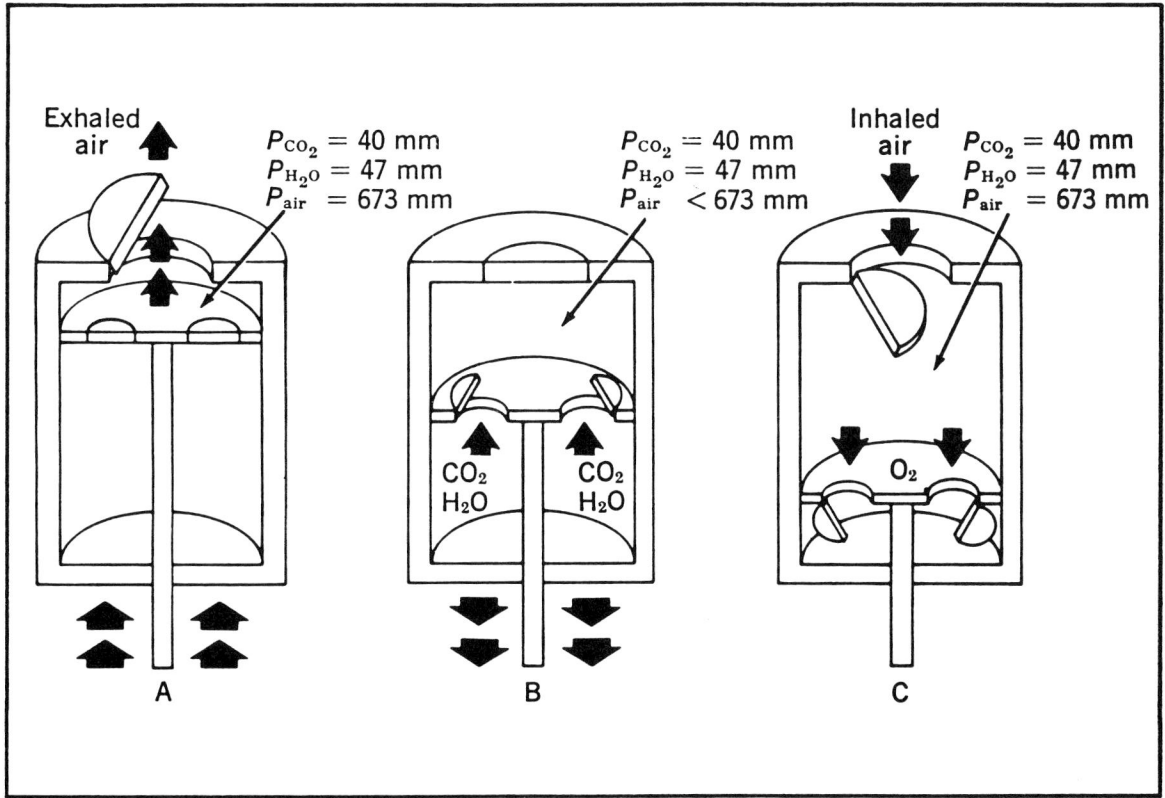

Figure 9.2 Schematic representation of breathing showing the change in the volume of the lungs during inhalation and exhalation [1].

pressure of the gas. As applied to this situation, Henry's law may be expressed as

$$P_{O_2} = K_{O_2,\text{blood}} X_{O_2} \qquad (9.1)$$

where

- P_{O_2} = the partial pressure of O_2 in equilibrium with the blood
- $K_{O_2,\text{blood}}$ = the Henry's law constant for O_2 in the blood
- X_{O_2} = the mole fraction of O_2 dissolved in the blood.

Oxygen, almost as rapidly as it dissolves, combines with the hemoglobin present in the red blood cells. This reaction effectively removes the oxygen from solution, allowing additional oxygen to be dissolved. The overall process may be represented as follows (Hb represents hemoglobin):

$$O_2(g) \leftrightarrows O_2(\text{soln}) \qquad (9.2)$$
$$O_2(\text{soln}) + \text{Hb}(\text{soln}) \leftrightarrows \text{HbO}_2(\text{soln}) \qquad (9.3)$$

Both of these reactions are readily reversible, and the position of each at equilibrium is determined by the partial pressure of oxygen in contact with the blood. As the blood leaves the lungs, it has an oxygen partial pressure (sometimes, in the medical profession, referred to as the oxygen "tension") of approximately 100 mm Hg, which is sufficient to produce 97% to 98% saturation of the hemoglobin as shown in Figure 9.3. Oxygen tension refers to the amount of oxygen dissolved in a solution, in this case the blood. Because the amount of oxygen is determined by its partial pressure in accordance with Henry's law, it is customary to express the concentration in terms of the partial pressure of oxygen with which the solution would be in equilibrium.

The transfer of oxygen from blood into body tissues occurs by diffusion. Arterial blood, having an oxygen partial pressure of about 100 mm Hg, flows into capillaries permeating the body tissues. The fluid bathing these tissues has an oxygen partial pressure of only 20 to 50 mm Hg. Consequently, oxygen dissolved in the blood diffuses into this tissue fluid. Loss of dissolved oxygen from the blood causes a reversal of reaction (equation 9.3), thus maintaining a supply of dissolved oxygen in the blood. By the time the blood has passed through the capillaries, 25% to 30% of its initial oxygen content has been lost, and the oxygen partial pressure is reduced to about 40 to 50 mm Hg. Returning to the lungs, the blood is resupplied with oxygen and the cycle is repeated.

Even though it is not necessary to our present discussion, it might be noted that the transport of carbon dioxide from body tissues to the lungs is also effected by a pressure gradient. The values in Table 9.2 illustrate this point. Just as the major portion of oxygen in the blood is transported in combination with hemoglobin, so is the major portion

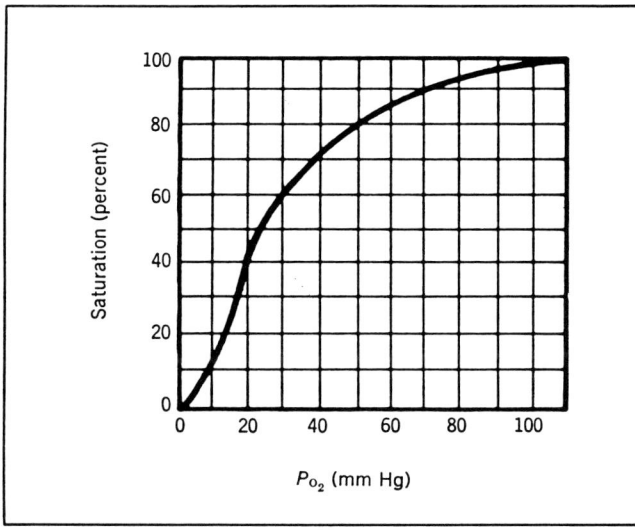

Figure 9.3 Percent saturation of hemoglobin with oxygen when exposed to different oxygen partial pressures at a constant CO_2 partial pressure of 40 mm Hg [5].

of carbon dioxide in the blood transported in a chemically combined form as the bicarbonate ion HCO_3^-.

Man requires a supply of oxygen at sufficient pressure to maintain a partial pressure in the lungs at the "normal" value of about 108 mm Hg. In a pure oxygen atmosphere, the minimum pressure that will provide the necessary amount of oxygen, taking into account the constant pressure of carbon dioxide and water vapor in the lungs, is about 40 + 47 + 108 = 195 mm Hg.

The partial pressure of oxygen in the lungs can vary within limits with little adverse effect. The extent of hemoglobin saturation can be reduced to about 90% before the symptoms associated with oxygen deficiency appear [3]. As shown in Figure 9.3, this means that the partial pressure of oxygen in the lungs can drop as low as 65 mm Hg with little apparent danger to the individual. The upper limit, however, is less well defined. It is known that breathing pure oxygen at pressures of 150 to 250 mm Hg for periods up to 14 days does not appear to produce any serious effects; but at a pressure of 450 mm Hg, man can survive for only about 1 day. Evidence suggests that the upper limit of oxygen partial pressure is in the range of 400 to 425 mm Hg [4].

Table 9.2 Partial Pressure of Carbon Dioxide in a Physiological System [6].

System Component	P_{CO_2}, mm Hg
Tissues	50–70
Venous blood	46
Arterial blood	40
Lungs	40
Expired air	20–30

Factors Impacting Pressure and Gas Control

A number of competing requirements influence the selection of cabin pressure and gas composition. Some of these factors are listed below:

1. An O_2 partial pressure that is neither hyperoxic nor hypoxic
2. An O_2 concentration that minimizes flammability of materials
3. A gas density that will provide adequate cooling to gas-cooled electronics
4. A total cabin/pressure suit pressure differential that is not conducive to dysbarism
5. Structure strength sufficient to contain the cabin pressure
6. Compensation for gas leakage

To demonstrate the way in which these factors interact, and the tradeoffs that are involved in emphasizing certain system parameters at the expense of others, let us examine a current pressure/gas control system. The Shuttle cabin atmosphere, at 760 mm Hg (1 atm) with 20% O_2 and 80% N_2, incorporates most of these factors; however, it does present a dysbarism problem when considered in combination with a 210 mm Hg (0.28 atm) pressure suit. This pressure differential requires 3 to 4 hours prebreathing with O_2 prior to decompression in the suit. The lengthy prebreathing requirement reduces the time available for EVA during a reasonable work day. To arrive at a more acceptable combination of cabin and pressure suit pressures, either the suit pressure must be increased or the cabin pressure reduced. If cabin pressure is reduced and O_2 pressure is held constant, the percentage concentration of O_2 increases, and thus flammability is increased. If cabin pressure is reduced and O_2 concentration is held constant, then O_2 partial pressure is reduced and hypoxia becomes a problem. If suit pressure is increased, then suit mobility becomes a problem. The ideal solution to this dilemma may ultimately be provided by the development of a high-pressure suit 413 mm Hg (0.54 atm) with good mobility. In the meantime, the compromise solution for Shuttle is to use a 525 mm Hg (0.69 atm) decompression step for 12 hours, breathe O_2 for 40 minutes, and then decompress to a 218 mm Hg (0.29 atm) pressure suit. This protocol involves a degree of reduction in gas cooling that is acceptable, an increase in flammability that is acceptable, some reduction in O_2 pressure (to 1,830 meters equivalent altitude) that will not result in clinical hypoxia, a prebreathing requirement that reduces EVA time, and some increase in suit pressure.

Life Support Systems

Maintenance of a suitable environment for human habitation requires the integration of a number of systems to provide overall life support. Not only is the atmosphere of concern, but also water and waste management and thermal control

Manned Space Flight

must be considered. Table 9.3 gives some representative data on the daily intake and output of an average human. This section discusses how the life support systems maintain a comfortable living environment.

Carbon Dioxide Removal

Carbon dioxide is one of the major contaminants encountered in manned spacecraft. It is formed as an end product of the oxidation of foodstuffs in man and is expired into the cabin atmosphere. The carbon dioxide output of an average individual is about 1 kg/day (2.2 lb/day). The minimum allowable concentration of carbon dioxide in a space cabin has been estimated to be approximately 3 parts per hundred [3] or a partial pressure of about 7.6 mm Hg. [4] Continuous exposure to concentrations in excess of this amount produces varying degrees of headaches, dizziness, confused thinking, and eventually death.

Removal of carbon dioxide from the cabin atmosphere on all flights to date (except Skylab) has been accomplished by its absorption with lithium hydroxide. Alkali metal hydroxides react readily with carbon dioxide in the presence of water vapor according to the following general equation (M represents any alkali metal):

$$2MOH(s) + CO_2(g) \rightarrow M_2CO_3(s) + H_2O(aq) \ . \quad (9.4)$$

A simple weight-weight calculation shows that to absorb the 1 kg (2.2 lb) of carbon dioxide released per man per day requires about 1.1 kg (2.4 lb) of LiOH, 1.8 kg (4.0 lb) of NaOH, or 2.5 kg (5.6 lb) of KOH. Thus, the use of LiOH gives the smallest weight penalty per kg (pound) of carbon dioxide absorbed.

$$1.0 \text{ kg } CO_2 \times \frac{1 \text{ mole } CO_2}{0.044 \text{ kg}} \times \frac{2 \text{ mole LiOH}}{1 \text{ mole } CO_2}$$
$$\times \frac{0.024 \text{ kg LiOH}}{1 \text{ mole LiOH}} = 1.1 \text{ kg of LiOH} \quad (9.5)$$

The LiOH/CO_2 reaction is essentially irreversible and is of little use if the carbon dioxide is to be used subsequently in a regenerative life support system. In such a system, the carbon dioxide must be removed from the cabin atmosphere and concentrated by some mechanism that will allow its easy recovery for subsequent treatment. One potential regenerative system uses a bed of synthetic zeolite (metal-ion alumino silicates, e.g., $NaAlSiO_4$) to absorb the carbon dioxide as cabin air is circulated through the bed (Figure 9.4). Zeolite, however, reacts readily with moisture losing its ability to absorb carbon dioxide. The airstream, therefore, must be predried to a dewpoint of about $-57°C$ by passage through a desiccant such as silica gel. The carbon dioxide absorbed on the zeolite can be desorbed easily by heating the zeolite to 340°C [5], by exposing it to a vacuum, or by using a combination of heat and vacuum.

A second regenerative system uses metal oxides which reversibly react with CO_2 as follows:

$$Ag_2O(s) + CO_2(g) \rightleftharpoons Ag_2CO_3(s) \quad (9.6)$$

The reaction of CO_2 with silver (I) oxide is 90% reversible in 4 hours at 180°C [1].

Oxygen Supply

Storage as Free Oxygen

Oxygen can be stored in a variety of forms, including high-pressure gas, supercritical fluid, cryogenic liquid, and oxygen-producing chemicals. Prestorage as a high-pressure gas was employed for the Mercury flights primarily because it is the simplest method. The gas, stored in heavy tanks under a pressure of approximately 3.9×10^5 mm Hg (510 atm) [14] is released at a rate sufficient to maintain the proper partial pressure of oxygen in the cabin. Although it is the simplest method, high-pressure storage also provides the greatest weight penalty because sturdy, pressure-resistant tanks are required.

Liquid storage has several advantages over gas storage. The greater density and lower vapor pressure of oxygen in condensed form reduce the size and strength of the tanks required. Among the disadvantages of liquid storage are the need for maintaining the temperature below the critical temperature of oxygen, 154.8 K, and the continuous presence of two phases in the storage tank as a result of vaporization of the liquid. Removal of oxygen from the container as a gas is desirable from the standpoint of regulating both temperature and pressure within the storage tank. Discharging O_2 in the form of a gas removes larger amounts of heat energy from the tank than would the discharge of O_2 as a liquid; thus, the capacity of the cooling system needed to maintain the O_2 in the form of a liquid is smaller. In a zero gravity environment, however, it is difficult to prevent discharge of the liquid.

As a compromise solution, the Gemini, Apollo, and Shuttle spacecraft have carried oxygen in the form of a supercritical fluid [11]. Stored as a supercritical fluid—that is, at a pressure greater than its critical pressure (3.78×10^4 mm Hg or 49.7 atm) and at a temperature above its critical temperature ($-118.4°C$)—the oxygen exists in one phase, behaving as a compressed liquid. Thus, the density advantages of a liquid are combined with the single-phase advantages of a compressed gas. The oxygen is dis-

Table 9.3 Daily Metabolic Values for a Human [1, 13]

	kg/person/day
Oxygen Consumption	0.6–1.0
CO_2 Production	0.7–1.2
Drinking Water	2.3–3.6
Wash Water	1.4–9.1
Urine	1.3–2.3
Feces	0.1
Food (dry weight)	0.5–0.9

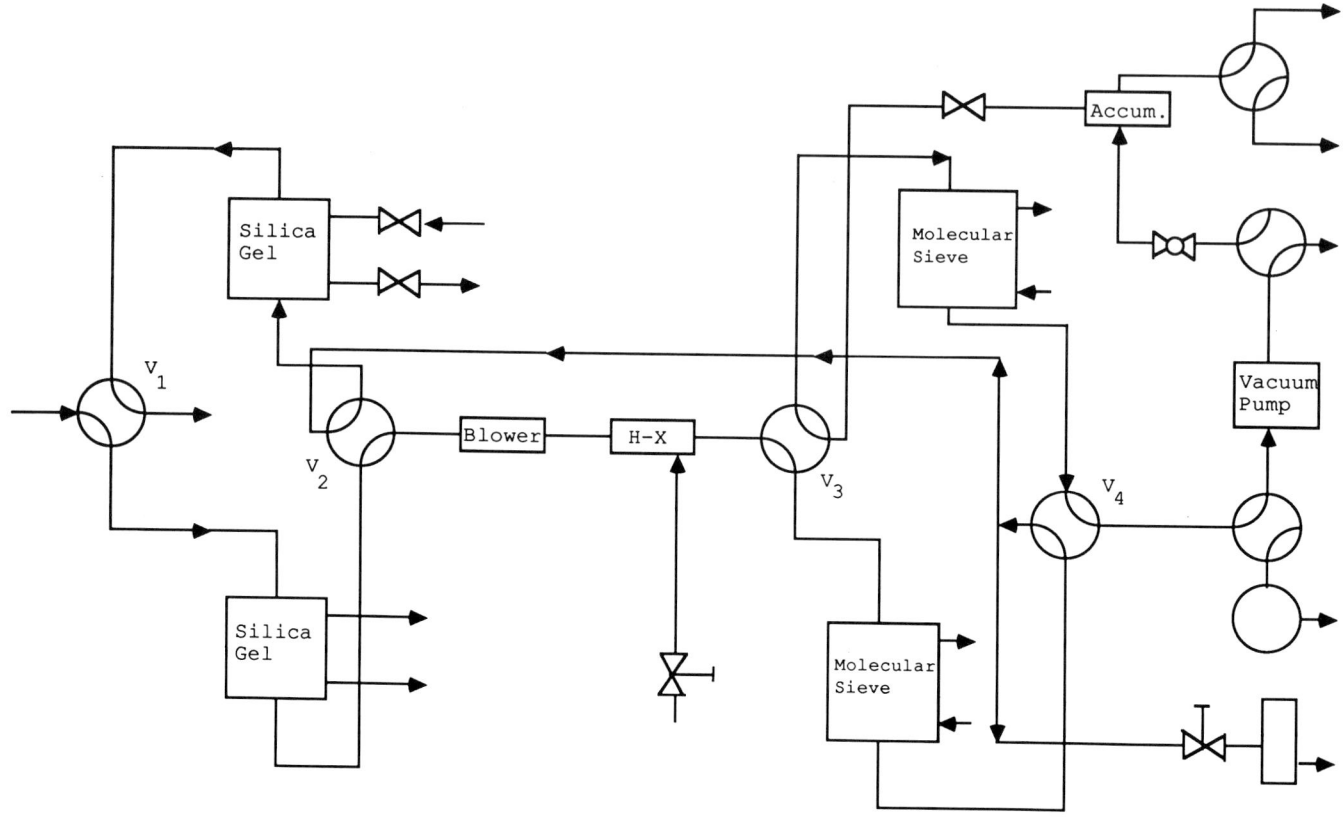

Figure 9.4 Schematic of a zeolite regenerative CO_2 absorption system [13].

charged from the tank at a constant pressure that is maintained by the regulated addition of heat to the storage system.

Chemical Regeneration

Numerous chemical methods of regenerating oxygen from waste products have been examined [15]. Most of these systems are based on the reduction of carbon dioxide and the direct or subsequent recovery of the oxygen contained in the carbon dioxide. The processes that appear most promising involve the reduction of carbon dioxide with molecular hydrogen to form water, which is subsequently electrolyzed to produce oxygen [16]. One variation of this process is known as the Sabatier or methanization reaction. The equation for this reaction is

$$CO_2(g) + 4H_2(g) \xrightarrow[200°-260°C]{Ni,Ru} CH_4(g) + 2H_2O(g) \ . \quad (9.7)$$

At a temperature of 200°C, about 99% of the carbon dioxide is converted. The methane from this reaction can be pyrolyzed, as follows

$$CH_4(g) \rightarrow C(s) + 2H_2(g) \quad (9.8)$$

and the resulting hydrogen can be recovered and then recycled. The water from the methanization reaction can be electrolyzed according to the reaction

$$2H_2O(l) \xrightarrow{e^-} 2H_2(g) + O_2(g) \quad (9.9)$$

which yields additional hydrogen and the desired oxygen. The sum of these three reactions gives the desired result; namely, the recovery of oxygen from carbon dioxide

$$CO_2(g) \rightarrow C(s) + O_2(g) \ . \quad (9.10)$$

A modification of the hydrogen-reduction process, known as the Bosch or carbonization reaction, reduces carbon dioxide to elemental carbon and water as follows:

$$CO_2(g) + 2H_2(g) \xrightarrow[590°-760°C]{Fe} C(s) + 2H_2O(g) \ . \quad (9.11)$$

Again the water is electrolyzed in a subsequent step to obtain the desired oxygen. Although it requires one less step than the Sabatier method, the Bosch process is less efficient, achieving generally less than 25% conversion of the carbon dioxide per pass through the reactor [14].

Storage in Chemically Combined Forms

Oxygen-releasing chemicals generally cannot compete with other methods of oxygen storage for long-duration missions, but they are being considered as a possible source of oxygen for spacesuit backpacks [4]. The compounds under investigation for this purpose are primarily peroxides, superoxides, chlorates, and perchlorates of alkali and alkaline earth metals. An illustrative example is the "chlorate candle," which has also been used on submarines and for emergency supplies of oxygen on aircraft. Sodium chlorate decomposes when heated according to the following reaction:

$$2NaClO_3(s) \xrightarrow{\Delta} 2NaCl(s) + 3O_2(g) \ . \quad (9.12)$$

Theoretically, 0.451 kg of oxygen is available per kilogram of sodium chlorate. Actual production is only about 40% of this theoretical maximum [1].

A combined carbon dioxide removal/oxygen sypply system, which has also been used successfully in submarines, employs potassium peroxide. The sequence of reactions may be represented as

$$2K_2O_2(s) + 2H_2O(g) \rightarrow 4KOH(s) + O_2(g) \quad (9.13)$$

and

$$4KOH(s) + 2CO_2(g) \rightarrow 2K_2CO_3(s) + 2H_2O(g) \ . \quad (9.14)$$

Water and Waste Management

A person requires about 3.1 kg or 6.9 lb per day of water for drinking and food preparation. A person produces about 3.5 kg (7.6 lb) per day of waste water, 1.5 kg (3.3 lb) in urine production and 1.9 kg (4.3 lb) in respiration and perspiration. For long duration trips, an additional 4.5 kg (10 lb) per person per day is needed for personal hygiene. A water management subsystem must be designed to recover water from waste and wash water. It must be reliable, efficient, light, and easy to integrate with other subsystems [17].

There have been many different proposals for water recovery. In vapor pyrolysis, waste water (urine, etc.) is basically distilled. Waste water is vaporized, with the vapor passing through a filter. The water vapor, along with some air, is heated and then passes into a pyrolysis (high heat) chamber, where ammonia and volatile organic compounds are oxidized. The water vapor is then condensed and separated from any residual gases. In other methods, waste heat from other spacecraft subsystems is used to vaporize waste water. Still other methods condense the water already in the vapor form from perspiration and respiration (a ''dehumidifier''). Depending on the exact method used, it is also often necessary to pass the recovered water through charcoal filters and/or ultraviolet light to eliminate impurities and microorganisms [17].

The state of the art in waste collection systems (WCS) is found on the Space Shuttle. It is assumed future systems will be simply modifications of the WCS. Rapid air flow past the user forces the waste into separate liquid and solid waste containers. On the Shuttle, wastes are dried, chemically treated, and stored until serviced on the ground. On interplanetary flights, urine water would be recovered. Fecal material would likely be dried by vacuum dehydration (evacuating evaporated fecal water to space), then possibly incinerated, with the remaining material either stored or dumped into space. An incineration device is also useful in disposing of other waste materials such as food containers.

Thermal Control

Early spaceflights used passive thermal control, receiving adequate temperature control through coatings and insulation. Larger systems such as the Space Shuttle need active thermal control systems to collect excess heat from the crew compartment and electronic systems. On the Space Shuttle, circulating water collects heat from the crew cabin and transfers its heat to freon coolant loops which circulate through radiator panels on the payload bay doors. Endothermic processess, such as vaporization of water and ammonia, are also used [17].

Evolutionary System Development

The Mercury spacecraft had a 255 mm Hg (0.34 atm) O_2 atmosphere supplied from a store of pressurized oxygen. CO_2 was controlled by a lithium hydroxide absorber in the environmental control loop. Temperature control was accomplished through cooling provided by a sublimator heat exchanger, which was supplied by a tank of cooling water augmented by condensate water from the ECS water separator. The sublimator vented water vapor overboard and cooling resulted from the change of state. The crew stayed in their pressure suits and the ECS system supplied both the pressure suit and the cabin [9].

The Gemini spacecraft retained the 255 mm Hg (0.34 atm) O_2 atmosphere used in Mercury; however, the primary source of oxygen was a liquid O_2 tank, with secondary oxygen supplies stored as high-pressure O_2. CO_2 was again controlled by the use of lithium hydroxide absorber. The primary means of heat rejection in the Gemini vehicle was a spacecraft radiator which radiated the heat to space. Heat loss was controlled by the flow of coolant to the radiator and to ECS heat exchangers. A secondary system permitted sublimation of water to the space vacuum if required for additional cooling. In the later Gemini flights, extensive periods were spent outside the pressure suit. Gemini was, incidentally, the first American spacecraft to support a pressure suit in space. The ECS system supplied the pressure suit through an umbilical connection [10].

The Apollo spacecraft atmosphere control systems were similar to the Gemini systems, but improved and more elaborate. The Apollo spacecraft system involved three modules: a command module ECS system, a lunar landing module ECS system, and a service module which carried consumables to support the command module system. The atmosphere was 255 mm Hg (0.34 atm) O_2 supplied from a cryogenic O_2 supply in the service module. Lithium hydroxide was used to absorb CO_2. Cabin temperature was maintained at 24°C ± 2.8°(75°F + 5°), with relative humidity limited to the range of 40% to 70%. The primary system for heat rejection was again a space radiator. The use of the radiators was facilitated by a slow, controlled roll of the command and service module called the passive thermal control flight mode. An evaporator was installed to provide additional cooling, but it was not used after Apollo 11 except for launch, Earth orbit, and reentry [11].

The Skylab atmosphere control system incorporated some significant changes. Cabin pressure was 255 mm Hg (0.34 atm), but to avoid minor chronic effects of hyperoxia over a long mission and possible interference of such effects with medical experiments, a two-gas environment was used. The atmospheric composition was 70% O_2 and 30% N_2, to provide an O_2 partial pressure just slightly higher than Earth-normal. There was automatic control of the two gases, but in practice, much of the control was accomplished manually to provide constant O_2 pressure during certain medical experiments. CO_2 absorption was accomplished with a regenerable molecular sieve system. This system allowed CO_2 to be flushed out of one bed by vacuum and heat and vented overboard while a second bed was absorbing CO_2. This regenerable system has obvious advantages for a long mission. However, a characteristic of this system was that it operated at a nominal CO_2 level of about 5 mm Hg, so that although it met the same 7.6 mm Hg CO_2 limit as earlier lithium hydroxide systems, the average CO_2 level was higher than on earlier spacecraft. The earlier systems had kept CO_2 near 1 mm Hg most of the time. The thermal control system for Skylab was primarily a passive system. The vehicle was carefully painted with paints of varying emissivity in specific patterns so that very little active control with radiators or evaporators was necessary. Several large panels of the exterior surface of Skylab were lost during launch, and later two separate types of shades were deployed to shade this bare area [12].

The Shuttle is the first American spacecraft to use a 760 mm Hg (1 atm) atmospheric pressure. Gas composition is 80% N_2 and 20% O_2, as on Earth. CO_2 absorption is accomplished with disposable lithium hydroxide cartridges, as in pre-Skylab flights, and thermal control is accomplished using radiators on the insides of the cargo bay doors [8].

Toxic Hazards in Space Operations

It has long been known that human exposure to trace levels of many gases can present a threat to health. Submarine operations and certain industrial workplace environments are the traditional spheres of concern with respect to toxic gas contamination. The closed-loop environment of spacecraft presented a new focus for such concerns. As early as the Apollo program, with the advent of longer space missions, steps were taken to provide adequate protection for crews by eliminating, or at least minimizing, crew exposures to possible harmful levels of trace contaminant gases and other hazardous material contained in the spacecraft cabin [18]. Accordingly, from the Apollo program onward each NASA manned spaceflight program has incorporated a toxicology program as an element of biomedical support. Such programs have had two primary objectives: (1) identification and control of sources of contaminant gases and (2) control or removal of these gases. This section summarizes the toxic hazard issues that are of concern for the Shuttle toxicology program and presents some initial results of Shuttle inflight toxicological sampling and analysis.

Whenever possible or appropriate, the chemicals described in this section include a threshold limit value (TLV), expressed in parts per million (ppm). The TLV specifies the concentration of a particular substance in air which is believed to present no danger of undesirable side effects; it is derived from observations of industrial workers under conditions of repeated daily exposure. TLVs are not poison thresholds, and can be exceeded for short periods under certain conditions. Nevertheless, TLVs do not take into account the conditions of space flight, during which crew members are exposed to potentially toxic substances continuously, rather than repeated 8 hour daily exposures.

Sources of Toxic Substances

Toxic substances can come from a number of different sources, including (1) leaks or spills from storage tanks, (2) volatile metabolic waste products of the crew, (3) particulate pollutants which are not easily removed from the air under conditions of weightlessness, (4) volatile components of spilled food, (5) leaks from environmental or flight control systems, (6) thermal reaction products produced by small electrical fires or contaminated removal systems, and (7) outgassing of cabin construction materials. All of the nonmetallic materials used in the interior of the Orbiter crew compartment are known to outgas contaminant compounds. Some specific sources are electrical insulation, paints, lubricants, adhesives, and the degradation of nonmetallic console and equipment structures. The heat produced by equipment operation increases outgassing. However, outgassing presents less of a problem in the Shuttle than in previous spacecraft because the atmospheric pressure is higher (760 mm Hg or 1 atm). Earlier spacecraft had reduced pressure atmospheres, a condition which accelerates outgassing. The environmental control and life support systems (ECLSS) of both the Space Shuttle and Spacelab have been designed so as to reduce trace contaminant gases to acceptable levels [19].

Fluids and Gases Aboard the Space Shuttle

A novel feature of the Space Shuttle in terms of U.S. manned spaceflight experience is its ground landing capability. This is significant because the transition from reaction jet to aerodynamic control and then to landing depends on spacecraft systems using fluids and gases with toxic potential. These materials are present in storage tanks and could present problems because of leaks during normal operation or from container rupture during a crash and rescue episode.

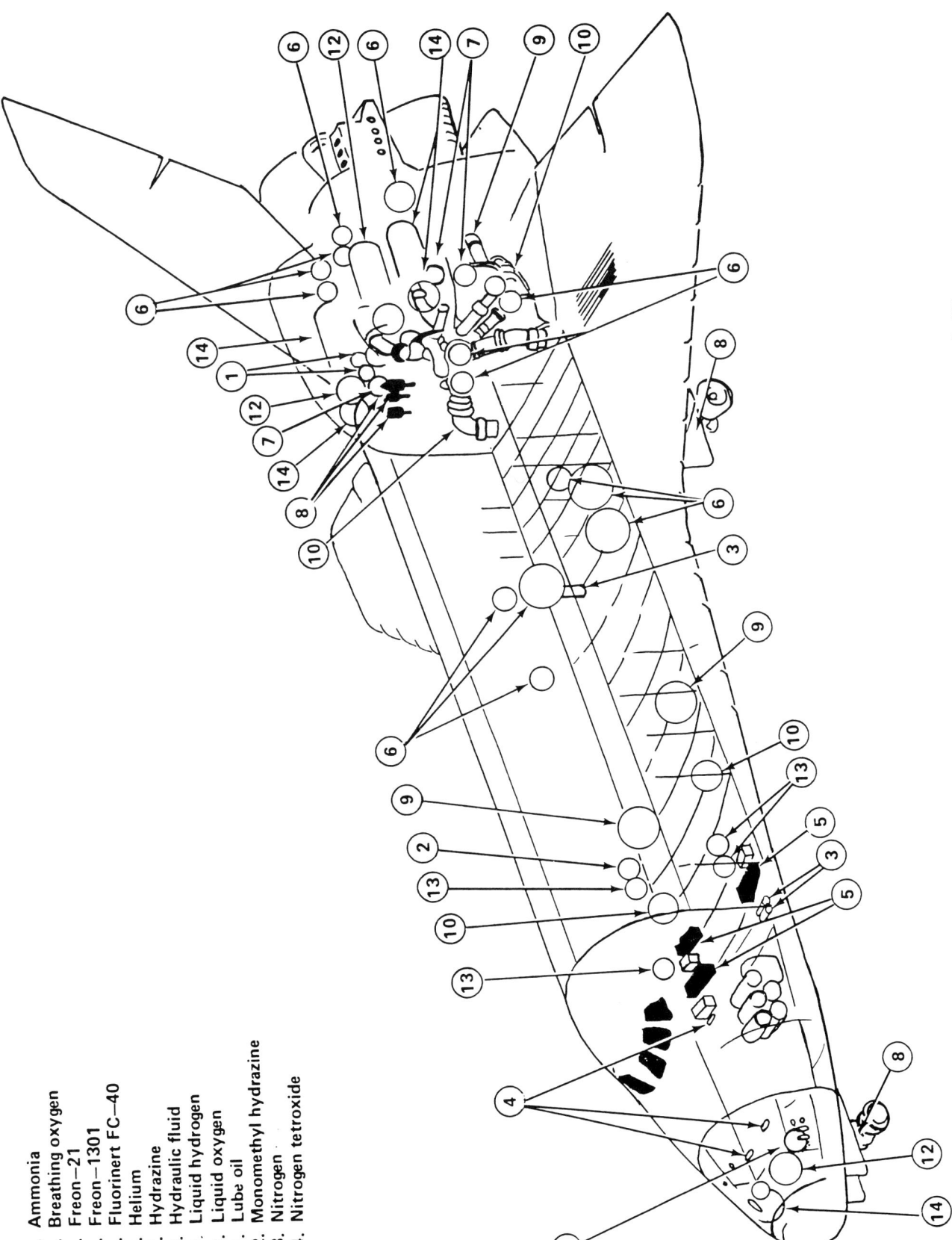

1. Ammonia
2. Breathing oxygen
3. Freon–21
4. Freon–1301
5. Fluorinert FC–40
6. Helium
7. Hydrazine
8. Hydraulic fluid
9. Liquid hydrogen
10. Liquid oxygen
11. Lube oil
12. Monomethyl hydrazine
13. Nitrogen
14. Nitrogen tetroxide

Figure 9.5 Location of potentially hazardous fluids and gases aboard the Space Shuttle [8].

Figure 9.5 shows the location of storage tanks on the Orbiter. The amount of material in each tank at landing depends upon whether or not the orbiter completed the full flight schedule. Exposure to each of these substances presents individual toxicological problems. A description of those which present the most significant potential health hazards follows.

Ammonia

Storage tanks are located in the aft-fuselage and are a part of Shuttle ECLSS. A colorless vapor with a pungent odor, ammonia is a powerful irritant to the eyes and to mucous memebranes of the upper respiratory tract. Symptoms of exposure include irritation of the eyes, conjunctivitis, swelling of the eyelids, irritation of the nose and throat, coughing, dyspnea, and vomiting. Liquid anhydrous ammonia produces severe burns on contact (TLV = 25 ppm, 18 mg/m^3).

Liquid and Gaseous Oxygen

The main health hazard associated with oxygen is fire. An oxygen-enriched atmosphere can be ignited by a spark. In the liquid state, oxygen can produce extensive frostbite "burns" due to low temperature on skin contact. Oxygen tanks are located in the mid-fuselage and are part of the ECLSS. There is no TLV; however, the upper limit is 100% for 6 hours at one atmosphere and the lower limit is 19%.

Freon 21

Storage tanks for this colorless, odorless, nonflammable gas are located in the mid- and aft-fuselage. Moderate concentrations can produce lightheadedness, giddiness, shortness of breath, and narcosis. The TLV is less than 1,000 ppm. Concentrations above this value may cause cardiac arrhythmia. If the exposed individual is having difficulty breathing, adrenalin or similar drugs should not be administered because of the possiblity of initiating irregular heartbeat.

Flourinert FC-40 (Fluorocarbon)

This fluorinated liquid is used as a dielectric coolant in the fuel cells of the electrical power system. At normal ground temperatures and pressures it presents little or no health hazard, but exposure to temperatures of 315°C (600°F) may produce toxic products.

Helium

Helium tanks are located at several locations within the Orbiter. The inert, nonflammable, colorless, and odorless gas acts as a simple asphyxiant in concentrations where the oxygen level is reduced to less than 15%.

Hydrazine (N_2H_4) and Monomethyl Hydrazine (MMH)

As part of the auxiliary power unit, hydrazine tanks are located in the aft-fuselage. At room temperature, hydrazine is a clear, oily liquid with an odor resembling ammonia. In the liquid or vapor form, it is extremely toxic. Liquid hydrazine produces severe burns on contact with the skin or eyes; it can penetrate the skin to cause systemic effects similar to those produced when swallowed or inhaled. The vapor causes local (irritation of the eye and respiratory tract) and systemic effects. Hydrazines produce effects on the central nervous system which result in convulsions and death. Rapid therapeutic treatment using barbiturates or barbiturates in combination suppresses the convulsive seizures and provides protection through the acute intoxication phase until the chemical has metabolized [20]. Repeated prolonged exposure to hydrazine may cause damage to the liver and kidneys. Pyridoxine, a form of vitamin B, prevents fatty liver changes from hydrazine. The TLV for hydrazine is 0.1 ppm; the TLV for MMH is 0.2 ppm. Table 9.4 presents the emergency exposure limits (EEL) for military and space operations for hydrazine and MMH.

Liquid Hydrogen

As part of the main propulsion system, liquid hydrogen tanks are located in the aft-fuselage. In the gaseous form, hydrogen acts as a simple asphyxiant. Liquid hydrogen produces serious "burns" on skin contact because of its low temperature.

Nitrogen

Tanks are located in the mid-fuselage and are part of the ECLSS. At room temperature, nitrogen is an inert, nontoxic, colorless, odorless, nonflammable gas. It acts as a simple asphyxiant when the oxygen level has been reduced to less than 15%.

Nitrogen Tetroxide

This highly toxic chemical is used in the reaction control system (RCS) and in the orbital maneuvering system (OMS). Skin contact with liquid nitrogen tetroxide results in a yellow stain if momentary, and severe chemcial burns if prolonged. In the eyes, the liquid can produce blindness. Inhalation results in irritation of the respiratory tract, and may cause pulmonary edema (TLV = 5 ppm). An unfortunate operational experience with exposure to this chemical occurred when the three U.S. crewmembers of the Apollo-Soyuz test project inhaled gaseous nitrogen tetroxide upon reentry. Some pulmonary edema occurred, forcing cancellation of many of the postflight medical tests [22].

Table 9.4 Emergency Exposure Limits for Hydrazine and MMH (Military and Space Operations) [21]

	10 min	30 min	60 min	24 hr
Hydrazine	30 ppm	20 ppm	10 ppm	—
MMH	90 ppm	30 ppm	15 ppm	1 ppm

Spacecraft Contaminants

A wide variety of contaminants has been found in the atmosphere of confined spaces during actual or simulated space flights. Analyses of samples taken from Apollo cabin atmospheres indicated the presence of some 300 different compounds [19]; approximately 400 compounds are outgassed in the Shuttle Orbiter cabin [24]. It has been very difficult to judge the toxic hazards of these contaminants because their concentrations are ususally not known. Many of these substances are commonly found in air, but in the confined atmosphere of the spacecraft they may build up to toxic levels.

Initial Results of Shuttle Missions

The Shuttle toxicology program provides for preflight and inflight atmospheric samplings and analyses during each operational mission. In the case of STS-1, the first operational flight test, atmospheric samples were obtained throughout the 54-hour flight. Four whole gas samples were taken during the mission: at the beginning; two evenly spaced periods in the middle; and at the end. Absorbed gas samples were taken on a continuous basis. In postflight chemical analysis of the samples, a total of 84 compounds were detected. Extrapolation of the data obtained in this mission to a 7-day mission assessment showed that the Orbiter's cabin environment was safe for a flight of that duration [19]. After the 54-hour STS-2 mission, analysis revealed a total of 99 compounds in the whole gas atmospheric samples [24].

Toxicological Aspects of Skylab

Toxicological support for the Space Shuttle benefited from the experience gained during the Skylab missions. Following the loss of Skylab 1's micrometeoroid shield, the interior wall insulation material of the orbital workshop overheated. This created a potential toxic hazard because the decomposition of the rigid polyurethane foam produced trace quantities of toluene diisocyanate, and because the overheating might have produced accelerated outgassing of all nonmetallic materials in the cabin. To protect the Skylab 2 crew from potentially toxic air contaminants, several measures were taken. Before the crew entered the space station cluster, a series of pressurization-depressurization cycles designed to discharge and dilute any contaminating gases were conducted. The crew also sampled the air for carbon monoxide and toluene diisocyanate by means of two types of gas analysis detector tubes. The analysis revealed no detectable toluene diisocyanate and an extrapolated 5 mg/m^3 level of carbon monoxide. The crew energized the ECLSS, which was designed to remove trace levels of contaminants by means of activated carbon filtration. Thirty minutes later the crew entered the space station, and the mission was completed with no more atmospheric trace gas problems.

The Skylab 4 crew took three samples of the atmosphere, on mission days 11, 46, and 77. Analysis of the three samples revealed the presence of more than 300 compounds in the Skylab atmosphere. The contaminants in each of the three samples were very similar, indicating that the amounts of individual trace contaminants were not rising throughout the mission. The gas generation rates of the contaminant sources and the removal rate by the ECLSS had attained a state of equilibrium.

The experience in Skylab demonstrated that certain specific toxic hazards can be ascertained and dealt with as they arise, and that a spacecraft's onboard ECLSS is capable of establishing an equilibrium between the generation of air contaminants and the removal rate, at least under normal operating conditions [24].

Radiation Exposure Issues

A major concern for space medicine is the radiation exposure experienced by space travelers. Over the past 25 years, we have developed a wealth of data describing the radiation environment in orbital space as well as through much of the rest of the solar system. Results from numerous space probes present a picture of heightened radiation levels, changing character of radiation, and even "radiation storms" as the level of solar activity waxes and wanes.

Radiation in space has been described as "the primary source of hazard for orbital and interplanetary space flight" [25]. The successful flights made by the United States and the former Soviet Union prove clearly that comparatively brief flights, when there are no solar flares and when flight trajectories are carefully planned, do not present a radiation hazard for space crews. However, as the space programs of both nations expand and as greater numbers of specialists spend more time in space, the problem of providing radiation protection grows in importance. Space medicine personnel must be knowledgeable concerning the extent of the danger; the means of providing an appropriate measure of protection; and, most important, the proper medical procedures for use in the event of a radiation emergency.

Measurement of Exposure

Consideration of radiation hazard and exposure risks for space crews requires an understanding of the units used for expressing radiation levels and the effect on biological systems. There are several interrelated units, which are defined as follows:

Roentgen

Radiation cannot be measured directly; therefore its magnitude is determined by the ionization produced from the passage of radiation through a medium. The roentgen refers to the ionization produced by the passage of x-ray or gamma radiation in air resulting in 0.001293 gram of air ions carrying one electrostatic unit of electricity of either sign. The roentgen as a measure is little used in describing exposure.

Rad

One rad (radiation absorbed dose) of any type of radiation corresponds to the absorption of 100 ergs per gram of any medium. This is the most common unit describing exposure. For biological systems, one rad is equivalent to 10^{-5} joules of energy absorbed per gram of tissue.

Rem

The rem (roentgen equivalent, man) is the most accurate unit for expressing exposure of man. One rem equals the absorbed dose of any ionizing radiation which produces the same biological effect in man as that resulting from the absorption of one roentgen of x-rays. The rad measure is qualified by an RBE (relative biological effectiveness) measure in order to arrive at an expression in rems. The RBE differs for various types of radiation, from a value of one for x-rays and gamma rays, to values as high as 15 to 20 for alpha particles at different energy levels. For the type of radiation encountered in space, particularly in low Earth orbit, 1 rem may be considered as roughly equal to 1 rad.

The biological effects of heavy particles, such as those found in galactic radiation, are not described adequately by a single dose parameter such as a rad. For these purposes, the linear energy transfer (LET) is used, which is the cosmic ray energy absorbed per micrometer of cells or living tissues penetrated. The LET is larger for higher energy, highly charged particles.

Investigations of radiation levels in space are accomplished with specially instrumented unmanned spacecraft which send information concerning radiation levels to ground stations on Earth. Radiation measuring devices also are carried on manned space vehicles. The purpose in this case is two-fold. First, there is a need to gather information concerning personnel exposures and to monitor the radiation hazard. Second, these missions add to the radiation mapping of near Earth space.

In the U.S. program, radiation is monitored in flight through use of two types of dosimeters, passive and active [26]. Precise radiation information is obtained from the passive dosimeters, which are composed of thermoluminescent dosimeter (TLD) chips, plastic sheets, and metal foils, each affected by different kinds and energies of radiation. These dosimeters are sealed units which must be processed in a laboratory following the flight. Active dosimeters are used as a means of determining in real-time the radiation danger being encountered. They may be read by a crewmember at any time. These are integrating dosimeters consisting of pen-sized ion chambers which measure three ranges of radiation exposure. The pocket dosimeter, low range (PDL), measures accurately in the range of zero to 100 rad. There also is included a contingency high rate dosimeter (HRD), provided for measurement of doses from zero to 600 rad. The combination of passive and active dosimeters allows accurate monitoring and recording of the unique radiation in space, including the electron, proton, and heavy cosmic rays encountered.

Human Response to Radiation

Human response to radiation can be quite varied, depending on the operation of a number of variables. Key variables affecting the response include:

1. Radiation Variables. The type of radiation, the possiblity of combined types, and the time-intensity relationship
2. Subject Variables. The location of exposure (whole body versus specific organs or tissues), the condition of the subject, the age, and complicating separate medical factors
3. Situational Variables. Stress factors such as weightlessness, acceleration, and body temperature

The bulk of the information in the medical literature on human response to radiation is drawn from clinical exposures and can be extrapolated to space flight only with caution. Also, little information is available concerning whole body exposures of healthy subjects and the effects of exposure to heavy ions. For these reasons, it is difficult to specify with precision the medical consequences of a particular spaceflight exposure situation [8].

It is characteristic of radiation exposures that all body organs and systems are not affected equally. Knowledge of the differential sensitivity of body structures is important both for protection purposes and for evaluation of clinical syndromes. Table 9.5 presents three groupings of body elements in terms of their relative radiosensitivity. The key characteristic of this most sensitive group is that all these tissues are growing. Tissue elements which are rapidly dividing are the most sensitive to any radiation exposure.

Tissues that show moderate sensitivity to radiation include the skin, the vascular endothelium, and internal structures such as the lung. These systems all have relatively

Table 9.5 Differential Radiosensitivity of Human Tissue [27]

High Sensitivity
- Lymphoid Tissue
- Bone Marrow Elements
- Gastrointestinal Epithelium
- Gonads (Testis and Ovary)
- Embryonic Tissues

Moderate Sensitivity
- Skin
- Vascular Endothelium
- Lung
- Kidney
- Liver
- Lens (Eye)

Low Sensitivity
- Central Nervous System
- Endocrine Glands (Except Gonads)
- Muscle
- Bone and Cartilage
- Connective Tissue

Manned Space Flight

Table 9.6 Radiation Exposure Limits Recommended for Spaceflight Crewmembers [28]

Constraint	Primary Reference Risk (Rem at 5 cm depth in tissue)	Ancillary Reference Risks			
		Bone Marrow* (Rem at 5 cm)	Skin (Rem at 0.1 mm)	Ocular Lens (Rem at 3 mm)	Testes (Rem at 3 cm)
1-year average		0.2	0.6	0.3	0.1
30-day maximum		25	75	37	13
Quarterly maximum**		34	105	52	18
Yearly maximum		75	225	112	38
Career limit	400	400	1,200	600	200

*Whole body exposure.
**May be allowed for two consecutive quarters followed by 6 months of restriction from further exposures.

slow growth characteristics. Of these, the lens of the eye is somewhat different in that it shows a particular sensitivity to neutron radiation, which can produce cataracts and lenticular opacities at lower levels of exposure than would be predicted. In general, by the time significant damage is received by any of these tissue systems, severe damage has occurred within the hematopoietic system.

Those tissues which show low radiosensitivity, such as the central nervous system, the endocrine glands, muscle, and bone and cartilage, reproduce very slowly and are, therefore, considered to be very stable.

Exposures During Space Flight

The Radiobiological Advisory Panel, Committee on Space Medicine of the National Academy of Sciences issued in 1970 a set of recommended radiation exposure limits for use by NASA in planning and developing protection for future manned space missions. These limits are presented in Table 9.6. The mission limit (30 days maximum for bone) was established at 25 rem in view of research which indicated this level of exposure would have an extremely low probability of episodes such as diarrhea, anorexia, and vomiting. The career limit of 400 rem was established on the basis of doubling the dose considered a risk for leukemia. Of course, this value of 400 rem refers to a career exposure and not an acute episode. Other intermediate values were calculated using the career dose as a reference.

The radiation levels encountered by both Soviet Union cosmonauts and United States astronauts during manned missions have been low. Table 9.7 presents the radiation exposures for crewmembers in 28 U.S. manned space missions. All radiation levels are seen to be acceptable. The highest exposure (7.8 rad) was recorded during the 84-day Skylab 4 mission. To put this risk into perspective, it has been stated that the Skylab 4 crewmen could fly a mission comparable to one 84-day flight per year for 50 years before exceeding the NASA career limit.

It is of interest that four nuclear devices were detonated by France at the Murora Test Site during the flight of Skylab 3. No radiation attributed to these tests was recorded in Skylab. Of the radiation which was received, however, it was calculated that galactic cosmic radiation contributed roughly 20% to 30%.

An indication of the risk associated with the exposures recorded during manned space missions can be obtained by comparing radiation levels encountered in space with those found in a number of other situations on Earth. Table 9.8 presents radiation exposures from such diverse sources as chest x rays and transcontinental jet flights. This table indicates that routine medical procedures such as an upper G.I. series or barium enema produce radiation exposures many times greater than that recorded during the two-day Space Shuttle flight of STS-1. Living in Kerala, India, for one year would be considerably more dangerous, due to the increased level of thorium in the local sands. Even there, however, the radiation presents little if any risk over the lifetime of a human.

Table 9.7 Measured Radiation Exposures in U.S. Manned Spaceflight Programs [30]

Mission	Mean Dose (Rad)	Mission	Mean Dose (Rad)
Mercury 9	0.027	Apollo 10	0.480
Gemini 3	0.020	Apollo 11	0.180
Gemini 4	0.045	Apollo 12	0.580
Gemini 5	0.177	Apollo 13	0.240
Gemini 6	0.025	Apollo 14	1.140
Gemini 7	0.150	Apollo 15	0.300
Gemini 8	0.010	Apollo 16	0.510
Gemini 9	0.018	Apollo 17	0.550
Gemini 10	0.840	Skylab 2	1.980
Gemini 11	0.025	Skylab 3	4.710
Gemini 12	0.015	Skylab 4	7.810
Apollo 7	0.160	STS-1	0.020
Apollo 8	0.160	STS-2	0.015
Apollo 9	0.200	STS-3	0.416

Table 9.8 Comparison of Radiation Exposure Under Various Conditions [30].

	Average Rad to Total Red Marrow
Transcontinental Round Trip, Jet	0.004
Chest X Ray (Photofluoroscopic)	0.045
Chest X Ray (Radiographic, Low Dose)	0.010
Living in Houston 1 Year	0.100
Living in Denver 1 Year	0.200
Upper G.I. Series	0.535
Barium Enema	0.875
Living in Kerala, India, 1 Year	5.——

Table 9.9 compares the radiation exposure of six astronauts from their space missions with the radiation these individuals were calculated to have received from all other sources, such as diagnostic x-ray procedures. With the exception of the two astronauts who had flown missions on the order of two and three months, radiation received from the more normal sources on Earth consistently exceeded that received in space. This low level of radiation during space flights has resulted from the combination of careful trajectory planning plus the good fortune to have avoided any unpredicted periods of heightened solar activity.

Were astronauts to encounter a sudden burst of radiation from the Sun, such as occurred in August 1972, or a nuclear detonation exploded at high altitude, the radiation picture would change dramatically. A nuclear explosion would be particularly dangerous. There would be prompt intense radiation in the form of x rays and neutron and gamma radiation. The spacecraft also might be unfortunate enough to fly through the debris cloud itself. Should such an event occur, crew exposures would be recorded and the data immediately trasmitted to ground stations. The mission management team, with medical consultation, then would decide whether circumstances would allow the crew to continue or if an immediate return to Earth was necessary.

A review of space crew exposures must include an interesting visual phenomenon, obviously related to space radiation, which was first noted in the Apollo program. Crewmembers of Apollo 11 reported seeing faint spots or flashes of light when the cabin was dark and their eyes had become dark-adapted. From these reports and more systematic studies on later Apollo flights, it was concluded that the light flashes resulted from high-energy, heavy cosmic rays penetrating the spacecraft structure and the crewmembers' eyes. The fact that dark adaptation is necessary for this experience indicates that the phenomenon occurs at the retina rather than through direct stimulation of the optic nerve.

Protection and Treatment Principles

The protection of spacecraft crews from radiation injury can follow any or all of four directions, as outlined by Grigor'yev (1976):

1. Passive physical protection—attenuation of radiation by increasing thickness of the skin of the spacecraft, arranging the equipment, creating shelter or shadow protection
2. Active protection—creating protection against charged particles with the aid of magnetic and electric fields
3. Local protection of critical organs and systems
4. Pharmaco-chemical antiradiation protection and methods of combined protection

The feasibility of each of these avenues of protection will be discussed briefly. Passive physical protection simply means increasing the thickness of a protective coating, generally the exterior of the spacecraft, in order to reduce the level of radiation from solar flares and from the Van Allen belts. However, the relationship is not straightforward. When the thickness of protection is increased to a certain limit, the dose from galactic cosmic radiation can lead to increased secondary radiation and thereby defeat the purpose of the protection system. The biggest problem with this approach, however, is the additional weight required for proper protection from spacecraft components such as onboard equipment, fuel reserves, etc.

Table 9.9 Representative Career Exposures for Astronauts [30]

Astronaut	Career Time	Space Exposure Time	Space Rad Dose	Diagnostic X-Ray and Other Rad Dose
X	15 years	9 days, 15 min	0.64	2.29
Y	10 years	12 days, 11 hrs, 4 min	0.18	2.01
Z	18 years	22 days, 2 hrs, 45 min, (4 flights)	1.83	2.76
A	15 years	28 days	1.66	2.82
B	15 years	59 days	3.71	2.36
C	11 years	84 days	8.01	2.65

It has been known, at least in theory, for a number of years that spacecraft can be protected from charged particles through the use of magnetic and electric fields developed around the vehicle. Such protection is promising and has a number of advantages, one being its comparatively low weight. In one approach, magnetic protection is based on the use of superconducting materials which have zero electrical resistance and are able to provide circulation of a current, once established, for a practically unlimited time. The energy consumption for magnetic protection of this type is not great and is determined basically by the power required for the cryogenic unit. Another advantage is that the magnetic field can be increased, by raising the working current level, when the amount of radiation exceeds predicted levels.

Local protection of critical organs and systems implies the availability of specific garments to be worn in the event of an anticipated increase in radiation. With dogs, it has been found that protective effect is highest with screening of the abdomen. Following radiation, when such abdominal screening is used, changes in peripheral blood, bone marrow, and immunobiological reactivity of the blood are not as severe as in other animals not similarly protected. Shielding a portion of the body, particularly if it is the abdominal area, appears therefore not only to increase survival rate, but also to reduce the severity and frequency of appearance of the primary syndromes of radiation injury.

Pharmaco-chemical protection is a topic which has been given considerable attention in recent years, principally for basic medical purposes. This is a broad field of endeavor. Saksonov notes that more than 15,000 different chemical substances, having widely dissimilar physicochemical properties and pharmacological effects, have been tested. These include vitamins, antibiotics, nitrites, cyanides, amino acids, alkaloids, flavonoids, polysaccharides, sulphur-containing substances, analeptics, narcotics, central nervous system stimulants, colline and acridine derivatives, local anesthetics, and others. Many substances have been found which, when given to test animals at a specific time before irradiation, result in some reduction of the damaging effect of the radiation. There are many problems, however. First, dose levels of the protective agent frequently must be quite high. Second, any number of side effects can be produced through the administration of the protective substance. For these reasons, the practical use of pharmacological or chemical substances for protection of space crews from radiation will not occur until some point in the distant future.

Summary

This chapter has discussed many of the important chemical aspects involved in manned space flight. In addition to the topics discussed in the previous chapters, spacecraft engineers encounter many additional design considerations for manned systems. These include cabin atmospheres, water and waste management, thermal control, toxic hazards, and radiation effects. The intention in this final chapter is to give the reader an appreciation of the tremendous complexity involved when humans go into space.

References

1. Lawrence, R. M., *Space Resources for Teachers: Chemistry*, Washington, D.C.: NASA EP-87, 1971.
2. Sanders, H. J., "Chemistry and the Atmosphere," *Chemical and Engineering News*, 1965, 44(13), 1A.
3. Konecei, E. B., "Manned Space Cabin Systems," in F. I. Ordway, *Advances in Space Science*, Vol 1, Academic Press Inc. (New York), 1959.
4. Welch, B. E., "Ecological Systems," in J. H. Brown, *Physiology of Man in Space*, Academic Press, Inc. (New York), 1963.
5. Anderson, A. L., *Essentials of Physiological Chemistry*, John Wiley & Sons, Inc. (New York), 1953.
6. Kleiner, I. S. and J. M. Orten, *Biochemistry*, 7th edition, C. V. Mosby Co. (St Louis), 1966.
7. Billingham, J., "Physiology of Man in the Space Environment," in P. E. Purser, M. A. Faget, and N. F. Smith, *Manned Spacecraft: Engineering Design and Operation*, Fairchild Publications, Inc. (New York), 1964.
8. Nicogossion, A. E. and J. F. Parker, *Space Physiology and Medicine*, NASA SP-447, 1982.
9. "Mercury Project Summary Including Results of the Fourth Manned Orbital Flight," NASA SP-45, 1963.
10. "Gemini Midprogram Conference," NASA SP-121, 1966.
11. Brady, J. C., et al., "Apollo Command and Service Module and Lunar Module Environmental Control Systems," in NASA SP-368, 1975.
12. Belew, L. F. (ed), *Skylab, Our First Space Station*, NASA SP-400, 1977.
13. Heitchue, R. D. *Space Systems Technology*, Reinhold Book Co., 1968.
14. Smylie, R. E. and M. R. Reumont, "Lift-Support Systems," in P. E. Purser, M. A. Faget, and N. F. Smith, *Manned Spacecraft: Engineering Design and Operation*, Fairchild Publications, Inc. (New York), 1964.
15. *Significant Achievements in Space Bio Science 1950 to 1964*, Washington, D.C.: NASA SP-92, 1966.
16. Smylie, J. W., "Evaluation and Application of Chemically Regenerative Atmospheric Control Systems," Office of Technical Services, U.S. Dept. of Commerce, NTIS N64-31979#.
17. Space Chemistry Handout, Chemistry 325, U.S. Air Force Academy, Fall 1987.
18. Rippstein, W. J., "The Role of Toxicology in the Apollo Space Program," in R. F. Johnston, L. F. Die-

tlein, & C. A. Berry (Eds.), *Biomedical Results of Apollo* (NASA SP-368), Washington, D.C.: National Aeronautics and Space Administration, 1975.

19. Rippstein, W. J., "Shuttle Toxicology," in S. L. Pool, P. C. Johnson, Jr., & J. A. Mason (Eds.), "STS-1 Medical Report (NASA TM-58240). Washington, D.C.: National Aeronautics and Space Administration, December 1981.

20. Azar, A., Thomas, A. A., & Shillito, F. H., "Pyridoxine and Phenobarbital as Treatment for Aerozine-50 Toxicity," *Aerospace Medicine*, 1940, 41, 1.

21. Back, K. C., Carter V. L., & Thomas, A. A., "Occupational Hazards of Missile Operations With Special Regard to the Hydrazine Propellants," *Aviation, Space, and Environmental Medicine*, 1978, 49, 591–598.

22. Nicogossian, A. E., LaPinta, C. K., Burchard, E. C., Hoffler, G. W., & Bartelloni, P. J., "Crew Health," in A. E. Nicogossian (Ed.), *The Apollo-Soyuz Test Project Medical Report* (NASA SP-411). Springfield, VA: National Technical Information Service, 1977.

23. Rippstein, W. J., "Shuttle Toxicology," in S. L. Pool, P. C. Johnson, Jr., & J. A. Mason (Eds.), "STS-2 Medical Report" (NASA TM-58245). Houston, TX: Lyndon B. Johnson Space Center, May 1982.

24. Rippstein, W. J., & Schneider, H. J., "Toxicological Aspects of the Skylab Program," in R. S. Johnston & L. F. Dietlein (Eds.), *Biomedical Results From Skylab* (NASA SP-377). Washington, D.C.: U.S. Government Printing Office, 1977.

25. Petrov, V. M., Kovalev, E. E., & Sakovich, V. A., "Radiation: Risk and Protection in Manned Space Flight," *Acta Astronautica*, 1981, 8(8–10), 1091–1098.

26. Barnes, C. M., "Radiological Health," in S. L. Pool, P. C. Johnson, Jr., and J. A. Mason (Eds.), "STS-1 Medical Report" (NASA TM-58240). Washington, D.C.: National Aeronautics and Space Administration. Scientific and Technical Information Branch, December 1981.

27. Tyler, P., "Health Effect of Ionizing Radiation," 53rd Annual Scientific Meeting of the Aerospace Medical Association, Bal Harbour, FL, 10–13 May 1982.

28. National Academy of Sciences (Radiobiological Advisory Panel, Committee on Space Medicine). "Radiation Protection Guides and Constraints for Space Mission and Vehicle-Design Studies Involving Nuclear Systems. Washington, D.C.: U.S. Government Printing Office, 1970.

29. Bailey, J. V., Hoffman, R. A., & English, R. A., "Radiological Protection and Medical Dosimetry for the Skylab Crewmen," in R. S. Johnston & L. F. Dietlein (Eds.), *Biomedical Results From Skylab* (NASA SP-377), Washington, D.C.: U.S. Government Printing Office, 1977.

30. Nachtwey, S., "Radiation Exposure, Detection, and Protection," 53rd Annual Scientific Meeting of the Aerospace Medical Association, Bal Harbour, FL, 10–13 May 1982.

31. Osborne, W. Z., Pinsky, L. S., & Bailey, J. V., "Apollo Light Flash Investigations," in R. S. Johnston, L. F. Dietlein, and C. A. Berry (Managing Eds.), *Biomedical Results of Apollo* (NASA SP-368). Washington, D.C.: U.S. Government Printing Office, 1975.

32. Grigor'yev, Yu.G., "Radiation Safety of Space Flights" (NASA TT F-16, 853). Translated by SCITRAN under contract NASW-2791. Washington D.C.: National Aeronautics and Space Administration, 1976.

33. Johnson, R. D., & Holbrow, C. (Eds.), *Space Settlements—A Design Study* (NASA SP-413), Washington, D.C.: U.S. Government Printing Office, 1977.

34. Saksonov, P. P., "Protection Against Radiation (Biological, Pharmacological, Chemical, Physical)," in M. Calvin and O. G. Gazenko (General Eds.), *Foundations of Space Biology and Medicine* (Vol. III), Washington, D.C.: U.S. Government Printing Office, 1975.

Discussion Questions

1. Describe the process by which oxygen is transported from the lungs to body tissue. Include in your discussion the key equilibrium reactions.

2. Cabin atmosphere has been a subject of great debate throughout the history of the manned space program. Discuss the issues which must be analyzed when selecting atmospheric composition for a spacecraft cabin.

3. Future space missions of extended durations will require an oxygen generation capability. Discuss some of the designs being considered in order to meet this need.

4. In the previous chapters, a number of different materials have been discussed which could represent potential health hazards to crew members. Identify those materials representing the greatest risk (and the conditions under which this could occur) and discuss precautions that must be taken.

5. The schedule for manned space missions is partly based upon the level of solar activity.
   a. Why is it dangerous to have crews in space during increased levels of solar activity?
   b. Do you think this is a concern of Space Station designers?

6. a. What parts of the body are most sensitive to ionizing radiation?
   b. What protection is necessary to prevent harmful exposure levels?
   c. What alternatives are being considered for radiation protection and treatment?

7. Arguments as to the ultimate benefits of manned versus unmanned missions have been ongoing for years. These arguments are especially important in light of funding cutbacks to the space program and the lower cost of unmanned

spacecraft. Discuss the types of missions appropriate for manned spacecraft and unmanned spacecraft.

Problems

1. A hyperbaric chamber can simulate atmospheric conditions at various altitudes and is used to familiarize aircrews with the effects of hypoxia.
   a. Calculate the partial pressure of oxygen within the lungs at an altitude of 6.1 km (20,000 ft) (Neglect dissolution of oxygen into blood).
   b. Why do runners often do "altitude training" to prepare for a big race?

2. During the course of a Shuttle mission, astronauts must change the lithium hydroxide cartridges used in the atmosphere filtration system.
   a. Calculate the mass of LiOH necessary for CO_2 removal for a crew of five during a typical seven day mission.
   b. What other compound might you suggest for CO_2 removal instead of LiOH?

3. Table 9.3 lists the daily metabolic values for a human.
   a. What is the total mass of expendables in and out?
   b. Why isn't the output equal to the input?

4. An astronaut experiences a radiation exposure of 5 rem.
   a. This individual feels the effect of the absorption of how much x-ray energy?
   b. Will more or fewer alpha particles be required to have the same radiological effect on the human?

Appendix A
Erosion Yields of Various Materials Exposed to Atomic Oxygen in Low Earth Orbit

Material	Erosion Yield (10^{-24} cm^3/Atom)	Reference	Material	Erosion Yield (10^{-24} cm^3/Atom)	Reference
Aluminum (150 Å)	0.0	1	Graphite Epoxy:		
Aluminum-coated Kapton	0.01	2	1034 C	2.1	10
			5208/T300	2.6	10
Aluminum-coated Kapton	0.1	2	GSFC Green	0.0	1
Al$_2$O$_3$	<0.025	3	HOS-875 (bare and preox)	0.0	1
Al$_2$O$_3$ (700 Å) on Kapton H	<0.02	4	Indium Tin Oxide	0.002	15,16
Apiezon grease 2 µm	>0.625	5	Indium Tin Oxide/ Kapton (aluminized)	0.01	2
Aquadag E (graphite in an aqueous binder)	1.23	6	Iridium film	0.0007	17
			Lead	0.0	1
Carbon	1.2	7,1,8,9	Magnesium	0.0	1
Carbon (various forms)	0.9–1.7	10	Magnesium Fluoride on glass	0.007	15,16
Carbon/Kapton 100XAC37	1.5	11	Molybdenum (1,000 Å)	0.0056	4
401-C10 (flat black)	0.30	12	Molybdenum (1,000 Å)	0.006	15,16
Chromium (123 Å)	partially eroded	14	Molybdenum	0.0	1
Chromium (125 Å) on Kapton H	0.006	15,16	Mylar	3.4	10
			Mylar	2.3	15,19
Copper (bulk)	0.0	17	Mylar	3.9	15,19,9
Copper (1,000 Å) on sapphire	0.007	15,16	Mylar	1.5–3.9	15
			Mylar A	3.7	18
Copper (1,000 Å)	0.0064	14	Mylar A	3.4	21,6
Diamond	0.021	17	Mylar A	3.6	6
Electrodag 402 (silver in a silicone binder)	0.057	6	Mylar D	3.0	6
			Mylar D	2.9	21
Electrodag 106 (graphite in an epoxy binder)	1.17	6	Mylar with Antiox	heavily attacked	22
			Nichrome (100 Å)	0.0	1
			Nickel film	0.0	17
Epoxy	1.7	10,16	Nickel	0.0	8
Fluoropolymers:			Niobium film	0.0	17,1
FEP Kapton	0.03	18	Osmium	0.026	10
Kapton F	<0.05	6	Osmium	heavily attacked	20
Teflon, FEP	0.037	5	Osmium (bulk)	0.314	17
Teflon, FEP	<0.05	10	Parylene, 2.5 µm	eroded away	22
Teflon, TFE	<0.05	10,6	Platinum	0.0	1
Teflon, FEP and TFE	0.0 and 0.2	15,19	Platinum	appears resistant	20
Teflon, FEP and TFE	0.1	15	Platinum film	0.0	17
Teflon	0.109	19	Polybenzimidazole	1.5	10,7
Teflon	0.5	15	Polycarbonate	6.0	8
Teflon	0.03	15	Polycarbonate resin	2.9	17
Teflon	<0.03	9	Polyester—7% Poly-ane/93% Polyimide	0.6	10
Gold (bulk)	0.0	17			
Gold	appears resistant	20	Polyester	heavily attacked	10,22

Material	Erosion Yield (10^{-24} cm^3/Atom)	Reference	Material	Erosion Yield (10^{-24} cm^3/Atom)	Reference
Polyester with Antiox	heavily attacked	10,22	Polyvinylidene Fluoride	0.6	9
Polyester (Pen-2,6)	2.9	23	Pyrone:		
Polyethylene	3.7	10,21,16,15	PMDA-DAB	2.5	23
Polyethylene	3.3	18,6	S-13-GLO, white	0.0	12
Polyimides:			SiO$_2$ (650 Å) on Kapton H	0.00103	4
BJPIPSX-9	0.28	23			
BJPIPSX-9	0.071	24	SiO$_x$/Kapton (aluminized)	0.01	2
BJPIPSX-11	0.56	23			
BJPIPSX-11	0.15	24	Silicones:		
BTDA-Benzidene	3.08	23	DCl-2577	0.055	21
BTDA-DAF	2.82	23	DCl-2755-coated Kapton	0.05	15
BTDA-DAF	0.8	24			
BTDA-mm-DD502	2.29	23	DCl-2775-coated Kapton	<0.5	15
BTDA-mm-MDA	3.12	23			
BTDA-pp-DABP	2.91	23	DC6-1104	0.0515	20
BTDA-pp-DABP	3.97	23	Grease 60 μm	intact but oxidized	25
Kapton (black)	1.4–2.2	15,12	RTV-560	0.443	21
Kapton (TV blanket)	2.0	15	RTV-615 (black, conductive)	0.0	20
Kapton (TV blanket)	2.04	19			
Kapton (OSS—1 blanket)	2.55	15	RTV-615 (clear)	0.0625	5
			RTV-670	0.0	1
Kapton (OSS—1 blanket)	2.5	15	RTV-S695	1.48	11
			RTV-3145	0.128	1
Kapton H	3.0	10,15,19 4,6,9	T-650-coated Kapton	<0.5	15
			Siloxane Polyimide (25% Sx)	0.3	7
Kapton H	2.4	15,19			
Kapton H	2.7	15,18	Siloxane Polyimide (7% Sx)	0.6	7
Kapton H	1.5–2.8	15			
Kapton H	2.0	18	Silver	10.5	5
Kapton H	3.1	18	Tantalum	appears resistant	20
ODPA-mm-DABP	3.53	23	Tedlar	3.2	10
PMDA-pp-DABP	3.82	23	Tedlar (clear)	1.3 and 3.2	15
PMDA-pp-MDA	3.17	23,24	Tedlar (clear)	3.2	18,6
PMDA-pp-ODA	4.66	23	Tedlar (white)	0.4 and 0.6	15
Polymethylmethacrylate	3.1	16	Tedlar (white)	0.05	15
25% Polysiloxane, 45% Polyimide	0.3	10	TiO$_2$, (1,000 Å)	0.0067	5
			Trophet 30 (bare and preox)	0.0	1
25% Polysiloxane-Polyimide	0.3	9	Tungsten	0.0	8
Polystyrene	1.7	10,16,9	Tungsten Carbide	0.0	8
Polysulfone	2.4	10,16	YB-71 (ZOT)	0.0	7

Appendixes

References

1. Marshall Space Flight Center
2. Smith, K. A. "Evaluation of oxygen interaction with materials (EOIM)—STS-8 atomic oxygen effects." AIAA-85-7021. November, 1985.
3. Durcanin, J. T., and Chalmers, D. R. "The definition of low earth orbital environment and its effect on thermal control materials." AIAA-87-1599. June, 1987.
4. Banks, B. A., Mirtich, M. J., Rutledge, S. K., and Nahra, H. K. "Protection of solar array blankets from attack by low earth orbital atomic oxygen." 18th IEEE Photovoltaic Specialists Conference. October, 1985.
5. Purvis, C. K., Ferguson, D. C. Snyder, D. B. Grier, N. T., Staskus, J. V., and Roche, J. C. "Environmental interactions considerations for space station and solar array design." Preliminary—December, 1986.
6. Visentine, J. T., Leger, L. G., Kuminecz, J. F., and Spiker, I. K. "NASA JSC STS-8 atomic oxygen effects experiment." AIAA 23rd Aerospace Sciences Meeting. January 1985.
7. Langley Research Center
8. University of Alabama at Huntsville
9. Coutler, D. R., Liang, R. H., Chung, S. H., Smith, K. O., and Gupta, A. "O-atom degradation mechanisms of materials." Taken from *Proceedings of NASA Workshop on Atomic Oxygen Effects*. June 1, 1987, p. 42.
10. Leger, L. G., Santos-Mason, B., Visentine, J. T., and Kuminecz, J. F. "Review of LEO flight experiments." *Proceedings on the NASA Workshop on Atomic Oxygen Effects*. November, 1986, p.6.
11. British Aerospace
12. Whitaker, A. F. "LEO atomic oxygen effects on spacecraft materials."
13. Martin Marietta
14. Lewis Research Center
15. Leger, L. J., Spiker, I. K., Kuminecz, J. F., Ballentine, T. J., and Visentine, J. T. "STS Flight 5 LEO effects experiment—Background description and thin film results." AIAA-83-2631-CP. October, 1983.
16. Jet Propulsion Laboratory
17. Gregory, J. C. "Interaction of hyperthermal atoms on surfaces in orbit: The University of Alabama Experiment." *Proceedings of the NASA Workshop on Atomic Oxygen Effects*. November, 1986, p. 31.
18. Leger, L. J., Visentine, J. T., and Kuminecz, J. F. "Low earth orbit."
19. Leger, L. J. "Oxygen atom reaction with shuttle materials at orbital altitudes—Data and experiment status." AIAA-83-0073. January, 1983.
20. Goddard Space Flight Center
21. Johnson Space Center
22. Washington University
23. Slemp, W. S., Santos-Mason, B., Bykes, G. F., Jr. and Witte, W. S., Jr. "Effects of STS-8 atomic oxygen exposure on composites, polymeric films and coatings." AIAA-85-0421. January, 1985.
24. Santos, B. "The dependence of atomic oxygen resistance on polyimide structures." (Preliminary results of STS-8). NASA Headquarters. January 23–24, 1984.
25. The Aerospace Corporation

Appendix B

Effects of Rocket Exhaust Contaminants on Selected Materials

The Expected Degradation of Selected Materials by Contaminants from Rocket Exhausts

		Kapton	Teflon	Glass/Quartz	Sapphire	MgF$_2$	Al	Ag	Au	Poly-Urethane	Silicone
	NH$_3$	None	None	Slight to None	None	?	None	None	None	Some	None
	N$_2$H$_4$	Moderate	None	Slight to Moderate for borosilicates	None	?	None	Slight to None	None	Some	Possible
Contaminant	HNO$_3$	Moderate	Slight to None	Slight to None	None	Dissolves	Slight to Moderate	Forms AgNO$_3$	None	Some	Some
	H$_2$O	None	None	None	None	None	None	None	None	None	None
	HCl	Slight to Moderate	None	None	Slight to None	None	Slight to Moderate	None	None	Some	Some
	H$_2$O$_2$	Slight to Moderate	None	None	None	?	None	?	?	Some	Some
	NO$_2$	?	None	None	None	?	None	Moderate	None	?	?
	Aniline	None	None	None	None	—	None	None	None	Some	Possible

From Shaw, C. "Surface Effects Evaluation Study," Air Force Rocket Propulsion Laboratory TR-86-093, February 1987.

INDEX

ablation, 168
absorptivity, 158
adsorption, 88
alpha, 123
ammonia, 186
amorphous, 19
angular momentum, 26
apoapsis, 25
Apollo, 93, 183
argument of perigee, 27
atmosphere, 37
atmospheric density, 43
atmospheric heating, 165
atomic mass, 2
atomic number, 2
atomic oxygen, 40, 45, 52, 75, 83, 107, 162, 195
atomic structure, 1, 73

ballistic coefficient, 30
batteries, 108
    Ag-Zn/Ag-Cd, 110
    Na-S, 113
    Ni-Cd, 111
    Ni-H_2, 109
beta, 123
binders, 148
black body radiation, 158
bond energies, 137
bonding
    covalent, 5, 103
    ionic, 4
Boyle's law, 6

cabin atmospheres, 177
carbon, 195
carbon dioxide removal, 181
career radiation exposures, 190
ceramics, 90
chamber temperature, 150
Charles's law, 6
Chemglaze paint, 96
chemical kinetics, 17
chemical reactions, 9
chemical regeneration, 182
chlorofluorocarbons (CFC), 40
coatings, 162
cold welding, 47
composites, 87
conduction, 155

contaminants, spacecraft, ix, 187, 198
convection, 155, 160
corrosion, 75
Coulomb's law, 1
covalent bonding, 5, 103
crosslinking, 19
crystal, 73

Dalton's law, 6
density scale height, 32
density wave, 49
dipole-dipole interactions, 7
dispersion forces. *See* London forces
drag force, 30

eccentricity, 25
electrochemistry, 20
ellipse, 25
emissitivity, 158–159
endothermic reaction, 14
energy
    conservation of, 24
    specific (battery), 109
    total specific, 24
enthalpy, 13, 135, 151
entropy, 15, 135
environmental effects, 107
epoxy, 90
equilibrium, 11
equilibrium constant, 12
exospheric temperature, 45
exothermic reaction, 14
extreme ultraviolet (EUV), 51

flight path angle, 26
fuel cell, 116
fuels. *See also* liquid fuels
    solid, 147

galactic cosmic radiation, 190
Galileo, 121, 127
gamma, 123
gases 6, 149
Gemini, 166, 183
geomagnetic activity, 48
Gibbs's free energy 16, 135
glasses, 90
graphite, 89
gravitational parameter, 23
Great Red Spot, 64

half-cell reduction potential, 21
heat pipe, 163
heat transfer, 155
heat transfer coefficient (convection), 166
Heisenberg's uncertainty principle, 1
helium, 186
hemoglobin, 180
Henry's law, 179
Hohmann transfer, 27
HOTOL, 173
hydrosphere, 7
hypergolic, 143

ideal gas law, 6, 43
inclination, 27
insulation, 162
Intelsat IV, 164
Intelsat V, 165
interdiffusion, 88
intermolecular forces, 7
interplanetary trajectories, 29
ionic bonding, 4

Jacchia '77, 42
Jupiter, 62

kapton, 90, 195, 198
King-Hele, 43

LDEF, vii, 83, 92
life support systems, 180
liquid fuels
    hydrazine, 138, 143, 186
    hydrocarbons, 141
    hydrogen, 138, 141, 186
    monomethyl hydrazine, 138, 145, 186
    unsymmetrical dimethylhydrazine (UDMH), 144
liquid oxidizers
    fluorine, 139
    hydrogen peroxide, 138, 141
    nitrogen tetroxide, 138, 141
    oxygen, 138, 139
liquid propellants, 137
lithosphere, 37
London forces, 7
longitude of ascending node, 27
louvers, 163

Magellan, 55
major axis, 25
Mariner, 57
Mars, 60
mass-energy conversion, 126
Mendeleev, 2
Mercury, 55, 183
mesosphere, 38
metals, 73
micrometeoroids, 96
mole, 2
monomethyl hydrazine, 138, 145, 186
multilayer insulation (MLI), 162
mylar, 195

National Aero-Space Plane (NASP), 173
Neptune, 66
nitrogen tetroxide, 186
n-type, 103
nuclear fission, 124
Nusselt number, 160

orbital decay, 30
orbital lifetime, 31, 59
orbital velocity, 27
orbits
   circular, 23
   direct, 27
   elliptical, 24
   geostationary, 28
   polar, 27
   retrograde, 27
oxidation, 75
oxidation-reduction reactions, 11
oxidizers 138, 147. *See also* liquid oxidizers
oxygen, partial pressure, 177
ozone, 38

particulate impacts, 93
patched conic maneuver, 30
periapsis, 25

periodic table, 1, 3
phase change materials (PCM), 128, 162
phase diagrams, 9
photovoltaics, 103
Planck's constant, 4
plane change, 29
planets
   inner/terrestrial, 55
   outer/Jovian, 55
Pluto, 67
polarization, 118
polymer, 18, 78,
   as binder, 148
power, 103
Prandtl number, 160
pressure scale height, 43
propellants. *See* liquid propellants; solid propellants
propulsion, 135
proton belt, 50
p-type, 103

rad, 188
radiation, 155
radiation effects, 73, 79, 187
radiators, 163
radioisotope fuels, 121
radioisotope thermoelectric generator (RTG), 120
rate law, 17
reaction rate, 17
reflectivity, 158
rem, 188
Reynolds number, 160
roentgen, 187

Sanger II, 173
Saturn, 66
semilatus rectum, 26
Shuttle thermal protection system, 169
silicon, 103, 196
Skylab, 187

SNAP-10A, 129
solar activity, 45
solar array design, 106
solar cell, 103
solar concentrators, 124
solar cycle, 44
Solar Max, 93
solid propellants, 146
solids, 8
SP-100, 125
space station, ix, 128
specific impulse, 136
Stanton number, 160
Stephan-Boltzmann constant, 158
stoichiometry, 10
stratosphere, 38
surface tension, 7, 163

teflon, 80, 195
thermal conductivity, 156
thermal control, 155
thermochemistry, 13, 135, 149
thermoelectric conversion, 122
thermosphere, 38
toxic substances, 184
transmissitivity, 158
trapped radiation, 50
troposphere, 38

UDMH. *See* liquid fuels
ultraviolet (UV), 81, 93, 158
universal gravitation, law of, 23
Upper Atmospheric Research Satellite (UARS), 41
Uranus, 66

vacuum effects, 74
Van Allen radiation belt, 50
Venus, 56
viscosity, 7, 138
vis viva equation, 24
Voyager, 67, 121